AN INTRODUCTION TO
BIOCERAMICS

ADVANCED SERIES IN CERAMICS

Editors-in-Chief: M McLaren and D E Niesz

Advanced Series in Ceramics – Vol. 1

AN INTRODUCTION TO
BIOCERAMICS

Editors

Larry L. Hench & June Wilson
University of Florida
Gainesville, Florida

World Scientific
Singapore • New Jersey • London • Hong Kong

Published by

World Scientific Publishing Co. Pte. Ltd.

P O Box 128, Farrer Road, Singapore 912805

USA office: Suite 1B, 1060 Main Street, River Edge, NJ 07661

UK office: 57 Shelton Street, Covent Garden, London WC2H 9HE

British Library Cataloguing-in-Publication Data
A catalogue record for this book is available from the British Library.

First published 1993
First reprint 1998
Second reprint 1999

AN INTRODUCTION TO BIOCERAMICS

ISBN 981-02-1400-6
ISBN 981-02-1626-2 (pbk)

Cover: Fluorochrome-labelled cancellous bone. (Photograph courtesy Dr. Tom Wronski, University of Florida)

Printed in Singapore.

Dedicated to Gerry Merwin (1947-1992) and Bill Hall (1922-1992), clinicians who pioneered the use of new biomaterials.

PREFACE

Since the 1970's, when it was first realized that the special properties of ceramic materials could be exploited to provide better materials for certain implant applications, the field has expanded enormously. Initial applications depended on the fact that smooth ceramic surfaces elicited very little tissue reaction and provided wear characteristics suitable for bearing surfaces (Chapters 2, 11). Resultant orthopaedic use has enjoyed twenty years' clinical success, notably in Europe.

Today, as well as those so-called inert bioceramics, materials have been developed which have properties which allow their use where bonding to soft or hard tissues is needed, where controlled degradation is required, where loads are to be bourne, where tissue is to be augmented, or where the special properties of ceramics can be allied with those of polymers or metals to provide implant materials with advantages over each.

In all of these applications and many others described in this text, the tissue reactions to, and properties of these bioceramics have been increasingly carefully studied so that they can be controlled and more importantly, predicted. This is the information which must be understood before they are applied clinically.

A recent assessment of the growth of the field of bioactive ceramics showed that the number of presentations on that subject at the first World Biomaterials Congress in 1980 formed 6% of the program. By the time of the fourth such congress in 1992 that figure was 23% of the whole (Fig. 1). In 1980 presentations came from 12 centers in 5 countries, in 1992 from 88 centers in 21 countries (Fig. 2). Research is international and clearly still growing.

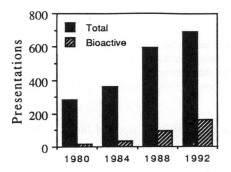

Fig. 1

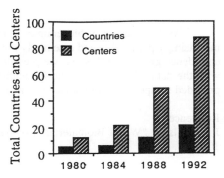

Fig. 2

Bioactive materials can be divided into two major areas, one contains bioactive glasses and glass ceramics (Chapters 3-8 and chapter 13) which develop biological hydroxyapatite at their surfaces after implantation and the other, calcium phosphate ceramics (Chapters 9-12) which are usually developed from chemical precursors. For an exception to this, see Chapter 10.

Materials from both groups have been used as powders and sometimes as solids in applications where mechanical requirements are low and as composites and coatings where mechanical requirements are high. Some have been designed specifically for high strength applications. (Chapters 5,6,8)

Coatings are discussed in Chapters 12-14. At the 1980 congress a single paper described the coating of bioactive glass on 'metal' (316L stainless steel). By 1992 a total of 37 presentations was made, 31 of which described coatings of hydroxyapatite on titanium or its alloy. The interest in coating stainless steel was mainly to provide non-cemented fixation in orthopaedics. This has now been supplanted by coatings on titanium, driven by its clinical success as a dental implant. Figure 3 shows changes in emphasis between 1980 and 1992.

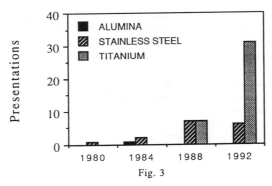

Fig. 3

As the behavior of bioceramics in both short and long-term applications becomes increasingly predictable and essentially reliable, their clinical application will increase as confidence grows. In this text we present the state of research world-wide at this time, with the data which provide the foundations of that research. We hope we have also provided signposts to those areas in which solutions to clinical needs are yet to be found.

Reference:
June Wilson, "World Biomaterials Congresses 1980-1992," *J. Applied Biom.* **4** (1993) 103-105.

June Wilson
Larry L. Hench
Gainesville, FL
May 6, 1993

CONTENTS

x

Chapter 1

INTRODUCTION

Larry L. Hench* and June Wilson**
*Advanced Materials Research Center
**Bioglass® Research Center
University of Florida
Gainesville, FL

OVERVIEW

Thousands of years ago humans discovered that clay could be irreversibly transformed by fire into ceramic pottery. Ceramic pots stored grains for long periods of time with minimal deterioration. Impervious ceramic vessels held water and were resistant to fire, which allowed new forms of cooking. This discovery was a large factor in the transformation of human culture from nomadic hunters to agrarian settlers. This cultural revolution led to a great improvement in the quality and length of life.

During the last forty years another revolution has occurred in the use of ceramics to improve the quality of life of humans. This revolution is the development of specially designed and fabricated ceramics for the repair and reconstruction of diseased, damaged or "worn out" parts of the body. Ceramics used for this purpose are called *bioceramics*. This book describes the principles involved in the use of ceramics in the body. Most clinical applications of bioceramics relate to the repair of the skeletal system, composed of bones, joints and teeth, and to augment both hard and soft tissues. Ceramics are also used to replace parts of the cardiovascular system, especially heart valves. Special formulations of glasses are also used therapeutically for the treatment of tumors.

Bioceramics are produced in a variety of forms and phases and serve many different functions in repair of the body, which are summarized in Fig. 1 and Table 1. In many applications ceramics are used in the form of bulk materials of a specific shape, called *implants, prostheses, or prosthetic devices*. Bioceramics are also used to fill space while the natural repair processes restore function. In other situations the ceramic is used as a coating on a substrate, or as a second phase in a composite, combining the characteristics of both into a new material with enhanced mechanical and biochemical properties.

Bioceramics are made in many different phases. They can be single crystals (sapphire), polycrystalline (alumina or hydroxyapatite), glass (Bioglass®), glass-ceramics (A/W glass-ceramic), or composites (polyethylene-hydroxyapatite). The phase or phases used depend on the properties and function required. For example, single crystal sapphire is used as a dental implant because of its high strength. A/W glass-ceramic is used to replace vertebrae because it has high strength and bonds to bone. Bioactive

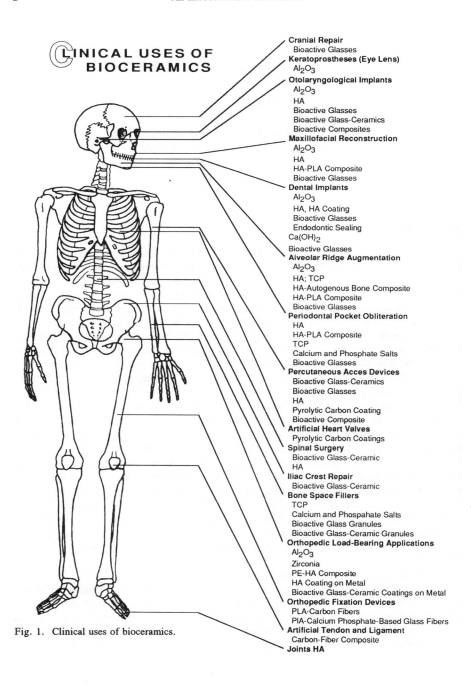

CLINICAL USES OF BIOCERAMICS

Cranial Repair
Bioactive Glasses
Keratoprostheses (Eye Lens)
Al_2O_3
Otolaryngological Implants
Al_2O_3
HA
Bioactive Glasses
Bioactive Glass-Ceramics
Bioactive Composites
Maxillofacial Reconstruction
Al_2O_3
HA
HA-PLA Composite
Bioactive Glasses
Dental Implants
Al_2O_3
HA, HA Coating
Bioactive Glasses
Endodontic Sealing
$Ca(OH)_2$
Bioactive Glasses
Aiveolar Ridge Augmentation
Al_2O_3
HA; TCP
HA-Autogenous Bone Composite
HA-PLA Composite
Bioactive Glasses
Periodontal Pocket Obliteration
HA
HA-PLA Composite
TCP
Calcium and Phosphate Salts
Bioactive Glasses
Percutaneous Acces Devices
Bioactive Glass-Ceramics
Bioactive Glasses
HA
Pyrolytic Carbon Coating
Bioactive Composite
Artificial Heart Valves
Pyrolytic Carbon Coatings
Spinal Surgery
Bioactive Glass-Ceramic
HA
Iliac Crest Repair
Bioactive Glass-Ceramic
Bone Space Fillers
TCP
Calcium and Phospahate Salts
Bioactive Glass Granules
Bioactive Glass-Ceramic Granules
Orthopedic Load-Bearing Applications
Al_2O_3
Zirconia
PE-HA Composite
HA Coating on Metal
Bioactive Glass-Ceramic Coatings on Metal
Orthopedic Fixation Devices
PLA-Carbon Fibers
PIA-Calcium Phosphate-Based Glass Fibers
Artificial Tendon and Ligament
Carbon-Fiber Composite
Joints HA

Fig. 1. Clinical uses of bioceramics.

Table 1. Form, Phase and Function of Bioceramics.

Form	Phase	Function
Powder	Polycrystalline Glass	Space-filling, therapeutic treatment, regeneration of tissues
Coating	Polycrystalline Glass Glass-Ceramic	Tissue bonding, thromboresistance, corrosion protection
Bulk	Single Crystal Polycrystalline Glass Glass-Ceramic Composite (Multi-Phase)	Replacement and augmentation of tissue, replace functioning parts

glasses have low strength but bond rapidly to bone so are used to augment the repair of boney defects.

Ceramics and glasses have been used for a long time outside the body for a variety of applications in the health care industry. Eye glasses, diagnostic instruments, chemical ware, thermometers, tissue culture flasks, chromatography columns, lasers and fibre optics for endoscopy are commonplace products in the multi-billion dollar industry. Ceramics are widely used in dentistry as restorative materials, gold porcelain crowns, glass-filled ionomer cements, endodontic treatments, dentures, etc. Such materials, called dental ceramics, are reviewed by Preston, 1988. However, use of ceramics *inside* the body as implants is relatively new; alumina hip implants have been used for just over 20 years. (See Hulbert et al., 1987, for a review of the history of bioceramics.)

This book is devoted to the use of ceramics as implants. Many compositions of ceramics have been tested for potential use in the body but few have reached human clinical application. Clinical success requires the simultaneous achievement of a stable interface with connective tissue and an appropriate, functional match of the mechanical behavior of the implant with the tissue to be replaced. Few materials satisfy this severe dual requirement for clinical use.

TYPES OF BIOCERAMICS-TISSUE INTERFACES

No material implanted in living tissues is inert; all materials elicit a response from the host tissue. The response occurs at the tissue-implant interface and depends upon many factors, listed in Table 2.

There are four general types of implant-tissue response, as summarized in Table 3. It is critical that any implant material avoid a toxic response that kills cells in the surrounding tissues or releases chemicals that can migrate within tissue fluids and cause

Table 2. Factors Affecting Implant-Tissue Interfacial Response.

Tissue Side	Implant Side
-Type of Tissue	-Composition of Implant
-Health of Tissue	-Phases in Implant
-Age of Tissue	-Phase Boundaries
-Blood Circulation in Tissue	-Surface Morphology
-Blood Circulation at Interface	-Surface Porosity
-Motion at Interface	-Chemical Reactions
-Closeness of Fit	-Closeness of Fit
-Mechanical Load	-Mechanical Load

Table 3. Consequences of Implant-Tissue Interactions.

Implant-Tissue Reaction	Consequence
Toxic	Tissue dies
Biologically nearly inert	Tissue forms a non-adherent fibrous capsule around the implant
Bioactive	Tissue forms an interfacial bond with the implant
Dissolution of implant	Tissue replaces implant

systemic damage to the patient (Black, 1984). One of the main reasons for the interest in ceramic implants is their lack of toxicity.

The most common response of tissues to an implant is formation of a non-adherent fibrous capsule. The fibrous tissue is formed in order to "wall off" or isolate the implant from the host. It is a protective mechanism and with time can lead to complete encapsulation of an implant within the fibrous layer. Metals and most polymers produce this type of interfacial response, the cellular mechanisms which influence this response are described in a later section.

Biologically inactive, nearly inert ceramics, such as alumina or zirconia, also develop fibrous capsules at their interface. The thickness of the fibrous layer depends on the factors listed in Table 2. The chemical inertness of alumina and zirconia results in a very thin fibrous layer under optimal conditions (Fig. 2). More chemically reactive metallic implants elicit thicker interfacial layers. However, it is important to remember that the thickness of an interfacial fibrous layer also depends upon motion and fit at the interface, as well as the other factors indicated in Table 2.

The third type of interfacial response, indicated in Table 3, is when a bond forms across the interface between implant and the tissue. This is termed a "bioactive" interface. The interfacial bond prevents motion between the two materials and mimics

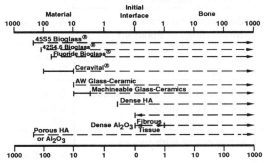

Fig. 2. Comparison of interfacial thickness of reaction layer of bioactive implants or fibrous tissue of inactive bioceramics in bone. (Reprinted from L. L. Hench, "Bioceramics: From Concept to Clinic," *J. Amer. Ceram. Soc.*, **74**[7] (1991) 1487-570, with permission.)

the type of interface that is formed when natural tissues repair themselves. This type of interface requires the material to have a controlled rate of chemical reactivity, as discussed in Chapters 3 and 5. An important characteristic of a bioactive interface is that it changes with time, as do natural tissues, which are in a state of dynamic equilibrium.

When the rate of change of a bioactive interface is sufficiently rapid the material "dissolves" or "resorbs" and is replaced by the surrounding tissues. Thus, a resorbable biomaterial must be of a composition that can be degraded chemically by body fluids or digested easily by macrophages (see below). The degradation products must be chemical compounds that are not toxic and can be easily disposed of without damage to cells.

TYPES OF BIOCERAMIC-TISSUE ATTACHMENTS

The mechanism of attachment of tissues to an implant is directly related to the tissue response at the implant interface. There are four types of bioceramics, each with a different type of tissue attachment, summarized in Table 4 with examples. The factors that influence the implant-tissue interfacial response listed in Table 2 also affect the type and stability of tissue attachment listed in Table 4.

The relative chemical activity of the different types of bioceramics is compared in Fig. 3. The relative reactivity shown in Fig. 3(a) correlates with the rate of formation of an interfacial bond of implants with bone (Fig. 3(b)). A type 1, nearly inert, implant does not form a bond with bone. A type 2, porous, implant forms a mechanical bond via ingrowth of bone into the pores. A type 3, bioactive, implant forms a bond with bone via chemical reactions at the interface. A type 4, resorbable, implant is replaced by bone.

Table 4. Types of Tissue Attachment of Bioceramic Prostheses.

Type of Implant	Type of Attachment	Example
(1) Nearly inert	Mechanical interlock (Morphological Fixation)	Al_2O_3, Zirconia
(2) Porous	Ingrowth of tissues into pores (Biological Fixation)	Hydroxyapatite (HA) HA coated porous metals
(3) Bioactive	Interfacial bonding with tissues (Bioactive Fixation)	Bioactive glasses Bioactive glass-ceramics HA
(4) Resorbable	Replacement with tissues	Tricalcium phosphate Bioactive glasses

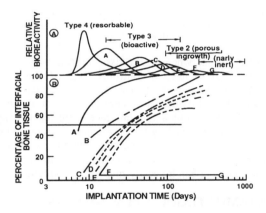

Fig. 3. Bioactivity spectrum for various bioceramic implants: (a) relative rate of bioreactivity and (b) time dependence of formation of bone bonding at an implant interface ((A) 45S5 Bioglass®, (B) KGS Ceravital®, (C) 55S4.3 Bioglass®, (D) A/W glass-ceramic, (E) HA, (F) KGX Ceravital®, and (G) Al_2O_3-Si_3N_4). (Reprinted from L. L. Hench, "Bioceramics: From Concept to Clinic," *J. Amer. Ceram. Soc.*, **74[7]** (1991) 1487-570, with permission.)

   The relative level of reactivity of an implant also influences the thickness of the interfacial layer between the material and the tissue (Fig. 2). A type 1, nearly inert, implant forms a non-adherent fibrous layer at the interface. A chemically stable material like alumina elicits a very thin capsule. Consequently, when alumina or zirconia implants are implanted with a tight mechanical fit and movement does not occur at the interface they are clinically successful.

   However, if a type 1, nearly inert, implant is loaded such that interfacial movement occurs, the fibrous capsule can become several hundred micrometers thick and the implant loosens very quickly. Loosening invariably leads to clinical failure for a variety of reasons which includes fracture of the implant or the bone adjacent to the implant.

Type 2 porous ceramics and HA coatings on porous metals were developed to prevent loosening of implants. The growth of bone into surface porosity provides a large interfacial area between the implant and its host. This method of attachment is often called *biological fixation*. It is capable of withstanding more complex stress states than type 1 implants which achieve only "morphological fixation".

A limitation of type 2 porous implants is the necessity for the pores to be at least 100 micrometers in diameter. This large pore size is needed so that capillaries can provide a blood supply to the ingrown connective tissues. Without blood and nutrition the bone will die. Vascular tissue does not appear in pores < 100 μm. If micromovement occurs at the interface of a porous implant the capillaries can be cut off, leading to tissue death, inflammation and destruction of interfacial stability.

When the porous implant is a metal, the large interfacial area can provide a focus for corrosion of the implant and loss of metal ions into the tissues, which may cause a variety of medical problems. Coating a porous metal implant with a bioactive ceramic, such as hydroxyapatite, diminishes some of these limitations. The HA coating also speeds the rate of bone growth into the pores. The coatings often dissolve with time which limits their effectiveness. The large size and volume fraction of porosity required for stable interfacial bone growth degrades the strength of the material. This limits the porous method of fixation to coatings or unloaded space fillers in tissues.

Resorbable implants (type 4 in Table 4) are designed to degrade gradually with time and be replaced with natural tissues. A very thin or non-existent interfacial thickness, Fig. 2, is the final result. This approach is the optimal solution to the problems of interfacial stability. It leads to the regeneration of tissues instead of their replacement. The difficulty is meeting the requirements of strength and short-term mechanical performance of an implant while regeneration of tissues is occurring. The resorption rates must be matched to the repair rates of body tissues (Fig. 3) which vary greatly depending on the factors listed in Table 2. Some materials dissolve too rapidly and some too slowly. Large quantities of material must be handled by cells so the constituents of a resorbable implant must be metabolically acceptable. This is a severe limitation on the compositions that can be used.

Successful examples of resorbable implants include specially formulated polymers. Resorbable sutures composed of poly(lactic acid)-poly(glycolic acid) are metabolized to carbon dioxide and water. Thus, they function for a time to hold tissues together during wound healing then dissolve and disappear. Tricalcium phosphate (TCP) ceramics degrade to calcium and phosphate salts and can be used for space filling of bone.

Bioactive implants (type 3 in Table 4) offer another approach to achieve interfacial attachment. The concept of bioactive fixation is intermediate between resorbable and bioinert behavior. A bioactive material undergoes chemical reactions in the body, but only at its surface. The surface reactions lead to bonding of tissues at the interface. Thus, a bioactive material is defined as: "a material that elicits a specific biological response at the interface of the material which results in the formation of a bond between the tissues and the material."

The bioactive concept has been expanded to include many bioactive materials with a wide range of bonding rates and thickness of interfacial bonding layers, Figs. 2 and 3. They include bioactive glasses such as Bioglass® bioactive glass-ceramics such as A/W glass-ceramic; dense synthetic hydroxyapatite (HA), bioactive composites such as polyethylene-HA, and bioactive coatings such as HA on porous titanium alloy. All of these materials form an interfacial bond with bone. The time dependence of bonding, the strength of the bond, the mechanism of bonding, the thickness of the bonding zone, and the mechanical strength and fracture toughness differ for the various materials.

No bioactive material is optimal for all applications. It is essential to match the form, phases and properties of a bioactive implant with its rate of bonding and its function in the body (Table 1). Relatively small changes in composition can affect whether a bioceramic is nearly inert, resorbable or bioactive. These compositional effects are described in Chapter 3. It was discovered in 1981 that certain bioactive glass compositions will bond to soft connective tissues as well as bone. The compositions that bond to soft tissues have the highest rates of surface reaction of all the bioactive materials.

A common characteristic of all bioactive implants is the formation of a hydroxy-carbonate apatite (HCA) layer on their surface when implanted. The HCA phase is equivalent in composition and structure to the mineral phase of bone. The HCA layer grows as polycrystalline agglomerates. Collagen fibrils are incorporated within the agglomerates thereby binding the inorganic implant surface to the organic constituents of tissues. Thus, the interface between a bioactive implant and bone is nearly identical to the naturally occurring interfaces between bone and tendons and ligaments. The stress gradients across a bioactive interface are a closer match to natural stress gradients than those across the interface of type 1 or type 2 implants.

TISSUE RESPONSE TO IMPLANTS

To understand the way in which tissues respond to an implant it is necessary to understand the nature of the tissue at the interface and the significance of any alterations seen there. The significance of such changes will vary with the material and will be governed both by their severity and by their persistence, a transient change or a continuing one may both appear to be identical shortly after implantation.

The act of implantation evokes tissue changes from the surgery and the persistence and resolution of those changes may or may not be independent of the implant material and its properties. Some damage is inevitable on implantation in all but a few situations. Only when these materials are delivered by injection is the effect produced at a point distant from that at which it enters the body.

This introduction will discuss the inflammatory response, which is the tissue reaction to any form of damage. In this context damage may be due to surgery, material properties or mechanical damage due to wear particles. To understand the inflammatory response requires some knowledge of the normal tissue architecture and function and in addition, certain frequently [and sometimes loosely] used terms will be defined.

Every organ in the body is made up from a combination, in varying proportions, of four tissue types.

1. Epithelium
2. Muscle
3. Nervous tissue
4. Connective tissue

Epithelial tissues cover and line organs throughout the body and can also secrete a wide variety of substances either directly into the system through ducts or into the blood stream. Glands are made up of such secretary epithelium.

Muscle tissue is found wherever movement is required. In the skeleton, muscle is under voluntary control, elsewhere such as in the cardiovascular, digestive and respiratory systems it is controlled biochemically.

Nervous tissue is specialized to transmit signals between the outside world, the brain and all of the body system.

Connective tissue, the fourth group, is well named since its constituent tissues connect and service all of the others. It includes blood supply to and from the organs. No organ in the body is without a connective tissue component and it is with connective tissue that the ceramic biomaterials, which are the subject of this book, interact.

For more information and details of the appearance of these tissue types see Ham or any similar histological textbook.

An inflammatory response will always be present immediately after surgery while the damaged tissue, blood clot, and bacteria introduced at that time are removed. The reddening and swelling which can be seen in inflammation mark the increase in blood supply (and its consequences) produced by the chemicals released by damaged tissues. In with the blood supply arrive the cells involved in the repair process. These include many cells known as phagocytes, for their ability to ingest, sometimes digest, and remove foreign material. It is the presence of these phagocytes at any time other than immediately post-implantation, which can indicate problems with a material or an implant.

All phagocytic cells begin life in the blood as one of the white cells or leucocytes (see Fig. 4).

The cells are distinguished by their size, shape, and staining characteristics (Ham 1969). They migrate from the blood into the tissues to deal with foreign material. The most numerous are the neutrophils and a massive increase in their numbers signifies, amongst other things, infection, since they ingest bacteria. All of the granular cells have lobed nuclei and may be termed polymorphs or "PMN" for short. They may also be termed "microphages". The non-granular cells have round nuclei and different functions. The lymphocytes are the cells which produce antibodies and an increase in lymphocytes

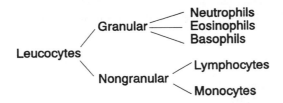

Fig. 4.   Classification of white blood cells.

and certain of the granulocytes can indicate an allergic response.   Monocytes in the blood are the source of the connective tissue phagocytes, the macrophages.   Monocytes migrate from the circulation into the connective tissue (where they are re-named histiocytes) and when needed move through the connective tissue to ingest foreign material which is too large to be dealt with by polymorphs.   Where that material is tissue debris the enzymes secreted by macrophages are sufficient to digest the material with relatively little harm to the cell.   However, when such debris is derived from an implant material or when the foreign body has attracted phagocytes because of its surface characteristics, the situation can be quite different.   Not only may the cell be unable to digest the material, it may be killed by it and thus release the material to be repeatedly ingested in a vain attempt to eliminate it, at the same time accompanied by increasing amounts of dead macrophage tissue.   The enzymes produced by these activated macrophages influence the fibroblasts, which produce the collagen to form the fibrous capsule around an implant.   For as long as phagocytic activity continues the capsule will become thicker.   When, a particle, or more notably a surface, is of a size that it can not be encompassed by a macrophage acting alone, then the giant cell appears.   Giant cells form when macrophages coalesce to produce a phagocyte large enough to deal with large particles or to attempt to deal with rough surfaces.   However, the characteristics of giant cells are similar, in many ways, to those of macrophages.   They do not themselves reproduce and the presence of giant cells at an interface some time after implantation can indicate a persistent stimulus.

Table 1 lists factors which can affect interfacial response and it should now be clear that all of these are mediated by the cells involved in the inflammatory response discussed above.   Any defect on the tissue side produced by age or disease will affect it, any damage to implant or tissue as a consequence of roughness, porosity or relative movement will affect it, the loading in use will affect it and the nature of the material and chemical reactions will also affect it.   Any material which, in its intended use, produces few, if any, of those factors which produce these tissue responses can be termed biocompatible.

A biocompatible material is one which possesses the ability to perform with an appropriate host response in a specific application.   This definition, arrived at by consensus in 1986, (see Williams 1987) emphasizes that biocompatibility is not lack of toxicity, but a requirement that a material performs appropriately. It is essential to recognize that every application of a material enforces different conditions and thus it may or may not be biocompatible in different applications.

The tissue response to nearly inert ceramics (type 1 implants) therefore is not dependant on chemistry so much as fit. If movement at the interface is minimal the phagocytic response will be transient and the thin capsule will be in place and quiescent shortly after implantation. With more chemically reactive materials, such as some metals, the reactive phase is extended and the capsule will therefore have more time to thicken before equilibrium is achieved. In the response to bioactive interfaces (type 3 implants) the capsule formation is minimal because of the removal of the influence of interfacial movement by the bonding mechanism. The reaction to resorbable implants (type 4) will persist until the components have been removed, for this type of reaction materials' properties will be the controlling factor in tissue response. Where porous materials or rough surfaces are concerned, (type 2 implants) those which depend on mechanical interlock, with or without bioactivity, for the tissue reaction, all factors are important and the tissue reaction most complex since almost all of the factors in Table 2 come into play, not only during the initial stabilization process but also during the long term. Because of the difficulty of achieving permanent stability within the pores under loaded conditions, breakdown within the pores is a potential problem and repair within the pores is difficult. These are a significant factor in development of these materials, discussed in Chapter 10.

TYPES OF BONE AT BIOCERAMIC INTERFACES

Most bioceramic implants are in contact with bone. Thus, it is important to understand that there are various types of bone in the body. Bone is a living material composed of cells and a blood supply encased in a strong, interwoven composite structure. There are three major components to the acellular structure of bone: collagen, which is flexible and very tough; hydroxycarbonate apatite, bone mineral, which is the reinforcing phase of the composite; and bone matrix or ground substance, which performs various cellular support functions. The three components are organized into a three dimensional system that has maximum strength and toughness along the lines of applied stress. See Ham, 1969, or Vaughn, 1974 for a description of the growth and structure of bone.

Two of the various types of bone are of most concern in the use of bioceramics. They are cancellous bone and cortical bone. Cancellous bone, also called trabecular or spongy bone, is less dense than cortical bone. It occurs across the ends of the long bones and is like a honeycomb in cross section, as shown on cover. Because of its lower density, cancellous bone has a lower modulus of elasticity and higher strain to failure than cortical bone, Table 5 and Fig. 5. Both types of bone have higher moduli of elasticity than soft connective tissues, such as tendons and ligaments (Table 5). The difference in stiffness (elastic modulus) between the various types of connective tissues ensures a smooth gradient in mechanical stress across a bone, between bones and between muscles and bones.

Table 5. Mechanical Properties of Skeletal Tissues.

Property	Cortical Bone	Cancellous Bone	Articular Cartilage	Tendon
Compressive Strength (MPa)	100-230[1,2]	2-12[3]		
Flexural, Tensile Strength (MPA)	50-150[1,2]	10-20[3]	10-40[4]	80-120[9]
Strain to Failure	1-3[6]	5-7[6]	15-50%[5]	10%[9]
Young's (Tensile) Modulus (GPa)	7-30[1,2,6]	0.5-0.05[3,6]	0.001-0.01[6]	1[6]
Fracture Toughness (K_{1c}) (MPa m$^{1/2}$)	2-12[2]			
Compressive Stiffness (N mm^{-1})			20-60[7]	
Compressive Creep Modulus (MPa)			4-15[8]	
Tensile Stiffness (MPa)			50-225[4]	

1.   F. G. Evans and A. King, *Biomedical Studies of the Musculoskeletal System* (Charles C. Thomas, Springfield, IL, 1961) pp. 49-53.
2.   W. Bonfield, "Elasticity and Viscoelasticity of Cortical Bone," in *Natural and Living Biomaterials*, eds. G. W. Hastings and P. Ducheyne (CRC Press, Boca Raton, FL, 1984) pp. 43-60.
3.   R. Van Audekercke and M. Martens, "Mechanical Properties of Cancellous Bone," in *Natural and Living Biomaterials*, eds. G. W. Hastings and P. Ducheyne (CRC Press, Boca Raton, FL, 1984) pp. 89-98.
4.   G. E. Kempson, "Relationship Between the Tensile Properties of Articular Cartilage from the Human Knee and Age," *Annals of the Rheumatic Diseases* **41** (1982) 508-511.
5.   D. L. Bader, G. E. Kempson, A. J. Barrett and W. Webb, "The Effects of Leukocyte Elastose on the Mechanical Properties of Adult Human Cartilage in Tension," *Biochim. et Biophys. Acta* **677** (1981) 103-108.
6.   J. Black, *Orthopedic Biomaterials in Research and Practice* (Churchill Livingston, New York, 1988).
7.   D. L. Bader, G. E. Kempson, J. Egan, W. Golbey and A. J. Barrett, "The Effects of Selective Matrix Degradation on the Short Term Compressive Properties of Adult Human Articular Cartilage," *Biochim. et Biophys. Acta*, **116** (1992) 147-52.
8.   S. Roberts, B. Weightman, J. Urban and D. Chappell, "Mechanical and Biochemical Properties of Human Articular Cartilage in Osteoarthritic Femoral Heads and in Autopsy Specimens," *J. Bone Jt. Surg.* **68B** (1986) 278-288.
9.   D. L. Butler, E. S. Grood, F. R. Noyes, R. F. Zernicke and K. Bracket, "Effects of Structure and Strain Measurement Techniques on the Material Properties of Young Human Tendons and Fascia," *J. Biomech.* **17**, 579-596 (1984)

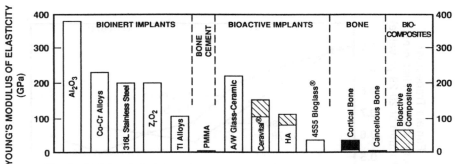

Fig. 5. Modulus of elasticity (GPa) for prosthetic materials compared with bone.

Bone at the interface with an implant is often structurally weak because of disease or ageing. Figure 6A shows the progressive loss of volume of bone with age. The decrease in bone area leads to a decrease in strength, Fig. 6B. See Revell, 1990, for a discussion of the pathology of bone, the effects of age and disease on the structure and rate of repair of bone.

The quality of bone at an implant-bone interface can deteriorate even further due to the presence of the implant or the method of fixation. Localized death of bone can occur, especially if bone cement, poly(methyl methacrylate) (PMMA), is used to provide mechanical attachment of the device. The local rise in temperature when the monomer cross-links to form the polymer is sufficient to kill bone cells to a depth of nearly a millimeter.

Another problem, called stress shielding, occurs when the implant prevents the bone from being properly loaded. The higher modulus of elasticity of the implant results in its carrying nearly all the load. Figure 5 compares the modulus of elasticity of the materials used for load bearing implants with the values of cortical bone and cancellous bone. The elastic modulus of cortical bone ranges between 7 and 25 GPa, depending upon age, location of the bone and direction of measurement (bone is anisotropic). This modulus is 10 to 50 times lower than that of alumina. Cancellous bone has a modulus that is several hundreds of times less than that of alumina.

The clinical problem arises because bone must be loaded in tension to remain healthy. Stress shielding weakens bone in the region where the applied load is lowest or in compression. Bone that is unloaded or is loaded in compression will undergo a biological change that leads to bone resorbtion.

The interface between a stress shielded bone and an implant deteriorates as the bone is weakened. Loosening and or fracture of the bone, the interface, or the implant will result. The presence of wear debris that often occurs in artificial hip and knee joints accelerates the weakening of the stress-shielded bone, because the increased cellular activity involved in removal of the foreign wear particles also attacks and destroys bone. The combination of stress shielding, wear debris and motion at an interface is especially damaging and usually leads to failure.

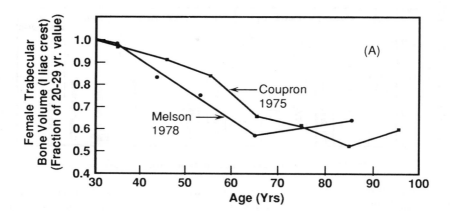

Fig. 6A. Effect of age on female trabecular bone volume of the Iliac crest. Data plotted from Coupron et al., 1975 and Melson, 1978.

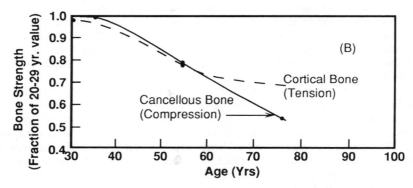

Fig. 6B. Effect of age on strength of bone. Data plotted from Yamada in Evans, 1970.

Elimination of stress shielding is one of the primary motivations for the development of bioceramic composites, discussed in Chapters 15 and 16. The elastic modulus of a two-phase composite can be matched to that of bone, as shown in Fig. 5. If one of the phases is a bioactive material the composite can also form a bioactive bond with bone, thereby eliminating two of the primary causes for implant failure, interfacial loosening and stress shielding.

TYPES OF PROCESSING AND MICROSTRUCTURE OF BIOCERAMICS

Bioceramic materials can be classified into eight categories based upon processing method used and the microstructure produced; i.e., the distribution of phases developed in the material (Table 6). The differences in microstructure of the eight categories are primarily due to the different starting materials and thermal processing steps involved in making the materials. Chapter 18 discusses the sequence of processing steps used in making bioceramics and the characterization methods required to ensure reproducibility of properties of the final product.

Table 6.

Type of Ceramic Processing	Example
1. Glass	45S5 Bioglass®
2. Cast or rapidly solidified polycrystalline ceramic	HA coating
3. Polycrystalline glass-ceramic	Ceravital®
4. Liquid-phase sintered (vitrified) ceramic	Glass-HA
5. Solid-state sintered ceramic	Alumina, zirconia
6. Hot pressed ceramic or glass-ceramic	A/W glass-ceramic
7. Sol-gel glass or ceramic	52S bioactive gel-glass
8. Multi-phase composite	PE-HA

Figure 7 summarizes the time-temperature profiles used in processing the ceramics listed in categories 1-6 in Table 6. The thermal processing of sol-gel glasses and ceramics involves much lower temperatures and different types of processing methods, as shown in Fig. 8. Processing of composites differ for each type of composite, as discussed in Chapters 15 and 16.

For the reader unfamiliar with ceramic processing, some of the concepts relating thermal processes with microstructural development follow. For detailed treatment of the theory and practice of ceramic processing consult Reed, Onoda and Hench, or Kingery, Bowen and Uhlmann.

The objective of ceramic processing is to make a specific form of the material that will perform a specific function (Table 1). This requires making a solid object, a coating or particulates (powders). There are two ways of making a specific shape; casting from the liquid state (types 1,2,3 in Table 6) or pre-forming the shape from fine-grained particulates followed by consolidation (types 4, 5, 6 in Table 6).

When a shape is made from powders it is called *forming*. The powders are usually mixed with water and an organic binder to achieve a plastic mass that can be cast, injected, extruded or pressed into a mold of the desired shape. The formed piece

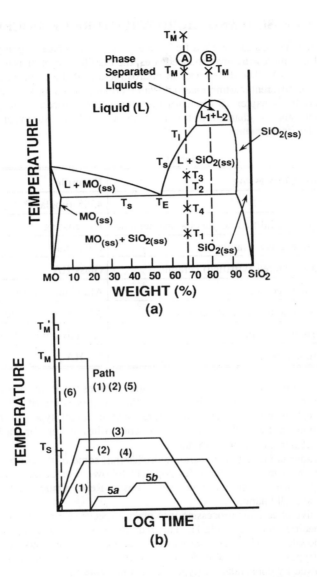

Fig. 7. (a) Composition A: Microstructure: (1) glass; (2) cast polycrystalline (large-grained); (3) liquid-phase-sintered (vitrified); (4) solid-state sintered; (5) polycrystalline glass-ceramic; (6) polycrystalline coating from T_m'. (b) Composition B: (1) Phase-separated glass. (2)-(5) Same as (a). (ss) = solid solution, T_s = solidus line.

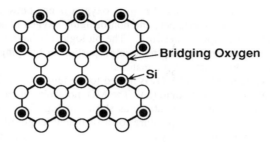

Fig. 8a. Schematic structure of a crystalline silicate. All Si(O$_4$) tetrahedra are bonded together by -Si-O-Si (siloxane) bonds.

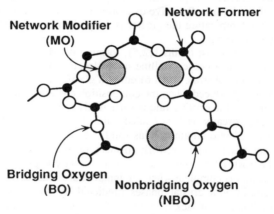

Fig. 8b. Schematic structure of a random glass network composed of network modifiers (MO) and network formers (SiO$_2$). Some of the Si are bonded to each other by bridging oxygen (BO) bonds and others are coordinated with non-bridging oxygen (NBO) bonds to network modifying ions.

is called *green ware*. Subsequently, the temperature is raised to evaporate the water *(drying)* and the binder is burned out, resulting in *bisque ware*. At a very much higher temperature the ware is densified during firing. After cooling to ambient temperature, one or more finishing steps may be applied, such as grinding and polishing, as illustrated in Fig. 1 of Chapter 18. The result is a finished product with desired properties. The properties depend upon the composition of the material, the phases developed during thermal processing and the microstructure of the material.

Phase equilibrium diagrams provide the basis for understanding the relationships between thermal processing schedules and the phases and microstructures produced. Figure 7a is a binary phase equilibrium diagram consisting of SiO$_2$ (silica), a network-forming oxide, and some arbitrary network modifier oxide, MO. MO can be Na$_2$O, K$_2$O, CaO, MgO, etc. Schematic structures of a glass, with a random network, and a crystal, with an ordered network, are shown in Fig. 8. There are two types of bonds in the glass or crystal network, bridging oxygen bonds between neighboring Si atoms, which hold the network together, and non-bridging oxygen bonds between Si and modifier atoms, which disrupt the network. The biological behavior of glasses, glass-ceramics and ceramics depends on the relative proportion of bridging oxygen bonds to non-bridging bonds in the phases of the material.

When a mixture of MO and SiO_2 is heated to the temperature TM in Fig. 7a the entire mass will melt and become liquid (L). The MO molecules break the Si-O-Si bonds of SiO_2 and lower the melting temperature, as shown in Fig. 7a. The liquid becomes homogeneous when held at this temperature for a sufficient length of time. In order to ensure homogeneity, melting is usually done several hundreds of degrees above TM. In a very rapid process such as plasma spray coating of HA (Chapter 12) or flame spray coating of bioactive glasses (Chapter 13), melting occurs but there is insufficient time for homogenization of the liquid. Selective evaporation of constituents of the melt can also occur, the higher the temperature the greater is the probability of this happening, leading to an inhomogeneous product.

When the liquid is cast (paths 1,2,5), forming the shape of the object during the casting, either a glass or a polycrystalline microstructure will result. When the liquid is rapidly cooled onto a substrate (path 6) either a glass or a polycrystalline coating will be formed. A glass is produced when the composition contains a sufficient concentration of network formers and the cooling rate is sufficiently rapid (path 1 or 6). The viscosity of the melt increases greatly as it is cooled until at T_1, the glass transition point, the material is transformed into an amorphous solid; i.e., a glass.

If there are insufficient network formers or the cooling rate is too slow, a polycrystalline microstructure will result. The crystals begin growing from T_1 and below. Crystallization is complete when the temperature reaches T_2. The final material consists of the equilibrium crystal phases predicted by the phase diagram (path 2). However, the combination of lack of network formers and very rapid cooling, such as occurs in plasma spray coating of hydroxyapatite, often produces a mixture of crystal phases which may or may not be equilibrium phases (see Chapters 9 and 12 and Bergeron and Risbud, 1984).

When the MO and SiO_2 powders are first formed into the shape of the desired object and fired at a temperature T_3, a liquid-phase-sintered structure will result (path 3). Before firing, the material will contain 10-40% porosity depending on the forming process used. During heating a liquid begins to form at grain boundaries at the eutectic temperature, T_2. The liquid dissolves the interface, penetrates between the grains, fills the pores and draws the grains together by capillary attraction (Fig. 9a). These effects decrease the volume of the compact. Since the mass remains unchanged but only rearranged, the density increases. The liquid content and composition can be predicted from the phase diagram for long firing times. However, in most ceramic processing, liquid formation does not proceed to equilibrium, due to the slowness of the reactions.

The microstructure resulting from liquid-phase sintering or vitrification, as it is commonly called, consists of grains from the original powder compact surrounded by a liquid phase formed during firing at T_3. As the compact is cooled from T_3 to T_2, the

(a) Liquid Phase Sintering

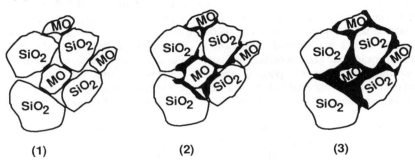

     (1)   (2)   (3)

Fig. 9a. Steps in Liquid Phase Sintering: (1) Liquid begins to form at MO-SiO$_2$ grain boundaries at T_E, (2) Liquid dissolves MO and SiO$_2$, (3) Liquid fills the pores and pulls the grains together into a dense object.

(b) Solid State Sintering

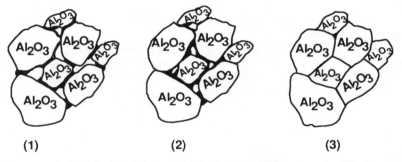

     (1)   (2)   (3)

Fig. 9b. Steps in Solid State Sintering: (1) Necks form at particle contacts by diffusion or creep, (2) Necks grow to close pore channels and particles rearrange to eliminate pores, (3) Pores are replaced by new grain boundaries.

solidus temperature T_S, the liquid phase crystallizes into a fine-grained matrix surrounding the original grains. If the liquid contains a sufficient concentration of network formers, the liquid will be quenched into a glassy matrix which surrounds the original grains. Hot-pressing of ceramics or glass-ceramics, such as A/W glass-ceramic, produces grain boundary reactions, densification and a final microstructure similar to that obtained by vitrification.

    A powder compact can be densified without the presence of a liquid phase by a process called solid-state sintering. This is the process used to make medical-grade alumina and zirconia. Prevention of formation of grain boundary phases that are susceptible to grain boundary corrosion is the main advantage of solid state sintering. In solid-state sintering, solid material is moved to areas of contact between particles. The driving force is reduction of surface energy gradients. A fully dense compact has no internal solid-vapor interfaces and therefore is lower energy than a porous material. Sintering occurs by thermal activation of molecules at the solid-pore interface; mechanisms include grain boundary diffusion, volume diffusion, surface diffusion, creep, or various combinations depending upon the temperature, sintering atmosphere, and composition of the material. Because long-range migration of atoms is necessary, solid-state sintering temperatures are usually greater than one-half of the melting point of the material: $T > T_L/2$ (path 2).

During solid-state sintering the material moves to eliminate the pores and open channels that exist between the grains of the compact (Fig. 9b). As the pores and open channels are closed during heat treatment the crystals become tightly bonded together at their grain boundaries and the density, strength, toughness, and corrosion resistance of the material increases greatly. The microstructure of a ceramic made by solid-state sintering consists of grains bonded together with a small amount of residual porosity.

The relative rate of densification during solid-state sintering is slower than that of liquid phase sintering because material transport is slower in a solid than in a liquid. However, it is possible to solid-state sinter single component materials such as pure alumina, since liquid development is not necessary. Consequently, when optimal mechanical and chemical performance is required, as it is in prostheses, solid-state sintering becomes essential.

Control of grain size during sintering is critical if properties are to be consistently superior. Excessive grain growth is always a potential hazard because of the high temperatures involved. Grain growth inhibitors can be used but they remain in the grain boundaries and may degrade grain boundary resistance to body fluids. Optimization of medical-grade ceramics requires a systematic characterization effort, described in Chapter 18.

Another class of microstructures is termed *glass-ceramics* because the object is formed as a glass but ends up as a polycrystalline ceramic. The transformation of the glass into a ceramic occurs in two steps. First, the glass is heat treated at a temperature in the range of 450-700°C (path 5a) to produce a large concentration of nuclei from which crystals can grow. When sufficient nuclei are present to ensure that a fine-grained microstructure will be obtained, the temperature of the object is raised to 600-900°C which promotes crystal growth (path 5b). Crystals grow from the nuclei until they impinge and 100% crystallization is achieved. The resulting microstructure is non-porous and fine-grained. The crystals are randomly oriented and can be very small and have a very uniform size distribution. The crystal phases may or may not correspond to the equilibrium crystal phases predicted by the phase diagram. When phase separation occurs, composition B in Fig. 7, a non-porous glass-in-glass microstructure can be obtained. Use of these concepts makes it possible to produce a very broad range of glass-ceramics, as described by Hohland and Vogel in Chapter 8.

SOL-GEL PROCESSING

Sol-gel processing is a chemically based method for producing ceramics, glass, glass-ceramics and composites at much lower temperatures than the traditional processing methods described above. Iler, Brinker and Scherer, and Hench and West describe the history, theory, processing details and applications of sol-gel processing. The sol-gel method was used by Jarcho, and many others subsequently, to make hydroxyapatite ceramics. The method has been recently used to make a new generation of bioactive gel-glasses (Li et al.) and offers promise for tailoring the composition of bioactive materials to match the requirements of specific disease states, as discussed in Chapter 21. Three methods can be used to make sol-gel materials:

Method 1:   Gelation of colloidal powders
Method 2:   Hypercritical Drying
Method 3:   Controlled hydrolysis and condensation of metal alkoxide precursors followed by drying at ambient pressure.

All three methods involve creation of a three-dimensional, interconnected network, termed a *gel*, from a suspension of very small, colloidal particles, called a sol. Colloids are solid particles with diameters < 100 nm. A sol is a dispersion of colloidal particles in a liquid. Milk is an example of a sol. A gel can be formed from an array of discrete colloidal particles by changing the pH of the sol (Method 1). The gel network can also be formed from the hydrolysis and condensation of liquid metal-alkoxide precursors (Methods 2 and 3), illustrated in Fig. 10. An example of a metal-alkoxide precursor used to provide the -Si-O-Si- network of bioactive gel-glasses is $Si(OR)_4$, where R is CH_3, C_2H_5, or C_3H_7. Other metal ions can also be used in addition to Si, such as, Ca, P, Ti, etc.

Seven steps are involved in making gel-glasses or ceramics by the sol-gel method. The first step shown in Fig. 10 is mixing the precursors which forms the low viscosity sol. As the network interconnects develop the viscosity increases greatly. Prior to completion of the network formation the sol can be applied as a coating, be pulled into a fiber, impregnated into a composite, formed into powders, or cast into a mold with a precise shape and surface features (Step 2). Gelation (Step 3) occurs in the mold or on the surface of a substrate forming a solid object or a surface coating. Powders can be made with very highly controlled size distributions.

The 3-D gel network is completely filled with pore liquid. Aging (Step 4) involves holding the gel in its pore liquid for several hours at 25-80°C. This leads to localized solution and reprecipitation of the solid network which increases the thickness of the interparticle necks, and the density and strength of the gel.

The pore liquid is removed during drying (Step 5). Colloidal gels are easily dried since their pore size is large, > 100 nm. Alkoxide-based gels have very small pores (1-10 nm) and thus large capillary stresses can arise during drying. Hypercritical drying at elevated temperature and pressure, above the pore liquid-solid critical point, avoids the solid-liquid interface and eliminates drying stresses (Method 2). Gels made

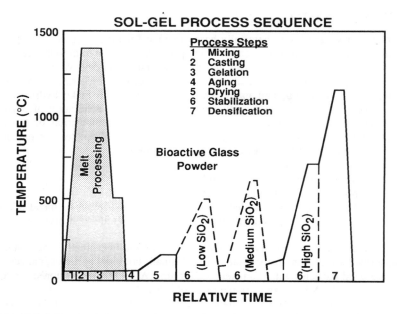

Fig. 10. Processing steps in making bioactive gel glasses by the sol-gel method. Note lower temperatures compared with melt processing of a low SiO₂ bioactive glass.

by this method, called *aerogels*, have very low densities and strengths. They are used for optical applications but as yet have no biomaterial's applications.

Gels dried under ambient temperature and relatively low temperatures are termed *xerogels* (Method 3). The generic term *gel* usually applies to a xerogel. Careful control of the hydrolysis and condensation rates in Step 1 by use of acid or base catalysts is required to produce the very narrow pore size distributions in xerogels needed to reduce drying stress gradients. A gel is dried (Step 5) when the physically adsorbed water is completely eliminated from the pores. This requires heating at controlled rates at temperatures of 120-180°C. The surface area of gels made by Method 3 is very large, 200-800 m^2/g. The pore sizes can be varied from 1.0-12 nm.

Chemical stabilization of a dried gel, Step 6 in Fig. 10, is necessary to control the environmental stability of the material. Thermal treatment in the range of 500-900°C desorbs surface silanols (Si-OH) and eliminates 3-membered silica rings from the gel. These surface chemical features are important in controlling the rate of HCA formation on the gel-glasses and their bioactivity. Stabilization also increases the density, strength and hardness of the gels and converts the network to a glass with network properties similar to melt-derived glasses.

Densification of alkoxide-derived gel-glasses is completed in the range of 900-1150°C depending upon composition. Gel-glasses with moderate (45-69%) SiO₂ content and high $CaO-P_2O_5$ content, for example, have all pores eliminated by the end of a 900°C treatment whereas gel-glasses with high SiO_2 content require 1000-1150°C. Hydroxyls and adsorbed water must be removed from the gels prior to closure of pores or bloating and inhomogeneous microstructures will result. A very important advantage of the sol-gel process is the ability to control the surface chemistry of the material by these thermal treatments.

READING LIST

L. L. Hench and E C. Ethridge, *Biomaterials: An Interfacial Approach* (Academic Press, New York, 1982).

S. F. Hulbert, J. C. Bokros, L. L. Hench, J. Wilson, and G. Heimke in *High Tech Ceramics*, ed. P. Vincenzini (Elsevier Science Pub. B.V., Amsterdam, 1987) pp 189-213.

Biocompatibility of Clinical Implant Materials, ed. D. Williams (CRC Press, Boca Raton, FL, 1981) Vol. II, pp 3-42.

J. Black, "Systemic Effects of Biomaterials," *Biomaterials* 5 (1984) 11.

A. W. Ham, *Histology* (J. B. Lippincott Co., Philadelphia, PA, 1969).

Bioceramics: Material Characteristics Versus In Vivo Behavior, ed. P. Ducheyne and J. Lemons, J., (Annals of New York Acad. Sci., New York, 1988) Vol. 523.

Ceramics in Clinical Applications, ed. P. Vincenzini (Elsevier, New York, 1987) p 297.

J. D. Preston, "Properties in Dental Ceramics," in *Proceedings of the IV Int. Symp. on Dental Materials* (Quintessa Pub. Co., Chicago, 1988).

M. Jarcho, "Calcium Phosphate Ceramics as Hard Tissue Prosthetics," *Clin. Orthop. Relat. Res.*, **157** (1981) 259.

S. F. Hulbert, F. W. Cooke, J. J. Klawitter, R. B. Leonard, B. W. Sauer, D. D. Moyle, and H. B. Skinner, "Attachment of Prostheses to the Musculo-Skeletal System by Tissue Ingrowth and Mechanical Interlocking," *Biomed. Mater. Symp.* 4 (1973) 1-23.

S. F. Hulbert, J. R. Matthews, J. J. Klawitter, B. W. Sauer, and R. B. Leonard, "Effect of Stress on Tissue Ingrowth into Porous Aluminum Oxide. Biomed.," *Biomed. Mater. Symp.* 5 (1974) 85-97.

T. Nakamura, T. Yamamuro, S. Higashi, T. Kokubo, and S. Ito, "A New Glass-Ceramic for Bone Replacement: Evaluation of its Bonding to Bone Tissue," *J. Biomed. Mater. Res.*, **19** (1985) 685.

T. Kokubo, S. Ito, S. Sakka, and T. Yamamuro, "Formation of a High-Strength Bioactive Glass- Ceramic in the System $MgO-CaO-SiO_2-P_2O_5$," *J. Mater. Sci.* **21** (1986) 536.

L. L. Hench, *J. Amer. Ceram. Soc.* 1991, 74(7), 1487-510.

L. L. Hench, "Bioactive Ceramics," in *Bioceramics: Materials Characteristics Versus In-Vivo Behavior* eds. P. Ducheyne and J. Lemons (Annals of New York Acad. Sci., New York, 1988) Vol. 523, pp 54.

L. L. Hench, L. L. R. J. Splinter, W. C. Allen, and T. K. Greenlee, Jr., *J. Biomed. Maters. Res.*, **2(1)** (1972) 117-141.

U. Gross,   R. Kinne, H. J. Schmitz, and V. Strunz, V. *CRC Critical Reviews in Biocompatibility* **4** (1988) 2.

Handbook on Bioactive Ceramics, Vol I: Bioactive Glasses and Glass-Ceramics; Vol. II: Calcium Phosphate and Hydroxylapatite Ceramics, eds. T. Yamamuro, L. L. Hench, L. L. and J. Wilson (CRC Press, Boca Raton, FL, 1990).

G. Onoda and L. L. Hench, *Ceramic Processing Before Firing* (J. Wiley and Sons, Inc., New York, 1978).

C. G. Bergeron and S. H. Risbud, *Introduction to Phase Equilibria in Ceramics* (The American Ceramic Society, Columbus, OH, 1984).

J. Reed, *Introduction to Ceramic Processing* (J. Wiley and Sons, Inc., New York, 1988).

Bioceramics, Volume 4, ed. W. Bonfield, G. W. Hastings, and K. E. Tanner (Butterworth-Heinemann Ltd., Guildford, England, 1991) pp 155-162.

Bioceramics 5, ed. T. Yamamuro (Kobunshi Kankokai, Inc., Kyoto, Japan, 1992)

The Bone-Biomaterial Interface, ed. J. E. Davies (University of Toronto Press, Toronto, Ontario, Canada, 1991).

R. Li, A. E. Clark, and L. L. Hench, *J. Appl. Biomaterials* **2** (1991) 231-239.

C. J. Brinker and G. W. Scherer, *Sol-Gel Science* (Academic Press, San Diego, CA, 1990).

L. L. Hench and J. K. West, "The Sol-Gel Process," *Chem. Rev.* **90** (1990) 33-72.

Biomaterials and Clinical Applications, eds. A. Pizzoferrato, P. G. Marchetti, A. Ravaglioli, and A.J.C. Lee (Elsevier, Amsterdam, 1987).

R. Li, A. E. Clark and L. L. Hench, "Effects of Structure and Surface Area on Bioactive Powders by Sol-Gel Process," in *Chemical Processing of Advanced Materials*, eds. L. L. Hench and J. K. West (J. Wiley and Sons, Inc., New York, 1992) pp. 627.

R. K. Iler, *The Chemistry of Silica* (J. Wiley and Sons, Inc., New York, 1979).

W. D. Kingery, K. Bowen and D. R. Uhlmann *Introduction to Ceramics* (J. Wiley and Sons, Inc., New York, 1976).

P. Revell, *Pathology of Bone* (Springer-Verlag, Berlin, New York, 1986).

Janet Vaughn, *The Physiology of Bone* (Clarendon Press, Oxford, 1981).

Chapter 2

THE USE OF ALUMINA AND ZIRCONIA IN SURGICAL IMPLANTS

S. F. Hulbert

Rose-Hulman Institute of Technology, Terre Haute, Indiana

INTRODUCTION

The use of ceramics in medicine has increased significantly during the past decade[1-5] and it is anticipated that the use of bioceramics will increase dramatically during the next.

All materials elicit a response from living tissues. Four types of responses are possible. (1) The material is toxic and the surrounding tissue dies. This material, obviously, should not be used as a biomaterial. (2) The material is non-toxic and biologically inactive and a fibrous tissue capsule of varying thickness forms around the material, or in the case of a bone implant the optical microscope may show direct apposition of bone to the material. Such a material is classified as a Nearly Inert Biomaterial. (3) The material is non-toxic and biologically active and an interfacial bond forms between the material and tissue. Such a material is referred to as a Surface-Active Biomaterial. (4) The material is non-toxic and dissolves with the surrounding tissue replacing the dissolved material. Such a material is classified as a Resorbable Biomaterial.

The potential of ceramics as biomaterials relies upon their compatibility with the physiological environment. Bioceramics are compatible because they are composed of ions commonly found in the physiological environment (calcium, potassium, magnesium, sodium, etc.) and of ions showing limited toxicity to body tissue (zirconium and titanium). This chapter deals with the two nearly inert ceramics most used in surgical implants, alumina and zirconia.

Inert bioceramics undergo little or no chemical change during long-term exposure to the physiological environment. Even in those cases where these bioceramics may undergo some long-term chemical or mechanical degradation, the concentration of degradation product in adjacent tissue is easily controlled by the body's natural regulatory mechanisms. Tissue response to immobilized inert bioceramics involves the formation of a very thin, several micrometers or less, fibrous membrane surrounding the implant material. Inert bioceramics may be attached to the physiological system through mechanical interlocking, by tissue ingrowth into undulating surfaces. The nearly inert ceramic most used for surgical implants is alumina.

ALUMINA CERAMICS AS IMPLANT MATERIALS

High-density, high purity ($>99.5\%$) Al_2O_3 (alumina) is used in load-bearing hip prostheses and dental implants because of its combination of excellent corrosion resistance, good biocompatibility, high wear resistance and high strength. Although some dental implants are single-crystal sapphire, most Al_2O_3 devices are very fine-grained polycrystalline α-Al_2O_3 produced by pressing and sintering at temperatures ranging from 1600-1800°C depending upon the properties of the raw material.

A very small amount of MgO ($<0.5\%$) is used as a grain growth inhibitor and is essential in order to achieve a fully dense sintered body with a fine grain microstructure. It is very important that the amount of SiO_2 and alkali oxides be below 0.1%, because they impede densification and promote grain growth. It is also essential that the amount of CaO be below 0.1% since its presence leads to a lowering of the static fatigue resistance.[6]

Strength, fatigue resistance, and fracture toughness of polycrystalline α-Al_2O_3 are a function of grain size and percentage of sintering aid, i.e., purity. Al_2O_3 with an average grain size of <4 μm and $>99.7\%$ purity exhibits good flexural strength and excellent compressive strength. These and other physical properties are summarized in Table 1 for a commercially available implant material, along with the International Standards Organization (ISO) requirements and the proposed new standards for alumina implants. Extensive testing has shown that alumina implants which meet or exceed ISO standards have excellent resistance to dynamic and static fatigue, and resist subcritical crack growth and impact failure.[7]

Table 1. Physical Characteristics of Al_2O_3 Bioceramics.

	Commercially Available High Alumina Ceramic Implants	ISO Standard 6474	Proposed New ISO Standards
Alumina Content (%by weight)	>99.7	≥ 99.51	
$SiO_2 + Na_2O\%$	<0.02	<0.1	
Density (g/cm^3)	3.98	$\geq$	≥ 3.94
Average Grain Size (μm)	3.6	<7	<4.5
Hardness (Vickers, HV)	2400	>2000	
Bending Strength (MP) (after testing in Ringer's solution)	595	>400	>450

Other typical properties of commercially available alumina implant materials are listed in Table 2.

Table 2. Other Typical Properties of Commercially Available Alumina Implants.

Surface finish Ra (μm)	0.02
Compressive strength (Mpa)	4000-4500
Young's Modulus (Gpa)	380-420
Fracture Toughness K_{1c} (MN/m$^{3/2}$)	4-6
Implant strength (Nc n/cm^2)	40-50

An increase in average grain size to >7 μm can decrease mechanical properties by about 20%. High concentration of sintering aids must be avoided because they remain in the grain boundaries and degrade fatigue resistance.

Methods exist for lifetime predictions and statistical design of proof tests for load bearing ceramics. Applications of these techniques show that specific prosthesis load limits can be set for an Al_2O_3 device based upon the flexural strength of the material and its use environment.[8] Load bearing lifetimes of 30 years at 12000 N loads 9a,b or 200 Mpa stresses (9c) have been predicted.[9] Figure 1 is an applied stress-probability of time to failure (SPT) diagram for medical grade alumina, based upon ref. 9c by Real et al. It shows that for 30 years survival with failure of no more than 1 in 100 components the maximum tensile stress that can be applied is limited to <200 Mpa. If stresses of 250 Mpa are applied to the ceramic component, within 3 years 4% of the implants are likely to fail and by 30 years 7% will probably fail. Use of SPT diagrams such as this together with finite element analyses of local stress distributions make it possible to design ceramic components that have very low probabilities of failure during the lifetime of the patient.

Results from aging and fatigue studies show that it is essential that Al_2O_3 implants be produced with the highest possible standards of quality assurance, especially if they are to be used as orthopedic prostheses in younger patients.

USE OF ALUMINA IN TOTAL HIP PROTHESES

Alumina ceramics are being used in hip and knee prostheses because of inertness, excellent biocompatibility and high wear resistance. It is estimated that one-half million hip prostheses to date have been implanted with alumina ball for the femoral head component and that the number is growing by at least a 100,000 per year.[10] Figure 2 shows three femoral components of total hip prostheses with alumina balls.

The main problem with present total hip systems is loosening of the acetabular component which is caused by wear debris. Numerous clinical studies indicates that

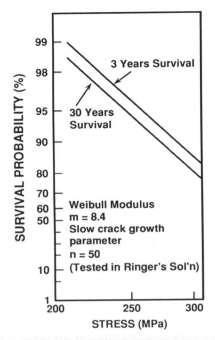

Fig. 1. Stress-probability of time to fracture diagram for medical grade alumina in Ringer's Solution. (Based upon M. W. Real et al., 1986, ref. 9c.)

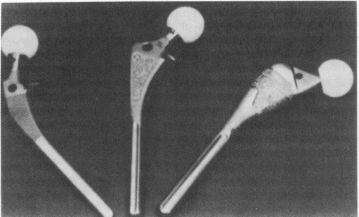

Fig. 2. Medical-grade alumina used as femoral balls in total hip replacement. Note three alternative types of metallic stems used for morphological fixation. (Photograph courtesy of J. Parr.) (Reprinted from "Bioceramics: From Concept to Clinic," L. L. Hench, *J. Am. Ceram. Soc.* **74[7]** (1991) 1487-1510, by permission of the American Ceramic Society.)

using femoral heads of alumina ceramic bearing against alumina cup sockets reduces wear debris by a factor of 10 or greater and by a factor of 2 or better when against ultrahigh molecular weight polyethylene (UHMWPE) cups.

The first clinical use of a total hip prosthesis with an alumina head and alumina socket was reported by Boutin[11] in 1971. In 1981 he reported on 1,330 cases. There were 4 socket and 6 ball fractures, 3 stem fractures and 7 incidents of severe wear.[12] Boutin and Blanquaert[13] reported wear rates ranging from 5-9 μm per year.

Dorlot, et al.,[14] in 1986 reported wear rates of less than 1 μm per year. Plitz, et al.,[15,16] reported significant wear rates with alumina on alumina total hip prostheses in 1982 and 1984.

In 1988, Dorlet, et al.,[17] reported on 20 retrieved ceramic-on-ceramic total hip prostheses. The average wear rate was 0.025 μm per year which is far less than that observed with UHMWPE bearing on metal or ceramic heads.

Winter, et al.,[18] reported on 10-14 year results using an alumina ball and alumina cup total hip placed in 100 patients between 1974 and 1979. Twenty-three patients could not be reached for follow-up. Twenty-five patients had to have revision systems. There were eight ball fractures. Eighty-percent of the 52 remaining prostheses were reported to have good clinical results.

Witvoet, et al.,[19] reported on 608 patients who received total hip prostheses consisting of a ceramic ball against a cemented alumina socket. Overall probability of survival was 88% after 8 years. The reason for failure was loosening at the cup-cement interface. Elastic modulus mismatch was assumed to be responsible for loosening. The survival rate of ceramic sockets was higher than with UHMWPE sockets in patients younger than 50 years of age.

In 1991, Sedel et al.,[20] reported on a ten-year clinical study of 187 cemented ceramic-to-ceramic total hip replacements. The major cause of failure was aseptic loosening of the acetabular component (15 failures). Fracture of the socket and/or the femoral head occurred in 5 patients. All of the mechanical failures occurred in components manufactured prior to 1979. Development of alumina ceramic which met or exceeded ISO standards has reduced the number of mechanical failures and in the case of the Sedel, et al., clinical completely eliminated mechanical failures. A ten year survival rate of 82.6% was reported. The outer diameter of the acetabular component was a major variable influencing the results. The outer diameter of the acetabular components must be at least 50 mm. Another major factor was the age of the patient. Sedel, et al.,[21] reported on alumina-to-alumina total hip prostheses involving 116 patients under 50 years of age, performed between April 1977 and August 1989. The survival analysis gives a 98.5% probability of retaining the prosthesis for more than 10 years.

In comparison, Harris, et al.,[22] report that 42% of all hip prostheses consisting of UHMWPE and metal-bearing surfaces require revision of the acetabular component within 10 years.

A number of clinical studies have compared wear rates of socket components of UHMWPE against alumina balls versus metal balls. There is considerable variation of data, but in each case the wear rate for systems with metal balls is much higher than with alumina balls.

Oonishi, et al.,[23] reported a study involving 956 total hip prostheses of alumina balls and UHMWPE cups performed between 1977 and 1988 and 117 prostheses involving metal heads and UHMWPE sockets performed between 1975 and 1981. The amount of wear was measured by x-ray techniques. The wear rate for alumina-UHMWPE combinations was 0.098 mm per year and 0.245 mm per year for metal-UHMWPE combinations.

Ohashi, et al.,[24] reported on 318 total hip prostheses done over a period of 13 years; 131 cases involved prostheses with metallic heads and 187 were prosthetics with alumina heads. UHMWPE acetabular components were used in all cases. The amount of wear was measured using x-ray techniques. The wear rate was 0.025 mm per year with alumina heads and 0.043 mm per year for metal heads.

Okumura[25] reported a study conducted between 1981 and 1988, involving 105 total hip prostheses, 73 hips using an alumina ball, the rest using metal balls articulated against UHMWPE sockets. Socket wear was measured using x-ray techniques. The observed socket wear with alumina balls was 0.08 mm per year while it was 0.14 mm per year with metal balls.

In 1990, Oonishi[26] reported a clinical study of total hip prostheses that showed alumina to UHMPE bearing surface combinations resulted in three times less wear than that observed with metal to UHMPE combinations. He also reported that laboratory knee simulation studies showed less than one tenth of the wear of alumina to UHMPE combinations.

In 1991 Asada, et al.,[27] compared the performance of a bipolar hip prothesis with a Co-Cr-Mo alloy device in beagle dogs. They concluded that an alumina ceramic was the better articulating material. One of the most important properties of articulating components is the wear and friction behavior over an extended period of time. Comparative tests in hip joint simulators have demonstrated tribologic superiority of high-purity alumina ceramic systems of metal-to-metal and metal-to-UHMWPE combinations.

Dörre[10] reports that sixteen years of clinical experiences have shown the annual average wear rates, shown in Table 3.

Table 3. Average Annual Wear Rates of Articulating Surfaces in Total Hip Prostheses.

Materials	Wear Rate (μm/yr)
Co-Cr-Mo Alloy/UHMWPE	200
Alumina/UHMWPE	20-130
Alumina/Alumina	2

In the case of alumina UHMWPE articulating surfaces, the 20 μm refer to observations of explanted components and include only wear measurements. The 130 μm refers to penetration measurements between head and acetabular components observed by means of radiographic analysis, it includes not only wear but also plastic flow.

The long term coefficient of friction of an alumina-alumina articulating surface decreases with time and approaches the value of a normal healthy joint, as illustrated in Fig. 3.

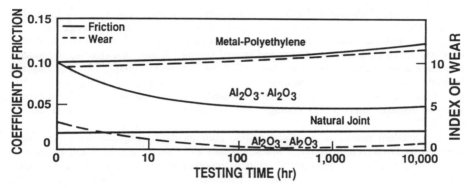

Fig. 3. Time dependence of (-) coefficient of friction and (---) index of wear of alumina-alumina versus metal-PE hip joint (*in vitro* testing). (Reprinted from "Bioceramics: From Concept to Clinic," L. L. Hench, *J. Am. Ceram. Soc.* **74**[7] (1991) 1487-1510, by permission of the American Ceramic Society.)

The outstanding frictional and wear properties of alumina ceramics are due to the materials' extremely low surface roughness and to their high surface energy which results in the fast and strong adsorption of biological molecules. These layers of adsorbed molecules provide a liquid-like covering which limits the direct contact of the articulating solid surfaces. The high surface energy of alumina is demonstrated through contact angle measurements. With the single phase fine grained structure of alumina the surface roughness is lower than for carbide-containing bi-phasic Co-Cr-Mo-alloys. Any roughness on alumina is inverse (due to pull out of grains) instead of the protruding asperities encountered with the hard metal carbides (M_7C_3). Having a concave surface roughness as apposed to protrusions is a major contributing factor to the outstanding wear properties of alumina ceramics.

A very important prerequisite for the superior tribologic results of alumina, particularly in the case of alumina/alumina articulation, is an extremely smooth polished surface with an average roughness of 0.01 μm and an extreme congruence of the sliding faces with a roundness deviation between 0.1 and 1 μm. The alumina ball and socket should be polished and used as a matched pair.

OTHER APPLICATIONS OF ALUMINA AS AN IMPLANT MATERIAL

Oonishi, et al.,[28] developed and tested clinically total knee prostheses consisting of an Al_2O_3 femoral component with a tibial component of UHMWPE.

Inoue, et al.,[29] reported a clinical study involving 52 knee prostheses consisting of an Al_2O_3 femoral component bearing on a UHMWPE tibia component. During the period of study, 1982-88, there were no revision cases. Volume 4 of *Orthopaedic Ceramic Implants*, published by the Japanese Society of Orthopaedic Ceramic Implants, contains 12 papers on total knee prostheses where one or both of the articulating surfaces are made of Al_2O_3.

Ceramic ankle joints,[30] elbows,[31] shoulders,[32] wrists,[33] and fingers[34] have also been tested clinically with success equal to or better than other material systems.

In 1964, Sandhaus [35] applied for a Swiss patent for an alumina ceramic dental implant. In 1975, after engineering and animal testing, Schulte and Heimke introduced a dense alumina ceramic dental implant called the Tübingen implant.[36,37] The implant is used both as an immediate implant after extraction or in edentulous regions. A 10-year follow-up of 610 Tübingen implants showed a success rate of 84.5%.[38] A recent study of approximately 1,300 Tübingen implants indicates a long-term success rate of 92.5%.[39]

Excellent results with polycrystalline alumina dental implants have also been reported by Driskell and Heller,[40] Mutschelknaus and Dörre[41] and Ehrl and Frenkel.[42]

Single-crystal alumina dental implants reported to have a bending strength three times that of polycrystalline alumina.[43] Yamagami[44] reported a study involving 402 patients with 617 single-crystal alumina dental implants. Using the NIH-Harvard Consensus Criteria, the cumulative survival rate was 97.3% for five years and 96.2% for ten years.

Alumina ceramics have been used in ENT and maxillofacial surgery. Plester and Jahnke[45-46] have developed and tested total and partial ossicular bone replacements made of alumina. Twelve years of clinical experience have shown a better success rate than polymeric components.[47]

In repairing the trachea due to pathological conditions or trauma, the cartilage which prevents the collapse of the trachea may need to be removed. This necessitates the insertion of some type of support structure to keep the trachea open. Zollner[48] et al., have designed and tested tracheal support rings made of alumina. The superior biocompatibility and strength of alumina account for better results over previously used polymeric rings.

Alumina implants have been used for various neurosurgical operations such as cranioplasty to repair bone defects in both the convexity and sub-occipital region and for reconstruction of the sellar floor and orbital wall.[49]

An eye prostheses consisting of a sapphire single crystal optical part and an alumina ceramic holding ring is used routinely in appropriate cases.[50]

Hormones, vaccines, and drugs can be administered either orally or by means of injections. These routes rarely deliver the chemicals at a steady dose-level. Administration of pharmaceuticals via implants in a controlled manner can deliver material at the desired level over long time periods and make it available in amounts which are effective yet not toxic. Buykx et al.,[51] have developed small porous collars of alumina to serve as reservoirs for an anti-inflammatory drug dexamethasone sodium phosphate, (DSP) in the immediate vicinity of the electrode of a cardiac pacemaker lead. Prototype collars of Al_2O_3 were fabricated containing 30% pores 7 μm in diameter and 30% pores 50 μm in diameter. The pores were filled with approximately 170 μg of DSP by impregnation, assembled into pacemaker leads with a 4 mm^2 Pt/Ir coated electrodes and tested in sheep, compared with leads having drug-eluting silicone collars, non drug-eluting ceramic collars, and conventional non drug-eluting "Laserdish" leads. The drug-eluting leads (both ceramic and silicone) by controlling inflammation at source showed a significant and lasting (52 weeks) reduction in the stimulation threshold voltage, and therefore may be valuable with a variety of pacemakers.

USE OF ZIRCONIA CERAMICS IN SURGICAL IMPLANTS

Alumina has outstanding biocompatibility and wear resistance, however, it exhibits moderate flexural strength and toughness. For this reason, the diameter of most alumina femoral head prostheses has been limited to 32 mm. Zirconia is also exceptionally inert in the physiological environment[52,53] and zirconia ceramics have an advantage over alumina ceramics of higher fracture toughness and higher flexural strength and lower Young's modulus.[54,55]

Zirconia ceramics suggested for surgical implants fall into two basic types: Tetragonal zirconia stabilized with yttria (TZP) and magnesium oxide partially stabilized zirconia (MG-PCZ). Properties of zirconia are compared with alumina ceramic implant materials in Table 4.

Table 4. Properties of Alumina Zirconia Ceramics Used in Surgical Implants.

Property	Unit	Al_2O_3	TZP	Mg-PSZ
Purity	%	>99.7	97	96.5
Y_2O_3/MgO	%	<.3	3 mol	3.4 wt
Density	g/cm^3	3.98	6.05	5.72
Grain Size (average)	μm	3.6	0.2-0.4	0.42
Bending Strength	Mpa	595	1000	800
Compressive Strength	Mpa	4250	2000	1850
Young's Modulus	GPA	400	150	208
Hardness	HV	2400	1200	1120
Fracture Toughness K_{lc}	MN/m$^{3/2}$	5	7	8

Zirconia may be suitable for bearing surfaces in total hip prostheses. However, there are three major controversies regarding zirconia. One is the reported strength reduction with time in physiological fluids. The second is its wear properties and third is the potential radioactivity of the material.

The deleterious martensitic transformation from tetragonal to monoclinic phase in yttria-doped zirconia due to aging in water and the accompanying reduction in toughness is well documented.[56] However, tests in simulated body fluids and in animals have shown only slight decreases in fracture strength and toughness.[57-59] The observed strength after two years is still much higher than the strength of alumina tested under similar conditions.[59]

Streicher, et al.,[54,55] recently reported on a detailed study of ceramic surfaces as wear partners for UHMWPE. Investigation was carried out on five grades of alumina and zirconia in a pin-on-disc test against UHMWPE for suitability as articulating components for total joint prostheses. The tests showed a difference in surface quality between the various grades of ceramics. The UHMWPE wear rate caused by the ceramic counterfaces was the lowest for alumina and 20% less than in combination with Co-Cr-Mo-alloy. Zirconia ceramics yielded unfavorable wear and friction results.

The reason for the inferior tribological behavior of zirconia ceramics against the non-polar UHMWPE is not understood. Alumina and zirconia ceramics have similar roughness and wetting characteristics. The only major difference in their surface properties is that alumina ceramics are much harder.

Sudanese, et al.,[60] evaluated zirconia wear resistance in a ceramic-ceramic coupling (ring on disc) test. The wear rate of zirconia-zirconia couplings was 5000 times that of alumina-alumina couplings. Zirconia ceramics should not be used for ceramic-ceramic articulating surfaces.

Zirconia is often accompanied by radioactive elements with a very long half-life, such as thorium and uranium. These elements are difficult and expensive to separate from zirconia. Sato,[61] recently observed 0.5 ppm of uranium 235 in a zirconia-based bioceramic.

There are two types of radiation of concern in zirconia ceramics: gamma and alpha. The gamma radioactivity of alumina, zirconia and Co-Cr-alloy femoral head prostheses has been measured.[62] Alumina was found to have the lowest gamma radioactivity and zirconia and the Co-Cr-alloy to be approximately the same. The gamma radioactivity for the Co-Cr-alloy and zirconia, were found to be of the same order of magnitude as the national ambient radioactivity in France.[62]

The data suggest that the level of gamma radiation in commercially available zirconia bioceramics is not a major concern. However, significant amounts of alpha-radiation have been observed with zirconia ceramics intended for surgical implants.[63] Alpha particles, because of their high ionization capacity, destroy soft and hard tissue cells. The alpha emission observed from zirconia ceramic femoral head prostheses is a concern. Although the activity is small, questions concerning the long-term effects of alpha radiation emission from zirconia ceramics must be answered.

SUMMARY

Alumina and zirconia ceramics are both exceptionally biocompatible due to their chemical stability in the physiological environment. However, the question of the long-term effects of alpha radiation from zirconia ceramics must be answered. Zirconia ceramics have a higher fracture strength, toughness and lower modulus of elasticity than alumina ceramics. The phenomena of slow crack growth, static and cyclic fracture, low toughness, stress corrosion, deterioration of toughness with time and sensitivity to tensile stresses are all serious concerns for both ceramics in high load bearing applications. Both alumina and zirconia ceramics undergo slight reduction in fracture strength with time in the physiological environment. [*]Alumina and zirconia bioceramics should be restricted to designs involving compressive loading or limited tensile loads.

The elastic modulus of nearly inert bioceramics is also a limitation on their use in the body. The Young's modulus (in Gpa) of cancellous bone has a range from 0.05 to 0.5 depending on location and age; cortical bone ranges from 7-25. In contrast, medical grade alumina ($>99.7\%$ Al_2O_3) has a Young's modulus of 380 to 420 Gpa and partially stabilized zirconia has a value of 150 to 208 Gpa. Thus, there is a modulus mismatch between cortical bone and an alumina implant in the range of 15-55X. The mismatch with cancellous bone is enormous, 760X to 7600X.[64]

A consequence of a mismatch in elastic modulus is that a bioceramic implant will shield a bone from mechanical loading, allowing nearly all the mechanical load to be carried by the implant. Living bone must be under a certain amount of load in order to remain healthy, if it is unloaded or is loaded in compression it will undergo biological changes which lead to resorption and weakening of the bone and deterioration of the implant-bone interface.[64] The high modulus of elasticity of alumina and zirconia limit their effectiveness as bone interface materials. Metal alloys suffer the same limitation. The answer to the modulus mismatch is a composite material. The high modulus of elasticity of alumina does not limit its ability to serve as an articulating surface.

Alumina is an excellent material for certain orthopedic applications, such as the ball in an artificial hip joint, because of its excellent biocompatability, low friction, high wear resistance, and high compressive strength. The tribological properties of alumina ceramics are far superior to zirconia ceramics. Alumina or alumina articulating surfaces in total joint replacements have the best tribological properties and, even more important, the best clinical results.

Table 5. Ranking of Articulating Surfaces for Total Joint Systems.

Superior	Al_2O_3 on Al_2O_3
Excellent	Al_2O_3 on UHMWPE
Good	Co-Cr-Mo on UHMWPE Ti-6Al-4V on UHMWPE
Poor	Metal on Metal

Other clinical applications of alumina include: knee prostheses, ankle prostheses, elbow prostheses, shoulder prostheses, wrist prostheses, finger prostheses, alveolar ridge and maxillofacial reconstruction, ossicular bone substitutes, keratoprostheses (corneal replacements), segmental bone replacements, controlled drug release systems, and dental implants.

As recently as twenty years ago, ceramics were widely ignored as potential biomaterials. Interest in bioceramics has increased dramatically over the past decade. Because of its outstanding biocompatibility and excellent tribological properties, alumina ceramics are widely used in total joint prothesis. It is anticipated that the use of alumina ceramics in otologic, maxillofacial and dental applications will continue to grow.

REFERENCE

1. S.F. Hulbert, J.C. Bokros, L.L. Hench, J. Wilson, and G. Heimke, "Ceramics in Clinical Applications, Past, Present and Future," in *High Tech Ceramics*, ed. P. Vincenzini (Proceedings of the World Congress on High Tech Ceramics, 6th CIMTEC, Milan, Italy, 1986) pp. 189-213.

2. L.L. Hench, and J. Wilson *Science* **226** (1984) 630-636.

3. S.F. Hulbert, L.L. Hench, D. Forbes, and L.S. Bowman, *Ceramics in Surgery* (1983) 3-25.

4. J.W. Boretos, *Ceramic Bulletin* **64** (1985) 1098-1100.

5. L.M. Sheppard, *Advanced Materials & Processes* **2(5)** (1986) 26-31.

6. G. Willmann, Die Bedeutung der ISO-Norm 6474 für Implantate aus Aluminiumoxid. *Zähnarztliche Praxis* **41** (1990) 286-290.

7. E. Dörre, and W. Dawihl, "Ceramic Hip Endoprostheses" in *Mechanical Properties of Biomaterials*, eds. G.W. Hastings, and D.F. Williams (J. Wiley & Sons, New York, 1980) pp. 113-127.

8. J.E. Ritter, Jr., D.C. Greenspan, R.A. Palmer, and L.L. Hench, "Use of Fracture Mechanics Theory in Lifetime Predictions for Alumina and Bioglass-coated Alumina", *J. Biomed. Mater. Res.* **13** (1979) 251-263.

9a. P. Christel, A. Meunier, J.M. Dorlot, J.M. Crolet, Witvolet, Jr., L. Sedel, and P. Bouritin, "Biomechanical Compatibility and Design of Ceramic Implants for Orthopedic Surgery" in *Bioceramics: Material Characteristics Versus In-Vivo Behavior*, Vol. 523, eds. P. Ducheyne, and J. Lemons (Annals of New York Academic of Science, 1988) p. 234.

9b. E. Dörre, W. Dawihl, V. Krohn, G. Altmeyer, and M. Semlitsch, "Do Ceramic Components of Hip Joints Maintain their Strength in Human Bodies?" in *Ceramics in Surgery*, ed. P. Vinenzini (Elsevier, Amsterdam 1983) p. 61.

9c. M.W. Real, D.R. Cooper, R. Morrell, R. Rawlings, B. Weightman, and R.W. Davidge, "Mechanical Assessment of Biograde Alumina," *Journal de Physique, Colloque C1*, **47** (1986) C11763-C11767.

10. E. Dörre, "Problems Concerning the Industrial Production of Alumina Ceramic Components for Hip Joint Prosthesis", *Bioceramics and the Human Body*, eds.

A. Ravaglioli, and A. Krajewski, Elsevier Applied Science (London and New York, 1991) pp 454-460.

11.   P. Boutin, "Arthroplastie totale de la hanche par prothese en alumine fritte", *Rev. Chir. Orthop.* **58** (1972) 229.

12.   P. Boutin, "T.H.R. Using Alumina-Alumina Sliding and a Metallic Stem: 1330 Cases and an 11-Year Follow-up," *Orthopaedic Ceramic Implants*, Vol. 1, eds. H. Oonishi, and Y. Ooi (Proceedings of Japanese Society of Orthopaedic Ceramic Implants, 1981).

13.   P. Boutin, D. Blanquaert, *Rev. Chir. Ortho.* **67** (1981) 279-287.

14.   J.M. Dorlot, P. Christel, L. Sedel, J. Witwoet, and P. Boutin, *Biological and Biomechanical Performances of Biomaterials*, eds. P. Christel, A. Meunier, and A.J.C. Lee, (Elsevier, Amsterdam, 1986) pp. 495-500.

15.   W. Plitz, H.U. Hoss, in *Biomaterials 1980*, eds. G.D. Winter, D.F. Gibbons, and H. Plenk (John Wiley, New York, 1982) pp. 187-196.

16.   A. Walter, W. Plitz, in *Biomaterials and Biomechanics 1983*, eds. P. Ducheyne, G. Van de Perre, and A.E. Aubert, (Elsevier, Amsterdam, 1984) pp. 55-60.

17.   J.M. Dorlot, P. Christel, A. Meunier, "Alumina Hip Prostheses: Long Term Behaviors," in *Proceedings of 1st International Symposium on Ceramics in Medicine*, eds. H. Oonishi, H. Aoki, and K. Sawai (Ishiyaku EuroAmerica, Inc., 1988) pp. 236-301.

18.   M. Winter, P. Griss, G. Scheller, and T. Moser, "10-14 Year Results of a Ceramic-Metal-Composite HIP Prostheses with a Cementless Socket," in *Proceedings of 2nd International Symposium on Ceramics in Medicine*, ed. G. Heimke (German Ceramic Society, 1989) pp. 436-444.

19.   J. Witvoet, P. Christel, L. Sedel, S. Herman, and D. Blanquaert, "Survivorship Analyses of Cemented Al_2O_3 Sockets," *Proceedings of 1st International Symposium on Ceramics in Medicine*, eds. H. Oonishi, H. Aoki, and K. Sawai (Ishiyaku EuroAmerica, Inc., 1988) pp. 314-319.

20.   L. Sedel, A. Meunier, R.S. Nizard, and J. Witvoet, "Ten Year Survivorship of Cemented Ceramic-Ceramic Total Hip Replacement," in *Bioceramics*, Vol. 4, eds. W. Bonfield, G.W. Hastings, and K.E. Tanner (Butterworth-Heinemann Ltd., London, UK, 1991) pp. 27-37.

21.   L. Sedel, P. Christel, A. Meunier, and P. Boutin, "Alumina/Bone Interface-- Experimental and Clinical Data," in *Proceedings of 1st International Symposium on Ceramics in Medicine*, eds. H. Oonishi, H. Aoki, and K. Sawai (Ishiyaku EuroAmerica, Inc., 1988) pp. 262-271.

22.   R.D. Mulroy, Jr., and W.H. Harris, "Improved Cementing Techniques: Effect on Femoral and Socket Fixation at 11 Years," *Transactions of the Society for Biomaterials*, **13** (1990) 147.

23.   H. Oonishi, H. Igaki, and Y. Takayama, "Comparison of Wear of UHMW Polyethylene Sliding Against Metal and Alumina in Total Hip Prostheses," in *Proceedings of 1st International Symposium on Ceramics in Medicine*, eds. H. Oonishi, H. Aoki, and K. Sawai (Ishiyaku EuroAmer., Inc., 1988) pp. 272-277.

24. T. Ohashi, S. Inoue, K. Kajikawa, K. Ibaragi, T. Tada, M. Oguchi, T. Arai, and
     K. Kondo, "The Clinical Wear Rate of Acetabular Component Accompanied with
     Alumina Ceramic Head," in *Proceedings of 1st International Symposium on
     Ceramics in Medicine*, H. Oonishi, H. Aoki, and K. Sawai (Ishiyaku
     EuroAmerica, Inc, 1988) pp. 278-283.

25. Okumura, T. Yamamuro, T. Kumar, T. Nakamura, and M. Oka, "Socket Wear
     in Total Hip Prosthesis with Alumina Ceramic Head," in *Proceedings of 1st
     International Symposium on Ceramics in Medicine*, H. Oonishi, H. Aoki, and K.
     Sawai (Ishiyaku EuroAmerica, Inc., 1988) pp. 284-289.

26. H. Oonishi, "Bioceramics in Orthopaedic Surgery--Our Clinical Experiences,"
     Abstracts 3rd International Symposium on Ceramics in Medicine, (1990).

27. K. Asada, Y. Yutani, H. Sakamoto, K. Yoshida, H. Sakane, H. Nakamua, and
     A. Shimazu, "Histological Study of Acetabular Clear Zone after Acetabular
     Reaming for Hip Arthroplasty using an Alumina Ceramic Endoprosthesis and a
     Metal One," in *Bioceramics*, Vol. 4, eds. W. Bonfield, G.W. Hastings, and K.E.
     Tanner, (Butterworth-Heinemann Ltd., 1991) pp. 47-54.

28. H. Oonishi, N. Okabe, T. Hamagchi, and T. Nabeshima, "Cementless Alumina
     Ceramic Total Knee Prosthesis," *Orthop. Ceramic Implants* 1 (1982) 157-160.

29. H. Inoue, Y. Yokoyama, G. Tanabe, "Follow-up Study of Alumina Ceramic
     Knee (KC-1 Type) Replacement," in *Proceedings of 1st International Symposium
     on Ceramics in Medicine*, eds. H. Oonishi, H. Aoki, and K. Sawai (Ishiyaku
     EuroAmerica, Inc., 1988) pp. 301-307.

30. Y. Ukon, H. Oonishi, N. Murata, T. Nabeshima, S. Kushitani, K. Tsuyama, S.
     Ootsuki, T. Yamamoto, Y. Wakimoto, and T. Hayashibara, "Clinical Experience
     of KOM Type Cementless Alumina Ceramic Artificial Ankle Joint," *Orthopaedic
     Ceramic Implants* 6 (1986) 125-129.

31. A. Murasawa, T. Hanyu, K. Nakazono, A. Hirano, T. Yamagishi, and M.
     Muraoki, "Indication and its Limit of Ceramic Prosthesis for Rheumatoid
     Elbows," *Orthopaedic Ceramic Implants* 5 (1985) 43-49.

32. H. Yoshikawa, H. Tokumaru, Y. Aoki, H. Nakashima, T. Kato, S. Saito, and
     H. Hamada, "Total Shoulder Replacement with Ceramic Prosthesis Following the
     Tikhoff-Linberg Procedure," *Orthopaedic Ceramic Implants* 1 (1983) 113-116.

33. M. Moritani, H. Hirako, H. Sakata, and R. Takenaka, "Ceramics-to-HDP Total
     Wrist Joint Prosthesis," *Orthopaedic Ceramic Implants* 2 (1982) 147-150.

34. K. Soi, S. Hattori, S. Kawai, N. Kuwata, T. Taguchi, and F. Kawakami,
     "Experimental and Clinical Studies of Alumina-Ceramic Finger Implants,"
     Orthopaedic, 2.

35. S. Sandhaus, "Bone Implants and Drills and Taps for Bone Surgery," British
     Patent 1083769, (1967).

36. W. Schulte, C.M. Busing, B. d'Hoedt, and G. Heimke, "Enosseous Implants of
     Aluminum Oxide Ceramics - a 5-Year Study in Human," in *Proceedings of
     International Congress on Implantology and Biomaterials in Stomatology*, ed.
     Kawahara, (Kyoto, Japan, 1980) pp. 157-167.

37.   W. Schulte, "The Intra-osseous Al_2O_3 (Frialit) Tübingen Implant, Development Status After Eight Years (1-3)", *Quintessence International* **15** (1984) 1-39.

38.   G. Heimke, B. d'Hoedt, and W. Schulte, "Ceramics in Dental Implantology," in *Biomaterials and Clinical Applications*, eds. A. Pizzoferrato, P.G. Marchetti, A. Ravaglioli and A.J.C. Lee (Elsevier Science Publishers, Amsterdam, 1987) pp. 93-104.

39.   W. Schulte, "The FRIALIT Tübingen Implant System," in *Osseo-Integrated Implants, Vol II, Implants in Oral and ENT Surgery*, ed. G. Heimke, (CRC Press, Boca Raton, Florida, 1990) pp. 12-32.

40.   T.D. Driskell, and A. L. Heller, "Clinical Use of Aluminum Oxide Endosseous Implants," *Oral Implantol.* **7** (1977) 1.

41.   E. Mutschelknaus, and E. Dörre, "Extension-Simplantate aus Aluminimumoxidkeramik," *Quintessenz* **28** (1977) 1-10.

42.   P.A. Ehrl, and G. Frenkel, "Experimental and Clinical Experiences with a Blade Vent-Abutment of Al_2O_3-Ceramic in the Shortended Dental Row-Situation of the Mandible," in *Dental Implants*, ed. G. Heimke (Hanser Verlag, Munchen, 1980) pp. 63-67.

43.   H. Kawahara, M. Hirabayashi, and T. Shikita, "Single Crystal Alumina for Dental Implants and Bone Screws," *J. Biomed. Mat. Res.* **14** (1980) 597-605.

44.   A. Yamagami, "Single-Crystal Alumina Dental Implant-13-Year Long Following-ups and Statistical Examination," in *Proceedings of 1st International Symposium on Ceramics in Medicine* eds. H. Oonishi, H. Aoki, and K. Sawai (Ishiyaku EuroAmerica, Inc., 1988) pp. 332-337.

45.   D. Plester and K. Jahnke, "Ceramic Implants in Otologic Surgery," *Am. J. Otology* **3** (1981) 104-108.

46.   K. Jahnke, "Otologic Surgery with Aluminum Oxide Ceramic Implants," in *Biomaterials in Otology*, ed. J. J. Grote (Martinus Nijhoff, The Hauge, 1984) pp. 205-209.

47.   K. Jahnke, M. Schrader, and J. Silberzahn, "Osseio-integrated Implants in Otorhinolaryngology," in *Osseo-integrated Implants, Vol II, Implants in Oral and ENT Surgery*, ed. G. Heimke (CRC Press, Boca Raton, Florida, 1990) p. 287.

48.   C. Zollner, H. Weerda, J. Strutz, "Aluminiumoxid-Keramik als Stutzgerust in der Trachealchirurgie," *Archives of Oto-Rhino-Laryngology*, Supplement II (Springer Verlag, 1983) pp. 214-216.

49.   H. Okudera, S. Kobayashi, and T. Takemae, "Neurosurgical Applications of Alumina Ceramic Implants," in *Bioceramics, Proceedings of the 1st Internat'l. Symp. on Ceramics in Medicine*, eds. H. Oonishi, H. Aoki, and K. Sawai (Ishiyaku EuroAmerica, Inc., 1988) pp. 320-325.

50.   F.M. Polack, and G. Heimke, "Ceramic Keratoprostheses, *Ophthalmology* **87** (1980) 693.

51.   W.J. Buykx, E. Drabaretk, K.D. Reeves, N. Anderson, R. Mathivanar, and M. Skalsky, "Development of Porous Ceramics for Drug Release and Other

Applications," *Abstracts, 3rd International Symposium on Ceramics in Medicine* (Terre Haute, IN, November 1990).

52.   R.C. Garvie, D. Urban, D.R. Kennedy, and J.C. McMeuer, "Biocompatibility of Magnesium Partially Stabilized Zirconia (Mg-PSZ Ceramics)," *J. Mat. Sci.* **19** (1984) 3224.

53.   T. Tateishi, and H. Yunoki, "Research and Development of Advanced Biomaterials and Application to the Artificial Hip Joint," *Bull. Mech. Engr. Lab.* **45** (1987) 1.

54.   R.M. Streicher, M. Semlitsch, and R. Schom, "Articulation of Ceramic Surfaces Against Polyethylene," in *Bioceramics and the Human Body*, eds. A. Ravaglioli and A. Krajewski (Elsevier Applied Science, London and New York, 1991) pp. 118-123.

55.   R.M. Streicher, Semlitsch, and R. Schon, "Ceramic Surfaces as Wear Partners for Polyethylene," *Bioceramics*, **Volume 4**, eds. W. Bonfield, G.W. Hastings, and K.E. Tanner (Butterworth-Heinemann Ltd., London, UK, 1991).

56.   T. Sato, and M. Shimada, "Transformation of Yttria-Doped Zirconia Polycrystals by Annealing in Water," *J. Am. Ceram. Soc.* **68** (1985) 356-359.

57.   P. Christel, A. Meunier, M. Heller, J.P. Torre, and C. Peille, N., "Mechanical Properties and Short-Term *In Vivo* Evaluation of Yttrium-Oxide-Partially-Stabilized Zirconia," *J. Biomed. Mat. Res.*, **23** (1989) 45-61.

58.   A. Mandrino, B. Moyer, A. Ben Abdallah, D. Treheurex and D. Orange, "Aluminas with Dispersoids. Tribologic Properties and *In Vivo* Aging," *Biomaterials* **11** (1990) 88-91.

59.   J.L. Drummond, "Aging Study of Magnesia Stabilized Zirconia," *Bioceramics*, Vol. 4, eds. J.E. Hulbert and S.F. Hulbert (Rose-Hulman Institute of Technology, Terre Haute, IN, 1992.

60.   A. Sudanese, A. Toni, G.L. Cattaneo, D. Ciaroni, T. Greggi, D. Dallari, G. Galli, and G. Guiunti, "Alumina vs. Zirconium Oxide: A Comparative Wear Test," *Proceedings of 1st International Symposium on Ceramics in Medicine*, eds. H. Oonishi, H. Aoki, and K. Sawai (Ishiyaku EuroAmer., Inc., 1988) pp. 237-240.

61.   Y. Sato, Aichi Medical University, Osaka, Private Communication, 3rd International Symposium on Ceramics in Medicine, (Terre Haute, Indiana, USA, November 1990).

62.   B. Cales, and C.N. Peille, "Radioactive Properties of Ceramic Hip Joint Heads," *Proceedings of 1st International Symposium on Ceramics in Medicine*, eds. H. Oonishi, H. Aoki, and K. Sawai, Ishiyaku EuroAmerica, Inc., (1988) 152-155.

63.   S. Cieur, R. Heindl, A. Robert, "Radioactivity of a Femoral Head of Zirconia Ceramics," *Bioceramics*, Vol. 3, eds. J. E. and S. F. Hulbert (Rose-Hulman Institute of Technology, Terre Haute, IN, 1991) pp. 367-371.

64.   L.L. Hench, "Bioceramics: From Concept to Clinic," *J. Am. Ceram. Soc.* **74(7)** (1991) 1487-1510.

Chapter 3

BIOACTIVE GLASSES

Larry L. Hench and *Örjan Andersson
University of Florida, Gainesville, Florida
*Abo Akademi, Turku, Finland

INTRODUCTION

It was discovered by Hench and colleagues in 1969 that bone can bond chemically to certain glass compositions. This group of glasses has become known as bioactive glasses based upon the following definition: *"A bioactive material is one that elicits a specific biological response at the interface of the material which results in the formation of a bond between the tissues and the material."* Bioactive glasses have numerous applications in the repair and reconstruction of diseased and damaged tissue, especially hard tissue (bone). Clinical applications are discussed in Chapter 4. One aspect that makes bioactive glasses different from other bioactive ceramics and glass-ceramics is the possibility of controlling a range of chemical properties and rate of bonding to tissues. The most reactive glass compositions develop a stable, bonded interface with soft tissues, as shown by Wilson and colleagues. It is possible to design glasses with properties specific to a particular clinical application. This is also possible with some glass-ceramics, but their heterogeneous microstructure restricts their versatility.

PROCESSING

Bioactive glasses are produced by conventional glass manufacturing methods (Chapter 1). Contamination of the glass must be avoided in order to retain the chemical reactivity of the material. Purity of raw materials must be assured. Analytical grade compounds are typically used for most components. Silica can be added in the form of high purity (flint quality) glass sand, since chemically prepared silicas are difficult to handle without adsorption of water and agglomeration. Choice of raw materials can affect properties of the glass. Andersson has shown that use of calcium phosphate compounds that contain crystal water result in glasses that crystallize more easily than if crystal water-free compounds are used. This effect is due to the dissolution of OH ions in the glass structure and the associated decrease of viscosity. See Chapter 13 for a discussion of the effects of viscosity on glass formation and crystallization. Preferential vaporization of fluxes will also affect glass viscosity and tendency to crystallize or phase separate as well as alter the final glass composition.

Weighing, mixing, melting, homogenizing and forming of the glass must be done without introducing impurities or losing volatile constituents, such as Na_2O or P_2O_5. Melting is usually done in the range of 1300-1450°C, depending on composition. The phase equilibrium diagram for the Na_2O-CaO-SiO_2 system shows a ternary eutectic very near the 45S5 glass composition (Table 1), which was the original basis for selecting this composition for investigation. There is a very steep liquidus as the composition increases in SiO_2 content which greatly affects the melting and homogenization behavior of the glass. Only platinum or platinum alloy crucibles or glass melter should be used to avoid contamination of the melt.

Bulk specimens can be formed by casting or injection molding in graphite or steel molds. Annealing is crucial, 450-550°C, because of the high coefficient of thermal expansion of the bioactive glass compositions. Each type of device must have its own annealing schedule established. Bioactive glasses are soft glasses and final shapes can be easily made by machining. Standard machine tools or dental handpieces can be used. Diamond cutting tools are preferred with copious irrigation, although dry grinding is also possible. If a granulated or powdered material is required, the melt can be rapidly quenched in water or air before grinding and sieving into the desired particle sizes. The glass frit (see Chapter 18) should be rapidly dried to avoid corrosion while in contact with water. Other processing methods used for bioactive glass coatings are described in Chapter 13. Composites made with bioactive glasses are discussed in Chapter 15.

COMPOSITIONS

The base components in most bioactive glasses are SiO_2, Na_2O, CaO, and P_2O_5 (Table 1). The first, and most well-studied composition, termed Bioglass® 45S5, contains 45% SiO_2, 24.5% Na_2O, 24.4% CaO and 6% P_2O_5, all in weight percent. The 45S5 composition in mole percent is given in Table 1, along with many other compositions investigated for surface reaction kinetics. Hench and co-workers have studied a series of glasses in this four-component system with a constant 6 weight percent P_2O_5 content. This work is summarized in the ternary SiO_2-Na_2O-CaO diagram shown in Fig. 1. The figure establishes the bioactive-bonding-boundary of compositions. In region A the glasses are bioactive and bond to bone. In the middle of this area a smaller region is indicated (broken line) within which soft tissue bonding also occurs. Glasses in region B behave as nearly-inert materials and are encapsulated by non-adherent fibrous tissue when implanted. Compositions in region C are resorbed within 10 to 30 days in tissue. In region D the compositions are not technically practical and have not been implanted. The boundary between region A and C depends upon the ratio of surface area of the glass to the effective solution volume of the tissue, as well as the glass composition. Fine glass powders resorb more quickly than bulk implants.

Partial substitution of CaO by CaF_2 does not significantly alter the bone-bonding behavior. The fluoride additions, however, reduce the rate of dissolution and affect the location of the A-C boundary in Fig. 1. Substitutions of MgO for CaO or K_2O for Na_2O also have little effect on bone bonding. B_2O_3 and Al_2O_3 have also been used in bioactive glasses to modify processing schedules (Chapter 13) and rates of surface dissolution.

®Registered trademark University of Florida, Gainesville, FL

Table 1. Glass Compositions by Mole %.

Designation	SiO_2	Na_2O	CaO	CaF_2	P_2O_5	B_2O_3	Al_2O_3
45S5.4F	46.1	24.4	16.2	10.8	2.6	0	0
45S5	46.1	24.4	26.9	0	2.6	0	0
#1(S63.5P6)	65.7	15.0	15.5	0	2.6	0.4	0.6
#9(S53P4)	53.9	22.6	21.8	0	1.7	0	0
#10(S45P7)	46.6	24.1	24.4	0	3.0	1.8	0
52S4.6	52.1	21.5	23.8	0	2.6	-	-
55S4.3	55.1	20.1	22.2	0	2.6	-	-
60S3.8	60.1	17.7	19.6	0	2.6	-	-
42SF	42.1	26.3	17.4	11.60	2.6	-	-
46SF	46.1	24.4	16.14	10.76	2.6	-	-
49SF	49.1	23.0	15.18	10.12	2.6	-	-
52SF	52.1	21.5	14.28	9.52	2.6	-	-
55SF	55.1	20.1	13.32	8.88	2.6	-	-
60SF	60.1	17.7	11.76	7.84	2.6	-	-
49S(gg)	50.	0	46.	0	4.	-	-
54S(gg)	55.	0	41.	0	4.	-	-
58S(gg)	60.	0	36.	0	4.	-	-
63S(gg)	65.	0	31.	0	4.	-	-
68S(gg)	70.	0	26.	0	4.	-	-
72S(gg)	75.	0	21.	0	4.	-	-
77S(gg)	80.	0	16.	0	4.	-	-
86S(gg)	90.	0	6.	0	4.	-	-

(gg) = gel-glass (see Li, Clark and Hench for details)

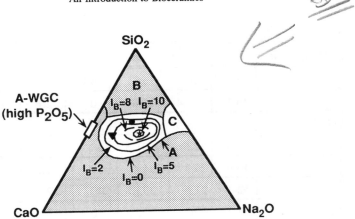

Fig. 1. Compositional dependence (in weight percent) of bone bonding and soft-tissue bonding of bioactive glasses and glass-ceramics. All compositions in region A have a constant 6 wt% of P_2O_5. A/W glass-ceramic has higher P_2O_5 content. Region E (soft-tissue bonding) is inside the dashed line where $I_B > 8$ ((*) 45S5 Bioglass®, (▼) Ceravital®, (●) 55S4.3 Bioglass®, and (---) soft-tissue bonding; $I_b = 100/t_{0.5bb}$). (Reprinted from L. L. Hench, "Bioceramics: From Concept to Clinic," *J. Amer. Ceram. Soc.*, **74**[7] (1991) 1487-570, with permission.)

Alumina is especially important in controlling glass surface durability and melting and forming characteristics. However, it is well established that Al_2O_3, in contrast to B_2O_3, can inhibit bone bonding. The amount of alumina that is tolerated depends on glass composition but is generally in the order of 1.0 to 1.5 weight percent. More alumina can be added to a glass with a high reactivity (high bioactivity) than to a glass which reacts more slowly. The dimensions of the bone-bonding boundary (Region A in Fig. 1) shrink as the percent of Al_2O_3 increases. Gross and co-workers have shown that the same effect occurs for other multi-valent cations, such as Ta_2O_5. Additions of more than 1.5 to 3% of multivalent ions usually make the glass inactive.

In a multi-component system like the SiO_2-Na_2O-CaO-P_2O_5-B_2O_3-Al_2O_3 system it is not possible to find a simple relationship between composition and tissue bonding that can be expressed in a two dimensional diagram, such as Fig. 1. Andersson et al. described the *in vivo* behavior of glasses in this complex system with a phenomenological model developed by regression analysis. The method predicts the *in vivo* behavior of glasses within certain compositional ranges. The prediction is based upon empirically determined factors and makes it possible to select glasses for specific applications without having to test them in animals. This method of compositional optimization is described in Chapter 13. It works because glass is an amorphous material and most properties are additive, within certain compositional limits.

The role of phosphate in bioactive glasses is interesting. Early on it was assumed that P_2O_5 was required for a glass to be bioactive. However, it is now known that phosphate-free glasses as well as glass-ceramics in which the phosphate is bound in a stable, relatively insoluble apatite phase are bioactive. Kokubo and co-workers have shown that the minimal melt-derived glass compositional system for bioactivity is CaO-SiO$_2$, with a compositional limit of about 60 mole percent. Li et al. have shown that gel-derived glasses in the Na$_2$O-CaO-SiO$_2$ are bioactive even up to 85 mole percent SiO$_2$ (Table 1). This very broad range of bioactive compositions makes it possible to tailor the reactivity of the glasses for various clinical applications, as discussed in Chapter 21. The role of phosphate in the glass appears only to aid in the nucleation of the calcium phosphate phase on the surface but is not a critical constituent because the surface will adsorb phosphate ions from solution.

PROPERTIES

The primary advantage of bioactive glasses is their rapid rate of surface reaction which leads to fast tissue bonding. Their primary disadvantage is mechanical weakness and low fracture toughness due to an amorphous two-dimensional glass network. Tensile bending strength of most of the compositions in Table 1 is in the range of 40-60 MPa, which make them unsuitable for load-bearing applications. Mechanical properties of bone are discussed in Chapter 1. For some applications low strength is offset by the glasses' low modulus of elasticity of 30-35 GPa. The importance of this value, which is close to that of cortical bone, is discussed in Chapter 1. The low strength does not influence the utility of bioactive glasses as a coating where interfacial strength between metal and the coating is the limiting factor. Low strength also has no effect on use of bioactive glasses as buried implants, in low-loaded or compressively loaded devices, in the form of powders, or as the bioactive phase in composites.

REACTION KINETICS

The basis of the bone bonding property of bioactive glasses is the chemical reactivity of the glass in body fluids. The surface chemical reactions result in the formation of an hydroxycarbonate apatite (HCA) layer to which bone can bond. Bonding occurs due to a sequence of reactions. On immersion of a bioactive glass in an aqueous solution three general processes occur: leaching, dissolution, and precipitation. Leaching is characterized by release, usually by cation exchange with H$^+$ or H$_3$O$^+$ ions, of alkali or alkaline earth elements. Ion exchange is easy because these cations are not part of the glass network, they only modify the network by forming non-bridging oxygen bonds (Chapter 1). The release of network-modifying ions is rapid for glasses in the bioactive compositional region (A in Fig. 1). This ion exchange process leads to an increase in interfacial pH, to values >7.4.

Network dissolution occurs concurrently by breaking of -S-O-Si-O-Si- bonds through the action of hydroxyl (OH) ions. Breakdown of the network occurs locally and releases silica into solution in the form of silicic acid [Si(OH)$_4$]. The rate of dissolution of silica depends very much on glass composition. The dissolution rate decreases greatly for compositions of >60% SiO$_2$ because of the larger number of bridging oxygen bonds in the glass structure. The hydrated silica (SiOH) formed on the glass surface by these reactions undergoes rearrangement by polycondensation of neighboring silanols, resulting in a silica-rich gel layer.

In the precipitation reaction, calcium and phosphate ions released from the glass together with those from solution form a calcia-phosphate-rich (CaP) layer on the surface. When formed *in vitro*, the CaP layer is mainly located on top of the silica gel, whereas *in vivo* it is formed within the gel layer. The calcium phosphate phase that accumulates in the gel surface is initially amorphous (a-CaP). It later crystallizes to a hydroxycarbonate apatite (HCA) structure by incorporating carbonate anions from solution within the a-CaP phase. The mechanism of nucleation and growth of the HCA layer appears to be the same *in vitro* and *in vivo* and is accelerated by the presence of hydrated silica.

Figure 2 shows the CaP and silica-rich layers formed on a 45S5 bioactive glass within one hour of implantation in a rat bone. The implant was removed as the bonding sequence was beginning. The data were obtained using Auger electron spectroscopy (AES) combined with Ar ion milling. The analysis is of 50 Angstrom "slices" of material, which combined together, yields a compositional profile of the reaction interface. The silica-rich layer has formed to a thickness of more than 12,000 Å (>1 μm). The CaP layer is already 0.8 μm thick after one hour of reaction. Biological molecules are bonded within the bilayer to a depth of 0.1 μm, as indicated by the C and N signals in the outer layer of the surface. It is important to note that the mixed organic-inorganic bonding occurs within a region that has Si as well as Ca and P present.

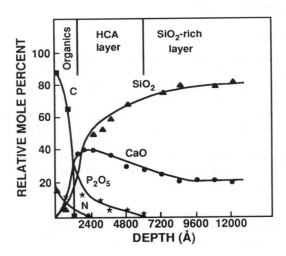

Fig. 2. Bilayer films formed on 45S5 Bioglass® after 1 h in rat bone, *in vivo* (1 Å = 10^{-1} nm). (Reprinted from L. L. Hench, "Bioceramics: From Concept to Clinic," *J. Amer. Ceram. Soc.*, **74**[7] (1991) 1487-570, with permission.)

Thus, the reactions on the implant side of the interface with a bioactive glass are:

Stage 1: Leaching and formation of silanols (SiOH).
Stage 2: Loss of soluble silica and formation of silanols.
Stage 3: Polycondensation of silanols to form a hydrated silica gel.
Stage 4: Formation of an amorphous calcium phosphate layer.
Stage 5: Crystallization of a hydroxycarbonate apatite layer.

Table 2 summarizes these five reaction stages in more detail.

Table 2. Reaction Stages of a Bioactive Glass Implant.

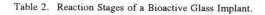

STAGE	
1	Rapid exchange of Na^+ or K^+ with H^+ or H_3O^+ from solution:
	$Si - O - Na^+ + H^+ + OH^- \rightarrow Si-OH^+ + Na^+ \ (solution) + OH^-$
	This stage is usually controlled by diffusion and exhibits a $t^{-1/2}$ dependence.
2	Loss of soluble silica in the form of $Si(OH)_4$ to the solution, resulting from breaking of Si-O-Si bonds and formation of Si-OH (silanols) at the glass solution interface:
	$Si - O - Si + H_2O \rightarrow Si - OH + OH - Si$
	This stage is usually controlled by interfacial reaction and exhibits a $t^{1.0}$ dependence.
3	Condensation and repolymerization of a SiO_2-rich layer on the surface depleted in alkalis and alkaline-earth cations:
	$$\underset{\underset{O}{\overset{\mid}{\mid}}}{\overset{O}{\overset{\mid}{\mid}}}O - Si - OH + HO - \underset{\underset{O}{\overset{\mid}{\mid}}}{\overset{O}{\overset{\mid}{\mid}}}Si - O \rightarrow O - \underset{\underset{O}{\overset{\mid}{\mid}}}{\overset{O}{\overset{\mid}{\mid}}}Si - O - \underset{\underset{O}{\overset{\mid}{\mid}}}{\overset{O}{\overset{\mid}{\mid}}}Si - O + H_2O$$
4	Migration of Ca^{2+} and PO_4^{3-} groups to the surface through the SiO_2-rich layer forming a CaO-P_2O_5-rich film on top of the SiO_2-rich layer, followed by growth of the amorphous CaO-P_2O_5-rich film by incorporation of soluble calcium and phosphates from solution.
5	Crystallization of the amorphous CaO-P_2O_5 film by incorporation of OH^-, CO_3^{2-} or F^- anions from solution to form a mixed hydroxyl, carbonate, fluorapatite layer.

For a bond with tissue to occur a layer of biologically active HCA must form. This appears to be the only common characteristic of all the known bioactive implants. The rate of tissue bonding appears to depend on the rate of HCA formation.

The kinetics of the reaction stages depend on the glass composition. Fourier Transform Infrared (FTIR) spectroscopy can be used to determine the reaction rates and mechanism of all five stages of reaction. Figure 3 shows the FTIR spectra (using a diffuse reflection stage) of a 45S5 Bioglass® implant after 0, 1 and 2 hours in tris-buffer solution at 37°C. The spectra are equivalent to those obtained using a simulated body

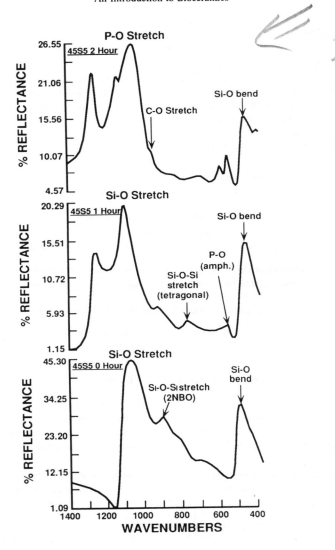

Fig. 3. FTIR spectra of a 45S5 Bioglass® implant after 0, 1, and 2 h in TBS at 37°C. (Reprinted from L. L. Hench, "Bioceramics: From Concept to Clinic," *J. Amer. Ceram. Soc.*, **74[7]** (1991) 1487-570, with permission.)

fluid (see Chapter 18).   The peak identifications are based upon previous assignments of
IR spectra (see Chapter 9).   The alkali-ion-hydronium ion exchange and network
dissolution (Stages 1 and 2 in Table 2) rapidly reduces the intensity of the Si-O-Na and
Si-O-Ca vibrational modes and replaces them with Si-OH bonds that have only one
nonbridging oxygen (NBO) ion.   Alkali content is depleted to a depth >0.5 μm within
a few minutes.   Clark et al. showed with AES that by 2 minutes alkali ion depletion
occurred to depths >0.1 μm.

As the stage 1 and 2 reactions continue, the single Si-OH NBO modes are
replaced by

$$\begin{array}{c} \text{H} \\ | \\ \text{O} \\ | \\ \text{—O—Si—OH} \\ | \\ \text{O} \\ | \end{array}$$

i.e; Si-2NBO stretching vibrations which are in the range of 930 cm^{-1}, decreasing to 880
cm^{-1}.   By 20 minutes, the Si-2NBO vibrations are largely replaced by a new mode
assigned to the Si-O-Si bond vibration between two adjacent SiO$_4$ tetrahedra (Fig. 4).

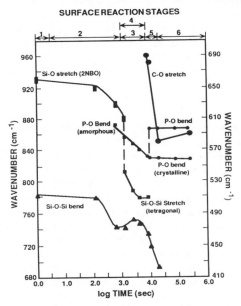

Fig. 4. Time-dependent changes in IR vibrations of the surface of 45S5 Bioglass® implant in a 37°C TBS.
(Reprinted from L. L. Hench, "Bioceramics: From Concept to Clinic," *J. Amer. Ceram. Soc.*, **74[7]**
(1991) 1487-570, with permission.)

This new vibrational mode corresponds to the formation of the silica-gel layer by the stage 3 (Table 2) polycondensation reaction between neighboring surface silanols. This mode decreases in frequency until it is hidden after 1 hour by the growing CaP-layer.

As early as 10 minutes, a P-O bending vibration associated with formation of an amorphous CaP layer appears. This is due to precipitation from solution (stage 4 in Table 2). Clark et al. showed, using AES, that by two minutes Ca and P enrichment occurred on the glass surface to a depth of approximately 20 nm. Ogino et al. showed, with AES, that by 1 hour the CaP layer grew to 200 nm in thickness.

Within 40 minutes (Fig. 4) the P-O bonding vibration is strong and exhibits a continually decreasing frequency as the CaP-rich layer builds. At about 1.5 ± 0.2 hrs, the P-O bending vibration associated with the amorphous calcium phosphate layer is replaced by two P-O modes (Fig. 3) assigned to crystalline apatite. Concurrent with the onset of apatite crystallization (stage 5 in Table 2) is the appearance of a C-O vibrational mode associated with the incorporation of carbonate anions in the apatite crystal lattice, as described by Kim et al. and discussed by LeGeros and LeGeros in Chapter 9. The crystals are nucleated and grow as hydroxycarbonate apatite (HCA), the same phase as biological HCA formed in mineralizing tissues. The reason for the equivalence is the similarity of nucleating mechanisms and physiological growth conditions.

The C-O mode decreases in wavenumber as the HCA layer grows. By 10 hours the HCA layer has grown to 4 μm in thickness, which is sufficient to dominate the FTIR spectra and mask most of the vibrational modes of the silica-gel layer or the bulk-glass substrate. By 100 hours the polycrystalline HCA layer is thick enough to yield X-ray diffraction (XRD) results, as discussed in Chapter 18. The primary 26 and 33 2Θ peaks of HCA are visible with considerable line broadening. By 2 weeks the FTIR spectra show three P-O vibrational modes and the XRD data are equivalent to biological HCA gown *in vivo*.

Thus, the bioactive glass implant surface provides a substrate that is favorable for the rapid nucleation and growth of biologically equivalent HCA (stage 5 in Table 2). Differences in the *in vivo* behavior of various glass compositions are due to the differences in the rate of stage 5, HCA formation.

Table 3, from Hench and LaTorre, summarizes a series of investigations of the effects of glass composition and solution composition on the kinetics of reaction stages 1-5. Filgueiras et al. showed that the sequence of surface reactions is independent of solution composition. However, the presence of Ca and P in a simulated body fluid (SBF) solution accelerates to a small extent the repolymerization of silica (stage 3) and formation of the amorphous calcium-phosphate (a-CaP) layer (stage 4). The major effect of solution composition is on the crystallization of HCA (stage 5). Figure 5 shows that the process of HCA crystallization is the same for the various glasses and the tris-buffer or SBF solution K-9, the only difference is the rate of crystallization. The rate of crystallization increases to a small extent in Ca- and P- containing SBF solutions (in 90 min rather than 120 min). However, Mg ions in SBF slow down formation of the a-CaP layer and greatly retard crystallization of HCA on the glass surface.

Table 3. Time for Onset of Reaction Stages 1,2,3,4 and 5 for Bioactive Glasses.

Composition	Time (Min) in Tris Buffer														
	1	2	10	20	40	60	90	120	150	360	720	1440	3600	4320	12000
45S5.4F(2)	1+2		3+4					5							
45S5 (old) (2)	1+2				3+4			5							
45S5 (new)	1+2		3+4					5							
Composition #1 (S63.5P6)	NO SPECTRAL CHANGES NOTED														
Composition #9 (S53P4)	1+2						3			4				5	
Composition #10 (2) (S45P7)	1+2			3+4						5					
42SF	1+2			3+4					5						
46SF	1+2			3+4				5							
49SF	1+2			3+4				5							
52SF	1+2			3+4					5						
55SF		1+2					3+4						5		
60SF					1+2					3					4
49S(Gel-Glass)			5												
54S(Gel-Glass)			5												
58S(Gel-Glass)			5												
63S(Gel-Glass)					1-5										
68S(Gel-Glass)					1-5										
72S(Gel-Glass)					1-5										
77S(Gel-Glass)												5			
86S(Gel-Glass)															5

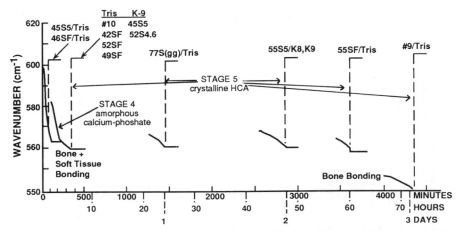

Fig. 5. Effect of glass composition on time for onset of crystallization of hydroxycarbonate apatite (HCA) on the surface in TRIS buffer or simulated body fluid (K-9) solutions.

Table 3 and Fig. 5 show that most of the effects of glass composition are on the time required for HCA crystallization. This finding makes it possible to summarize the relationship between surface reaction rates of bioactive glasses and their *in vivo* behavior. Figure 6, from Hench and LaTorre, shows the critical compositional relationship between *in vitro* and *in vivo* kinetics. For glasses with up to about 53 mole percent SiO_2 HCA crystallization occurs very rapidly on the glass surface, within 2 hours. These compositions develop a rapid bond with bone and also form an adherent, interdigitating collagen bond with soft tissues. Glasses with SiO_2 content between 53 and 58 mole percent SiO_2 require two to three days to form both the a-CaP layer and to crystallize HCA. Such glass compositions are bioactive, but they bond only to bone. When implanted in soft tissues the fibrous capsule formed around them is parallel to the interface and is non-adherent. Compositions with > 60% SiO_2 do not form a crystalline HCA layer even after four weeks in SBF. An amorphous calcium-phosphate layer forms but it does not crystallize to HCA. Such glasses are not bioactive and bond neither to bone nor soft tissues.

The *in vitro* kinetics studies described above were conducted in well specified conditions with the ratio of glass surface area (SA) to solution volume (V) fixed at 0.1 cm^{-1}. Changing the SA/V ratio changes the reaction kinetics. The SA/V ratio relevant to an implant is very difficult to estimate. It depends not only on surface area of the implant and the size of the cavity into which it is placed but also on tightness of fit, blood flow, metabolic rate of the tissues, inflammation, etc. Thus, the kinetics obtained from *in vitro* analyses are only an estimate of the reaction rates *in vivo*. However, the compositional effects observed *in vitro* appear to be equivalent *in vivo*.

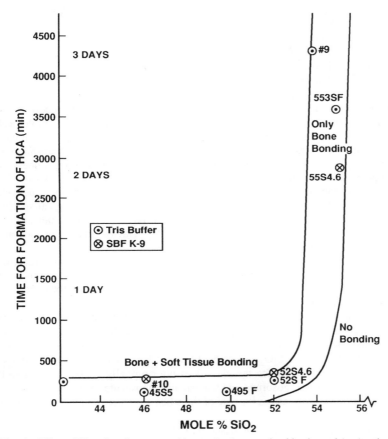

Fig. 6. Effect of bioactive glass composition on *in vitro* reaction kinetics and *in vivo* tissue response.

TISSUE BONDING

The five reaction stages that occur on the material side of the interface do not depend on the presence of tissues. They occur in distilled water, tris-buffer solutions or simulated body fluids. Bonding to tissues requires an additional series of reactions. They are less well defined and the factors that control the rates are not known. In some stages the biological processes involved are still under investigation and the materials factors cannot be determined until the biological understanding is complete. At present the sequence of events that appear to be associated with formation of a bond with tissues are:

Stage 6:   Adsorption of biological moieties in the SiO2-HCA layer
Stage 7:   Action of macrophages
Stage 8:   Attachment of stem cells
Stage 9:   Differentiation of stem cells
Stage 10:   Generation of matrix
Stage 11:   Mineralization of matrix

Rapid growth of HCA agglomerates on a bioactive glass surface incorporates collagen fibrils *in vitro* (Fig. 7) without the presence of cells, enzymes, or biological growth factors. The crystals appear to form around the collagen fibrils and form bonds with them on an ultrastructural level, similar to what has been observed by transmission electron microscopy of bone bonded to bioactive glass. The same bonding process of collagen incorporation within the growing gel layer has been observed in soft tissues by Wilson et al. and Wilson and Nolletti (Figs. 8 and 9). Thus, stage 6 appears to be well established with respect to collagen. However, it is likely that other biological species are adsorbed on the bioactive surface much earlier than collagen. These species are not yet known. There is also little information on stage 7. Davies and colleagues are investigating stages 8-10 using cell culture methods.

Within a week mineralizing bone appears at the interface of the more reactive bioactive glasses (stage 11). By four weeks the interface is completely bonded to bone without any intervening fibrous tissues. The time sequence of bone formation on bioactive glasses is reviewed by Hench and Clark and Gross et al.

Figure 10 shows a scanning electron micrograph of a cross-section of a bioactive glass (S46PO) after 8 weeks in rabbit tibia. In the back-scatter mode it is easy to distinguish the characteristic silica rich (dark) and calcium phosphate rich (bright) layers. The composition of the glass is SiO_2, 46.0; Na_2O, 26.0; CaO, 25.0; P_2O_5, 0.0; B_2O_3, 2.0 and Al_2O_3, 1.0 weight-%. Since the glass does not originally contain phosphate it is clearly seen that phosphate is absorbed and that calcium phosphate forms mainly within the silica-rich layer. This emphasizes the role of the silica structure in inducing the HCA formation.

The most important criterion for tissue bonding is the mechanical resistance of the tissue-implant interface. There is no consensus on the best test method to determine interfacial adherence of bioactive implants. Studies by the Florida group of bone bonding to bioactive glasses and glass-ceramics, primarily 45S5 composition, showed very high interfacial strength values using a variety of mechanical test methods. Most of the studies involved loaded prostheses, such as segmental bone replacements (Fig. 11) or femoral head prostheses. In most studies the interface did not fail, fracture occurred either in the implant, such as shown in Fig. 11, or in the bone distal to the implant, as discussed in Chapter 13.

Various push-out or pull-out tests have also been reported for bioactive glasses and glass-ceramics. Chapter 5 discusses the method used by the Kyoto group. The push-out test method used by Andersson and colleagues in Finland is illustrated in Fig. 12.

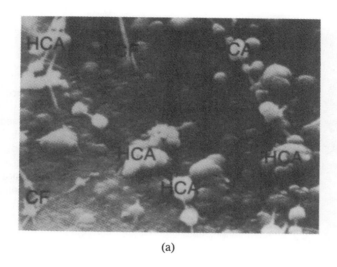

(a)

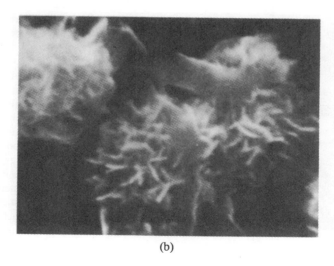

(b)

Fig. 7. (a) SEM micrograph of collagen fibrils incorporated within the HCA layer growing on a 45S5 Bioglass® substrate *in vitro*. (b) Close-up (~11300X) of the HCA crystals bonding to a collagen fibril. (Photographs courtesy of C. Pantano.)

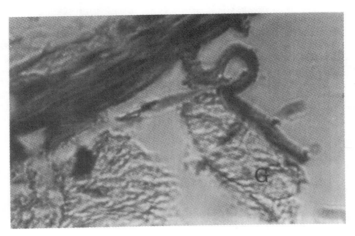

Fig. 8. Peeling of collagen, which remains adherent to the 45S5 bioactive-glass (G) decalcified section). (Original magnification X 250.)

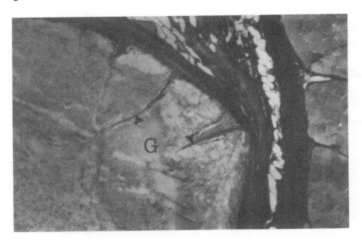

Fig. 9. Collagen fibers (▸) in cracks on the 45S5 bioactive-glass surface (G) undecalcified section). (Original magnification X 100.)

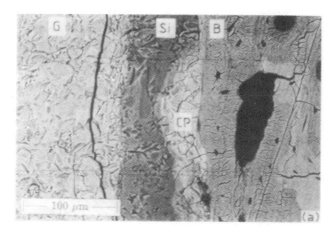

Fig. 10. Scanning electron micrograph of a cross-section of a phosphate-free bioactive glass (S46PO) after 8 weeks in rabbit tibia. G = bulk glass, Si = Si-rich layer, CP = Ca,P-rich layer, B = bone.

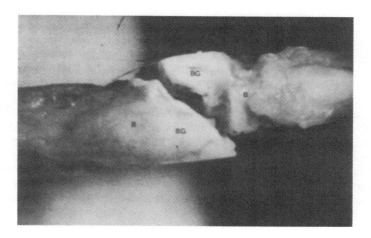

Fig. 11. Fracture of (BG) 45S5 Bioglass®-ceramic segmental bone replace in monkey due to impact torsional loading. Note (B) bonded interface. (Photograph courtesy G. Piotrowski.) (Reprinted from L. L. Hench, "Bioceramics: From Concept to Clinic," *J. Amer. Ceram. Soc.*, **74[7]** (1991) 1487-570, with permission.)

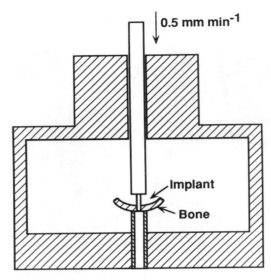

Fig. 12. Schematic illustration of a specimen in the push-out test fixture.

The conical implant is pushed out of the rabbit tibia at a cross-head speed of 0.5 mm/min. Prior to testing, the bone covering the base of the cone is removed by grinding. The base of the cone is also slightly ground in this process and a flat supporting surface is produced. The conical shape reduces the contribution from the surface roughness. It is difficult to measure the contact area since the bone grows along the surface of the implant and may be present only as a very thin layer. If the glass is sectioned along its axis after the test it is possible to estimate the thickness. With this method strength values of 15 to 25 MPa have been obtained for a number of bioactive glasses. For an inert glass the same test gave a value of 0.5 MPa or less, for glasses just outside the bioactivity border values of 2 to 3 MPa, and for smooth titanium 2 MPa. Thus, non-bonding biocompatible glasses behave as titanium. When testing bioactive glasses it is observed that the fracture line in some areas is within the silica-rich layer, in others within bone. Figure 13 shows a cross-section of glass S46PO after the push-out. It is clear that bone has adhered to the glass and that the fracture has occurred within the bone.

RATE OF BONDING

The rate of development of the interfacial bond between and implant and bone can be referred to as the level of bioactivity. Hench introduced an index of bioactivity as a measure of this. The index is given by $I_B = (100/t_{0.5bb})$, where $t_{0.5bb}$ is the time for more than 50% of the surface to be bonded to bone. In the ternary diagram for the

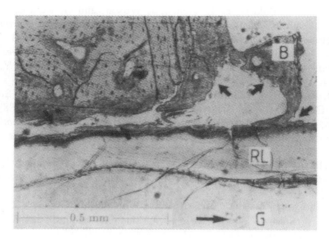

Fig. 13. Optical micrograph of a cross-section of glass S46PO after push-out. B = bone, RL = reaction layer, G = bulk glass. Direction of loading is indicated by long arrow and fracture line by short arrows.

compositional dependence of the bioactivity, iso-I_B contours have been indicated (Fig. 1). Thus, the closer to the bioactivity boundary a glass is, the slower is the rate of bonding. When the constant 6 weight-% P_2O_5 content in Fig. 1, I_B goes toward zero as the SiO_2 content is raised close to 60%. Bioactive implants with intermediate I_B values do not develop a stable bond with soft tissue. The broken line in Fig. 1 indicates the region within which the I_B values are sufficiently high for soft tissue bonding.

The thickness of the bonding zone is roughly proportional to the I_B value and the failure strength of a bioactive bond appears to be inversely proportional to the thickness of the zone. Thus, a very high I_B value gives a thick bonding zone and a low shear strength. Depending on whether rapid bonding or high shear strength is preferred, different compositions are optimal.

By subcutaneous implantation it has been found that if two specimens of the bioactive glass S53P4 come into contact with each other in the tissue they can fuse through their apatite surface layers. This phenomenon was observed at implantation for 24 hours. Thus, it seems that if a bioactive glass is accidently broken in the tissue it might self-repair. During this process the surfaces must stay in contact with each other. This emphasizes the importance of a rapid reaction such as in this case, however a similar phenomenon has been seen for A/W glass ceramic (see Chapter 5) which has a lower I_B value.

READING LIST

L. L. Hench and E. C. Ethridge, *Biomaterials: An Interfacial Approach* (Academic Press, New York, 1982).

S. F. Hulbert, J. C. Bokros, L. L. Hench, J. Wilson, and G. Heimke, "Ceramics in Clinical Applications, Past, Present and Future" in *High Tech Ceramics*, ed. P. Vincenzini (Elsevier Science Pub. B.V., Amsterdam, 1987) pp. 189-213 .

L. L. Hench, "Bioactive Ceramics," in *Bioceramics: Materials Characteristics Versus In Vivo Behavior*, Vol. 523, ed. P. Ducheyne and J. Lemons (Annals N.Y. Acad. Sci., 1988) pp. 54.

U. Gross, R. Kinne, H. J. Schmitz, V. Strunz, "The Response of Bone to Surface Active Glass/Glass-Ceramics," *CRC Critical Reviews in Biocompatibility* 4 (1988) 2.

L. L. Hench, "Cementless Fixation," in *Biomaterials and Clinical Applications*, eds. A. Pizzoferrato, P. G. Marchetti, A. Ravaglioli, and A.J.C. Lee (Elsevier, Amsterdam, 1987) p. 23.

L. L. Hench, R. J. Splinter, W. C. Allen, and T. K. Greenlee, Jr., "Bonding Mechanisms at the Interface of Ceramic Prosthetic Materials," *J. Biomed. Maters. Res.* 2[1] (1972) 117-141.

L. L. Hench and J. W. Wilson, "Surface-Active Biomaterials," *Science* 226 (1984) 630.

U. Gross and V. Strunz, "The Interface of Various Glasses and Glass-Ceramics with a Bony Implantation Bed," *J. Biomed. Mater. Res.* 19 (1985) 251.

J. Wilson, G. H. Pigott, F. J. Schoen, and L. L. Hench, "Toxicology and Biocompatibility of Bioglass," *J. Biomed. Mater. Res.* 15 (1981) 805.

T. Yamamuro, L. L. Hench, and J. Wilson, eds., *Handbook on Bioactive Ceramics, Vol I: Bioactive Glasses and Glass-Ceramics* (CRC Press, Boca Raton, FL, 1990).

June Wilson and D. Nolletti, "Bonding of Soft Tissues to Bioglass," in *Biological and Biomechanical Performance of Biomaterials*, eds. P. Christel, A. Meunier, and A.J.C. Lee. (Elsevier Science Publishers, Amsterdam, 1986) pp 99-104.

L. L. Hench, H. A. Paschall, W. C. Allen, and G. Piotrowski, "Interfacial Behavior of Ceramic Implants," *National Bureau of Standards Special Publication* 415 (1975) 19-35.

L. L. Hench and H. A. Paschall, "Histo-Chemical Responses at a Biomaterials Interface," *J. Biomed. Maters. Res.* 5[1] (1974) 49-64.

C. G. Pantano, D. E. Clark and L. L. Hench, *Corrosion of Glass* (The Glass Industry, New York, 1979).

L. L. Hench and D. E. Clark, "Physical Chemistry of Glass Surfaces," *J. Non-Crystalline Solids* 28 (1978) 83.

R. W. Douglas and T. M. El-Shamy, "Reactions of Glasses with Aqueous Solutions," *J. Amer. Ceram. Soc.* 50[1] (1967) 1-8.

C. M. Jantzen, "Prediction of Glass Durability as a Function of Environmental Conditions," *Maters. Res. Soc.* 125 (1988) 143-159.

B. C. Bunker, D. R. Tallant, T. J. Headley, G. L. Turner, and R. J. Kirkpatrick, "The Structure of Leached Sodium Borosilicate Glass," *Phys. and Chem. Glasses* 29[3] (1988) 106-120.

A. E. Clark, C. Y. Kim, J. K. West, J. Wilson and L. L. Hench, "Reactions of Fluoride and Nonfluoride Containing Bioactive Glasses" in *Handbook of Bioactive Ceramics, Vol I*, eds. T. Yamamuro, J. Wilson and L. L. Hench (CRC Press, Boca Raton, FL, 1990) p. 73.

M. Ogino, F. Ohuchi, and L. L. Hench, "Compositional Dependence of the Formation of Calcium Phosphate Films on Bioglass®," *J. Biomed. Maters. Res.* **14** (1980) 55-64.

A. E. Clark, C. G. Pantano and L. L. Hench, "Auger Spectroscopic Analysis of Bioglass® Corrosion Films," *J. Amer. Ceram. Soc.* **59[1-2]** (1976) 37-39.

Ö. H. Andersson, *The Bioactivity of Silicate Glass* (PhD Dissertation, Dept of Chemical Engineering, Abo Akademi, Finland, 1990).

C. Y. Kim, A. E. Clark and L. L. Hench, "Early Stages of Calcium-Phosphate Layer Formation in Bioglasses®," *J. Non-Cryst. Solids* **113** (1989) 195-202.

Ö. H. Andersson G. Liu, K. Kangasniemi and J. Juhanjoa, "Evaluation of the Acceptance of Glass in Bone," *J. Maters. Sci., Maters. in Med.* **3** (1992) 145-150.

D. M. Sanders, W. B. Person and L. L. Hench, "Quantitative Analysis of Glass Structure Using Infrared Reflection Spectra," *Appl. Spectroscopy* **28[3]** (1974) 247-255.

B. O. Fowler, "Infrared Studies of Apatites I Vibrational Assignments for Calcium Strontium and Barium Hydroxyapatite Utilizing Isotopic Substitution," *Inorg. Chem.* **13[1]** (1974) 194.

R. F. Le Geros, G. Bone and R. Le Geros, "Type of H_2O in Human Enamel and in Precipitated Apatites," *Calcif. Tissue Res.* **26** (1978) 111.

L. L. Hench, "Stability of Ceramics in the Physiological Environment," in *Fundamental Aspects of Biocompatibility, Vol 1*, ed. D. F. Williams (CRC Press, Boca Raton, FL, 1981) pp. 67-85.

C. Oktsuki, T. Kokubo, K. Takatsuka, and T. Yamamuro, "Compositional Dependence of Bioactivity of Glasses in the System $CaO-SiO_2-P_2O_5$ its *In Vitro* Evaluation," *Nippon Seramikkusu Kyokai Gakijutsu Ronbuski* **99[1]** (1991) 1-6.

R. Li, A. E. Clark, and L. L. Hench, "An Investigation of Bioactive Glass Powders by Sol-Gel Processing," *J. Appl. Biomaterials* **2** (1991) 231-239.

M. M. Walker, "An Investigation into the Bonding Mechanisms of Bioglass® (Masters Thesis, University of Florida, Gainesville, 1977).

J. E. Davies, "The Use of Cell and Tissue Culture to Investigate Bone Cell Reactions to Bioactive Materials," in *Handbook of Bioactive Ceramics, Vol I*, eds. T. Yamamuro, L. L. Hench and J. Wilson (CRC Press, Boca Raton, FL, 1990) p. 195.

J. E. Davies, ed., *Bone-Biomaterials Interface* (University of Toronto Press, Toronto, Ontario, 1991)

L. L. Hench and G. P. LaTorre, "Reaction Kinetics of Bioactive Ceramics Part IV: Effect of Glass and Solution Composition," in *Bioceramics 5*, eds. T. Yamamuro, T. Kokubo and T. Nakamura (Kobonshi Kankokai, Inc., Kyoto, Japan, 1992) pp. 67-74.

Jon K. West and Larry L. Hench, "Reaction Kinetics of Bioactive Ceramics Part V: Molecular Orbital Modeling of Bioactive Glass Surface Reactions," in *Bioceramics 5*, eds. T. Yamamuro, T. Kokubo and T. Nakamura (Kobonshi Kankokai, Inc., Kyoto, Japan, 1992) pp. 75-86.

Ö. H. Andersson and I. Kangasniemi, "Calcium Phosphate Formation at the Surface of Bioactive Glass *In Vitro*," *J. Biomed. Maters. Res.* **25** (1991) 1019-1030.

Ö. H. Andersson, K. H. Karlsson and I. Kangasniemi, "Calcium Phosphate Formation at the Surface of Bioactive Glass *In Vivo*," *J. Non-Cryst. Solids* **119** (1990) 290-296.

Ö. H. Andersson, G. Liu, K. H. Karlsson, J. Miettinen and J. Juhanoja, "*In Vivo* Behavior of Glasses in the SiO_2-Na_2O-CaO-P_2O_5-Al_2O_3-B_2O_3 System," *J. Maters. Sci., Maters. in Med.* **1** (1990) 219-227.

Ö. H. Andersson, "Glass Transition Temperature of Glasses in the SiO_2-Na_2O-CaO-P_2O_5-Al_2O_3-B_2O_3 System," *J. Maters. Sci., Maters. in Med.* **3** (1992) 326-328.

L. L. Hench, O. A. Andersson and G. P. LaTorre,"The Kinetics of Bioactive Ceramics, Part III: Surface Reactions for Bioactive Glasses Compared with an Inactive Glass," in *Bioceramics, Volume 4*, eds. W. Bonfield, G. W. Hastings and K. E. Tanner, (Butterworth-Heinemann Ltd., Guildford, England, 1991) pp. 155-162.

Chapter 4

BIOACTIVE GLASSES: CLINICAL APPLICATIONS

June Wilson, Antti Yli-Urpo* Risto-Pekka Happonen*

Bioglass® Research Center, University of Florida, Gainesville, Florida

*Institute of Dentistry, University of Turku, Finland,

INTRODUCTION

When it was first proposed that glass might be used as an implant material the concept was slow to find acceptance. The properties of glasses were, after all, well-known. They are brittle, generally fragile and could easily be broken or fractured accidentally, generating particles or pieces which could migrate and which were expected to damage tissues and blood vessels. Glass could be cast or molded but could not be easily drilled or carved by any but expert hands.

The bioactive materials, which happened to be glasses, had however, other valuable properties. They would develop at their surfaces, when in contact with body fluids and tissues, a reactive layer which, because of its gel-like structure, provided a compliant interface between bulk glass and tissue. It was found that the 45S5 formulation of Bioglass® and related compositions did not break-up when drilled with standard surgical drilling equipment. The glass was relatively soft and extremely suitable for microsurgical drilling techniques, opening up its use in certain otolaryngological applications (Fig. 1). However it could never be certain that particles would not appear

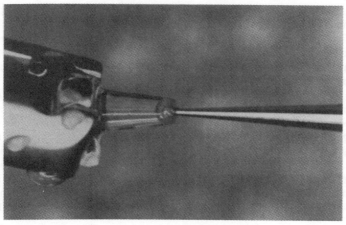

Fig. 1. The stem of a Bioglass® middle ear prosthesis is being drilled to fit over the residual stapes. The facet is approximately 0.5 mm wide.

after implantation of a solid device if it were to become damaged in some way and exhaustive tests were done to find out what the effect of "ground glass" of this nature might be in tissues. Fortunately the constituent chemicals, silicon, sodium, oxygen, calcium and phosphorous are all found in the body and at the concentrations derived from an implant did not disturb the adjacent tissues. Particles could be ingested by phagocytic cells, those cells found in blood and tissues which are programmed to pick up and digest all unusual particles, and did not affect the viability or behavior of those cells. Particles could be injected into the bloodstream and come to rest in the capillary bed, but did not cause any emboli or detectable effect on blood flow. Many other tests showed that these bioactive glasses in solid or particulate form were not toxic to any of the tissues or systems with which they were in contact. They were shown to bond to osseous tissues and to soft tissues through a compliant interface and these special properties defined the clinical applications for which they have been used in the last ten or more years.

Compositions of Bioglass®, notably 45S5 and that which contains some fluoride, and the glass designated #9 from Finland have been tested in many medical and dental applications. See Chapter 3 for the chemical compositions of these glasses and the mechanisms of bonding to tissues. Applications have required several different forms of the material.

1.   Solid shapes
2.   Particulates of various size ranges.
3.   Particulates combined with autologous bone particles.
4.   Particulates delivered via an injectable system.

It must be emphasized that clinical experience in man on many of the applications described below is still limited.

Solid Shapes

The glasses can be fabricated by casting into shaped implants for specific purposes where mechanical strength is of secondary importance.

The first successful use of 45S5 Bioglass® was as a replacement for the ossicles in the middle ear, as a treatment for the conductive hearing loss which develops when the sound waves impinging on the tympanic membrane do not reach the oval window in the inner ear. Conduction loss in the middle ear can result from trauma, chronic infection or be due to congenital abnormality. Replacement of one or more of the ossicles can restore the continuity of the conducting system. Materials previously used have included polymers, both porous and solid, and metals of various types. These materials engender a fibrous tissue reaction which effectively holds the implant in place. Scar tissue around an implant however will dampen, rather than transfer sound waves and implants which promote scar tissue become gradually less efficient. The major mode of failure is extrusion through the tympanic membrane. When metal or plastic implants are in continuous contact with the soft tissue of the eardrum they wear through and are lost through the hole. These two problems, immobilization by a means other than fibrous tissue and prevention of extrusion, are solved by the special properties of 45S5 Bioglass® which bonds to both hard and soft tissues. A bone bond can be achieved with the remainder of the stapes if it can be retained to protect the connection at the oval window. A collagenous bond mimicking the normal connection between ossicles can also be achieved if necessary. Most importantly the soft tissue bond between the implant and the tympanic membrane eliminates the movement at that interface which leads to extrusion. Success has been achieved by Bioglass®* devices even after a previously damaged tympanic membrane has been repaired with skin or muscle graft. A typical device is shown in Fig. 2. The stem of the implant may be shortened and shaped to fit the individual patient. These and similar devices have been in clinical use for many years.

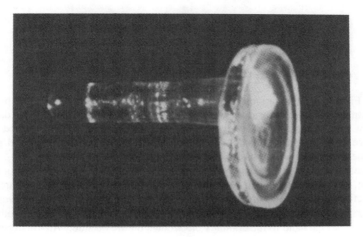

Fig. 2. A Bioglass® middle ear device.

In oral surgery, cone-shaped devices made from 45S5 Bioglass®, have been used to fill the defect in the jaw which is created when a tooth is removed (Fig. 3). Removal of one or more teeth produces changes in the jaw bone which are followed by gradual bone loss so that the normal shape of the bone which supports healthy teeth changes to a narrow "knife edge" ridge with reduced height which cannot comfortably support dentures. Without some means of preventing this bone loss, denture wearers are often destined to suffer increasing discomfort from ill-fitting dentures and in many cases may eventually become unable to wear dentures at all. These devices have now been in use for almost a decade and those made from bioactive glass have proven to be more successful than others which have been tried.

---

*Registered U.S. Trademark, University of Florida, Gainesville, FL, USA, 32611.

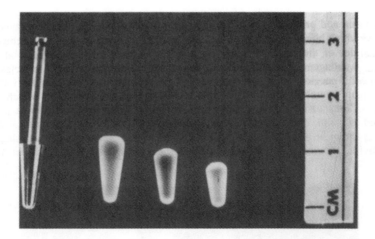

Fig. 3. Endosseous Ridge Maintenance Devices (ERMI) and matching dental burr.

In Finland an innovative use of solid, cast bioactive glass implants has been used in the treatment of facial injuries in which the bone which supports the eye is damaged. Such "blow out fractures" of the orbit have been treated by insertion of a curved implant of bioactive glass to restore the floor of the orbit (Fig. 4). A series of curved plates is cast and the most suitable one is selected by the surgeon at the operation.

Another use, in ENT surgery at present, but with the potential for much wider application, is that of providing a soft tissue seal for an implant which passes through the skin. Electrodes, which are an essential part of an extracochlear implant developed at the University of London to treat profound deafness, must be connected both to the cochlea (or inner ear) and to the complex electronics on the outside. Any material which passes through the skin and subcutaneous tissues without an effective seal can provide a channel along which bacteria, which are always present on the skin, can move to cause infection. This is particularly dangerous in any position close to the brain because of potentially fatal meningitis. The anchors which contain the electrodes in this implant are coated with 45S5 Bioglass® and are implanted in the cranium. The implant is placed so that part is in bone where the Bioglass® provides a bond to immobilize it. The part which passes through the skin bonds to the soft tissues and provides the essential seal. The soft tissue bond is protected from damage due to movement, by the bone bond and relatively thin layer of soft tissue overlying it. Such implants have been in place for several years with no significant problems.

Particulates of Various Size Ranges

Bioactive glass in the form of particulate has found application in the treatment of periodontal disease. In this condition the area of gum, and eventually bone, which surrounds the teeth becomes infected and in consequence the attachment of soft tissue

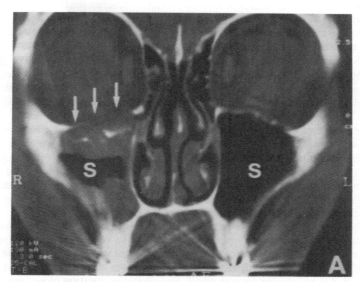

Fig. 4. A) Computer tomography (CT) of a 23-year old man demonstrates blow-out fracture of the right orbital floor (arrows) with herniation of orbital tissues into the maxillary sinus (S).

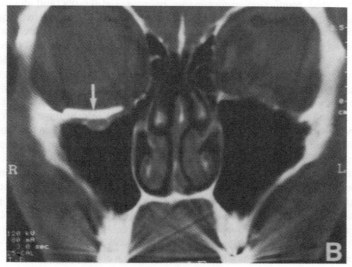

Fig 4. B) CT scan of the same patient taken 4 months after correction of the fracture using a curved bioactive glass implant (arrow). (Courtesy of Dr. Kalle Aitasalo, Department of Otolaryngology, University Central Hospital of Turku, Turku, Finland).

Intro

Soft tissue attachment

which seals and supports the teeth is broken and infection reaches the bone which is then destroyed (Fig. 5). As these supporting structures break down, the teeth are lost. Indeed more teeth are lost because of periodontal disease than for any other reason. The condition is treated by combinations of surgery and antibiotics but it is essential to replace lost bone and soft tissue connections if teeth are to be saved. Bioactive glasses have been shown to be effective in this use. In animal studies it was shown that bone and soft tissue connections were restored to a level close to that of normal teeth when a bioactive glass (45S5 Bioglass®) which could bond to hard and soft tissues was used. Results were much better than with hydroxylapatite materials previously used. These findings have been confirmed in clinical trials. Many periodontists use a membrane in conjunction with bone reconstructive materials to shape the space into which bone will grow. Studies in Finland are combining bioactive glass with such a membrane in a clinical study and comparing results with patients in which glass alone is used. The membrane in use is non-resorbable and must be removed at the end of treatment, necessitating a second operation. Bioactive glass with a resorbable membrane may prove an effective combination for some patients.

Another application which is being developed using 45S5 Bioglass® particulate, is for the treatment of patients with paralysis of one of the vocal cords. Such patients typically have had surgery, which may be for many different reasons, in which the nerve to the cord is cut. Sounds are produced when the vocal cords on each side of the larynx, which form a V shaped opening through which air is breathed, come together and vibrate as air is expelled from the lungs and passes over them. If one or both cords cannot move, proper sound production cannot occur because air escapes and the voice is "breathy". In addition the normal vocal cords protect the airway from aspiration of fluids and infection. Any treatment which can move the paralyzed cord close to the functional one so that closure can occur is effective in restoring voice and preventing infection. Such a treatment can be obtained by placing particulate Bioglass® behind the paralyzed cord. The material remains bonded in the soft tissues producing an augmentation which displaces the paralyzed cord to where it can function. The augmentation is more effective than that produced by an injectable material of PTFE which has been used for this purpose, because of the bonding to soft tissue which occurs and retains the bioactive glass at the site. The augmentation is due to the retained glass and not to the granulomatous reaction to an inert polymer which can disappear with migration of polymer beads from the site. Such migration has produced embolism distant from the site which has been damaging. Attempts to detect such embolism after administration of particulate Bioglass® have been unsuccessful. Such attempts have included light microscopy of adjacent tissues, drainage and other lymph nodes as well as chemical analysis of tissues in which emboli might conceivably be found. No sign, either direct or indirect, of migration of Bioglass® particles has been found and this, combined with specific administration of particulates *in vivo* and *in vitro* during toxicity testing suggests that migration and embolism will not be a problem.

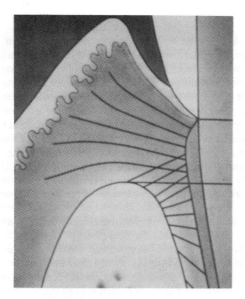

Fig. 5. A) shows the normal structure of the periodontium with insertion of the fibers of the periodontal ligament above the alveolar bone.

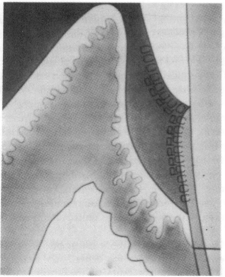

Fig. 5. B) shows the changes due to periodontitis. The fibers have been disrupted and the bone eroded. A successful treatment for this condition must restore the periodontal ligament as well as augment the alveolar bone (Figures courtesy of Dr. Samuel B. Low, Department of Periodontology, College of Dentistry, University of Florida, Gainesville, Florida).

Particulates with Autologous Bone

The success of particulate material in restoring bone in the periodontal area, where the surgical site preparation releases bone fragments and associated growth factors to combine with the bioactive glass, has led to a series of investigations where the glass has been mixed with autologous bone before implanting. The best material for restoring lost or damaged bone for reconstructive and cosmetic reasons is undoubtedly fresh autologous bone which carries none of the potential problems of rejection or viral contamination which are of concern when banked or freeze-dried bone are used. Such bone can be harvested locally or from a site such as the iliac crest and is always the material of choice. However, it is rarely available in large quantities and removal may leave a compromised ilium which may itself then need treatment. In addition there is significant morbidity associated with this procedure. Applications in which available autologous bone fragments are extended by mixing with bioactive glass are being tested in maxillofacial reconstruction. Granules, powders and small blocks of bioactive glass #9 have been combined, using a layering technique with available bone in patients with bone defects resulting from bone cysts and tumor removal. Several patients with chronic frontal sinus infections, chronic pansinusitis and mucocele have been successfully treated in this way A particular advantage seems to be the flexibility within the system, using blocks, powder, granules and bone as needed and as available. Successful results have been achieved in patients with conditions which have been considered intractable. For many reconstructive purposes the mechanical properties of the bone used are secondary in importance to the implant's ability to fill space and be stable. Some applications exist where improvement in mechanical properties as well as volume is desirable. Such an application is the reconstruction of the lower jaw, or mandible, after tumor removal or post-traumatic injury. The restored mandible must be both aesthetically pleasing and be able to support the stress of chewing. Microsurgical techniques have been developed which permit successful movement of a piece of bone, together with a blood supply. The bone chosen is usually rib, but sometimes fibula, and the aesthetic requirement of a reconstructed mandible can be met. Unfortunately such bone is rarely strong enough or big enough to satisfy mechanical requirements. In experiments using 45S5 Bioglass® particulates combined with autologous bone fragments, ribs have been augmented in dogs to up to 10x their original cross-sectional dimension. It is expected that this increase in volume will also supply more satisfactory mechanical properties when bone augmented in this way is used in load bearing situations. However this will require careful long term follow-up studies since all bone, autologous, freeze dried or augmented remodels under use conditions and the final remodelling will not take place until some time after transplantation.

A mixture of bioactive glass granules and autologous bone chips has been used also in situations where the amount of bone is insufficient for dental implants to be inserted. This is a rather common problem in posterior parts of the upper jaw (or maxilla), where, as a result of resorption of the alveolar crest, only a thin bone plate many be found in the floor or maxillary sinus. In such situations the mixture is grafted to the floor of the maxillary sinus around fixtures projecting into the sinus. Preliminary results of this so called "sinus lift" operation are promising (Fig. 6).

In many other applications, especially in orthopaedic surgery, but also in dental implantology, the deficiencies of bioactive glass materials in load bearing applications are overcome by using these materials as coatings on substrates which provide the mechanical properties. This area is at present being investigated extensively (see Chapter 13). Composites of bioactive glasses with other materials are also applied (see Chapters 15 and 16).

Particulates by Injection

A final group of applications are those which depend on the soft tissue adhesion of these materials and which are facilitated by the ability to introduce them by injection. For such applications the interrelationships between particle size and shape, characteristics of the vehicle and needle size and length provide many problems to be solved before bioactive glasses may be used in this way. When these problems can be overcome and the bioactive glass particles placed as required the treatments are effective. Preclinical experiments in animals have concentrated on the treatment of two urological conditions which occur in children and some adults.

The principle of tissue augmentation is as described for the movement of vocal cords, that is to augment specific areas of soft tissue safely and, as far as possible, permanently in a controlled fashion. In patients with incontinence due to decreased urethral resistance such as those with *spina bifida* and bladder exstrophy effective treatment results from tissue augmentation periurethrally which increases urethral resistance. Such treatment with a suspension of PTFE has been effective but, for reasons already given, a replacement material is needed. When tested in an animal model marked increases in urethral resistance were achieved and were maintained up to three months. The treatment is expected to be equally effective in children and clinical trials await the development of a satisfactory combination of materials and delivery system.

A second condition, ureteral reflux, is also amenable to treatment by specific localized areas of soft tissue augmentation. In this condition there is a failure of the mechanism which normally prevents a back flow (or reflux) of urine from the bladder to the ureter. Such reflux allows bacteria, found normally in the bladder, to reach the kidneys causing pyelonephritis and associated kidney damage. If reflux cannot be prevented, treatment must include a lifetime of antibiotic therapy which is undesirable. If the soft tissues beneath the ureter, at the point where it enters the bladder, can be augmented, the reflux can be controlled. Preclinical experiments in animals have shown that this is feasible. However clinical application requires that injection be made via an endoscope to avoid open surgery, particularly since the injection may need to be repeated if the effect is found to be insufficient or transitory. Clinical trials again await the development of a satisfactory combination of materials and delivery system.

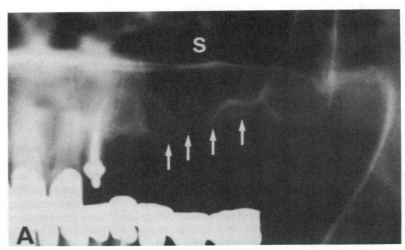

Fig. 6. A) Radiograph of a 69-year old woman shows that the amount of bone in the area of left maxillary sinus (S) is insufficient (arrows) to support dental implants. A sinus lift operation using mixture of bioactive glass granules and autologous bone chips was made to achieve more bone on the floor of the maxillary sinus. Implants were inserted 8 months after the sinus lift operation.

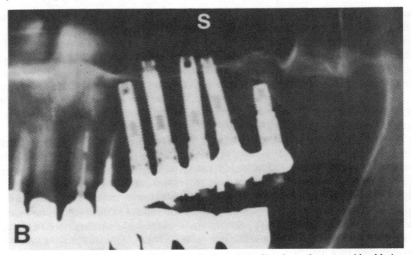

Fig. 6. B) Radiograph of the same area 6 months after insertion of implants shows considerable increase in the amount of bone on the floor of the maxillary sinus. Note good osseointegration of the fixtures. (Courtesy of Dr. Juha Peltola, Department of Oral Diseases, University Central Hospital of Turku, Turku, Finland).

SUMMARY

Bioactive glass materials are available in a range of compositions which are able to bond to soft tissues and/or bone. These materials vary in their reactivity and speed of bonding and in the ability of the bone to provide mechanical strength. They have been successfully applied as solids and particulates and may be combined with other materials, both natural and synthetic, to provide treatment for many disparate clinical conditions.

READING LIST

June Wilson and G. H. Pigott, "Toxicology and Biocompatibility of Bioglasses®," *J. Biomed. Mater. Res.* **15** (1981) 805-817.

June Wilson, G. E. Merwin, and L. L. Hench, "Machining Bioglass® Implants," *SAMPE Journal* **21[3]** (1985) 6-8.

June Wilson and David Nolletti, "Bonding of Soft Tissues to Bioglass®," in *Handbook of Bioactive Ceramics, Vol. I, Bioactive Glasses and Glass-Ceramics*, eds. T. Yamamuro, L. L. Hench and J. Wilson (CRC Press, Boca Raton, Florida, 1990) pp. 283-302.

June Wilson and S. B. Low, "Bioactive Ceramics for Periodontal Treatment: Comparative Studies in the Patus Monkey," *J. Applied Biomaterials* **3** (1992) 123-129.

L. L. Hench, June Wilson and G. Merwin, "Bioglass® Implants for Otology," in *Biomaterials in Otology*, ed. J. J. Grote (Martinus Nijhoff Publishers, The Hague-Bonton-London (1983) pp. 62-69.

H. R. Stanley, M. B. Hall, F. Colaizzi, and A. E. Clark, "Residual Alveolar Ridge Maintenance with a New Endosseous Implant Material," *J. Prosthetic Dentistry* **58[5]** (1987) 607-613.

June Wilson and G. E. Merwin, "Biomaterials for Facial Bone Augmentation: Comparative Studies," *J. Applied Biomaterials* **22[A2]** (1988) 159-177.

Chapter 5

A/W GLASS-CERAMIC: PROCESSING AND PROPERTIES

Tadashi Kokubo

Department of Materials Science, Kyoto University, Japan

COMPOSITION AND PROCESSING

Glass can be converted by a heat treatment into glass-crystal composites containing various kinds of crystalline phases with controlled sizes and contents. The resultant glass-ceramic can exhibit superior properties to the parent glass and to sintered crystalline ceramics. Generally, monophase bioactive ceramics such as Bioglass®-type glasses and sintered hydroxyapatite do not show as high a mechanical strength as human cortical bone. Natural bone is a composite in which an assembly of hydroxyapatite small crystal particles is effectively reinforced by organic collagen fibers. Kokubo et al attempted to prepare a similar composite by a process of crystallization of glass in 1982. In this attempt, ß-wollastonite ($CaO \cdot SiO_2$) consisting of a silicate chain structure was chosen as the reinforcing phase.[1]

A parent glass in the pseudoternary system $3CaO \cdot P_2O_5$-$CaO \cdot SiO_2$-$MgO \cdot CaO \cdot 2SiO_2$ was prepared by the conventional melt-quenching method. When the glass in a bulk form was heated up to 1050°C at a rate of 5°C/min., fine grained oxyapatite and fibrous ß-wollastonite precipitated. Large cracks were, however, formed in the middle part of the crystallized product, because the wollastonite precipitated only from the outer surfaces of the glass.[2] In the next preparation the parent glass was crushed into a fine powder 5 μm in average size, pressed into the desired form and then subjected to the same heat treatment. The cracks were eliminated, but small amount of micropores remained in the intergranular spaces of the glass powders, because the apatite and wollastonite precipitated before complete densification.[3] A small amount of CaF_2 was added to the composition of the parent glass, and glass powder of the composition MgO 4.6, CaO 44.7, SiO_2 34.0, P_2O_5 6.2, and CaF_2 0.5 wt% was subjected to the same treatment. As a result, the glass powders were fully densified at about 830°C, and then the oxyfluoroapatite ($Ca_{10}(PO_4)_6(O,F_2)$) and wollastonite precipitated successively at 870 and 900°C respectively, to give a crack- and pore-free dense and homogeneous glass-ceramic.[2]

Figure 1 shows a transmission electron micrograph of the resultant glass-ceramic. In this case, the wollastonite did not take the fibrous form. Both the apatite and wollastonite were homogeneously dispersed in a glassy matrix, taking the shape of a rice grain 50 to 100 nm in size. According to powder X-ray diffraction, the contents of apatite, wollastonite and residual glassy phase were 38, 34 and 28 wt%, respectively, and the composition of the residual glassy phase was estimated to be MgO 16.6, CaO 24.2 and SiO_2 59.2 wt%.[4] This glass-ceramic was named A/W after the names of the crystalline phases and is called Cerabone® A-W commercially.

75

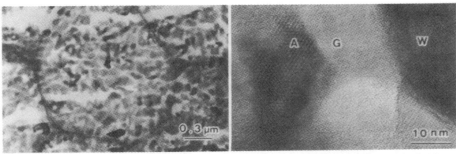

Fig. 1. Transmission electron micrograph of A/W glass-ceramic. A: Apatite, W: Wollastonite, G: Glassy phase.

MECHANICAL PROPERTIES

A/W glass-ceramic can easily be machined into various shapes, even into screws, by diamond tools. Figure 2 shows A/W glass-ceramic shaped into artificial vertebrae, intervertebral spacers, spinous process spacers and iliac spacers. Some physical properties of the glass-ceramic are shown in Table 1. The bending strength (215 MPa) of this glass-ceramic is almost twice that (115 MPa) of dense sintered hydroxyapatite and even higher than that (160 MPa) of human cortical bone in an air environment. The parent glass G and the glass-ceramic A, precipitating only the apatite, have bending strengths of 72 and 88 MPa, respectively. It is evident that the high bending strength of A/W glass-ceramic is due to the precipitation of the wollastonite as well as apatite. A/W glass-ceramic has a fracture toughness of 2.0 MPa·m$^{1/2}$ whereas glass G and the glass-ceramic A have only 0.8 and 1.2 MPa·m$^{1/2}$, respectively. This means that the high bending strength of A/W glass-ceramic is attributed to its high fracture toughness. A/W glass-ceramic has a fracture surface energy of 15.9 Jm^{-2}, whereas glass G and glass-

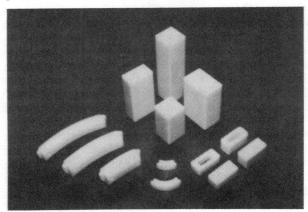

Fig. 2. Iliac spacers (left), artificial vertebrae (middle top), spinal spacers (middle bottom) and intervertebral spacers (right).

Table 1. Physical Properties of A/W Glass-Ceramic.

Density (g/cm^3)	3.07
Bending Strength (MPa)	215
Compressive Strength (MPa)	1080
Young's Modulus (GPa)	118
Vickers Hardness (HV)	680
Fracture Toughness (MPa$^{1/2}$)	2.0
Slow Crack Growth, n	33

ceramic A have only 3.3 and 6.4 Jm^{-2}, respectively. The high fracture toughness of A/W glass-ceramic is attributable to the high fracture surface energy. Glass G and glass-ceramic A show a fairly smooth fracture surface, whereas A/W glass-ceramic has a roughened fracture surface, as shown in Fig. 3. This means that the wollastonite effectively prevents straight propagation of the cracks, causing them to turn or branch out.[5] It is notable that the wollastonite exhibits such a reinforcing effect, even though it is not in a fibrous form.

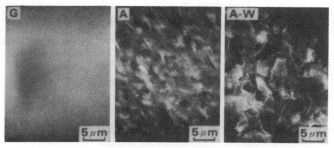

Fig. 3. Scanning electron micrographs of fracture surfaces of glass G, glass-ceramic A and A/W glass-ceramic.

When loaded in an aqueous body environment, this glass-ceramic shows a decrease in mechanical strength, i.e. fatigue, by slow crack growth due to stress corrosion, similar to other ceramics. The magnitude of its fatigue is, however, much lower than those of glass G and glass-ceramic A.[6] The decrease in bending strength of A/W glass-ceramic with decreasing stress rate in a simulated body fluid with ion concentrations nearly equal to those of the human blood plasma of pH 7.25 at 36.5°C (Table 2), is much lower than those of glass G and glass-ceramic A (Fig. 4). The parameter, n, of slow crack growth, which is derived from the dependence of the bending strength upon the stressing rate, is 33 for A/W glass-ceramic whereas it is 9 and 18 for glass G and glass-ceramic A, respectively. When a bending stress of 65 MPa is

Table 2. Ion Concentrations of Simulated Body Fluid (SBF) and Human Blood Plasma.

	Ion Concentration (mM)							
	Na^+	K^+	Ca^{2+}	Mg^{2+}	Cl^-	HCO_3^-	HPO_4^{2-}	SO_4^{2-}
SBF	142.0	5.0	2.5	1.5	147.8	4.2	1.0	0.5
Human Plasma*	142.0	5.0	2.5	1.5	103.0	27.0	1.0	0.5

Buffered at pH 7.25 with 50 mM tris (hydroxymethyl) aminomethane $((CH_2OM)_3CNH_2)$ and 45 mM hydrochoric acid (HCl).

*J. Gamble, *Chemical Anatomy, Physiology and Pathology of Extracellular Fluid,* 6 Ed. (Harvard University Press, Cambridge, 1967) p. 1.

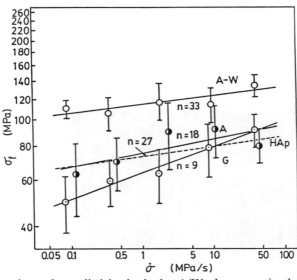

Fig. 4. Dependence of bending strength (σ_f) of glass G, glass-ceramic A, A/W glass-ceramic and dense sintered hydroxyapatite (HAp) in the simulated body fluid at 36.5°C upon stressing rate ($\dot{\sigma}$).

continuously applied in the body, A/W glass-ceramic should withstand it for over 10 years, whereas glass G, glass-ceramic A and dense sintered hydroxyapatite would survive only 1 min. The magnitude of the fatigue of A/W glass-ceramic can be further decreased by a surface modification such as Zr+ ion implantation.[7] Animal experiments have shown that A/W glass-ceramic maintains its high mechanical strength for long periods *in vivo*.[8]

SURFACE CHEMISTRY

A/W glass-ceramic is so tightly bonded to the living bone that the fracture under tensile stress does not usually occur at the interface between the glass-ceramic and bone, but within the bone. Its bioactivity is much higher than that of sintered hydroxyapatite. For example, its granular particles are covered with newly grown bone up to 90% of their surfaces within 4 weeks in rat tibia, whereas those of the sintered hydroxyapatite are only 60% covered even after 16 weeks, as shown in Fig. 5.[9] Thus high bioactivity of A/W glass-ceramic is attributed to a specific surface property.

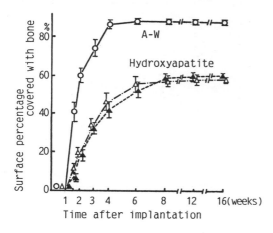

Fig. 5. Surface area of A/W glass-ceramic and two kinds of commercial sintered hydroxyapatite covered with newly grown bone as a function of time after implantation.

When A/W glass-ceramic is implanted into a bone defect, it forms a thin layer, rich in Ca and P, on its surface and bonds to the surrounding bone through this layer, as shown in Fig. 6.[10] This Ca,P-rich layer is identified as a layer of an apatite by X-ray microdiffraction as shown in Fig. 7.[11] With transmission electron microscopy, A/W glass-ceramic is seen closely connected to the living bone through this apatite layer, without a distinct boundary as shown in Fig 8.[12]

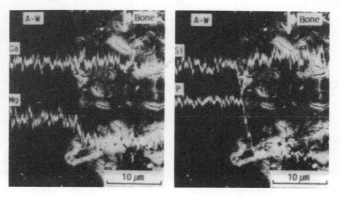

Fig. 6. Electron probe X-ray microanalysis of the interface of A/W glass-ceramic in rabbit tibia.

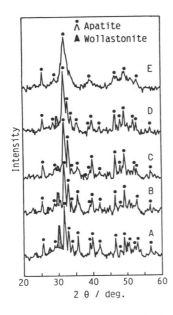

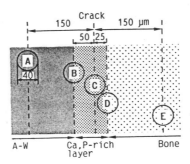

Fig. 7. X-ray microdiffraction of the interface of A/W glass-ceramic to the bone of sheep vertebra.

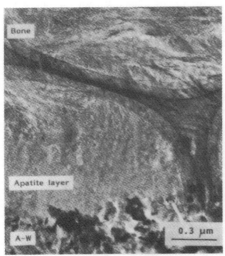

Fig. 8. Transmission electron micrograph of the interface of A/W glass-ceramic in rat tibia.

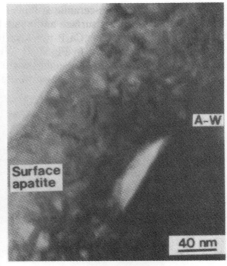

Fig. 9. Transmission electron micrograph of a cross section of A/W glass-ceramic soaked in simulated body fluid.

The same type of apatite layer is formed on the surface of A/W glass-ceramic even in a simulated body fluid with ion concentrations nearly equal to those of human blood plasma (Table 2), as shown in Fig. 9. According to thin-film X-ray diffraction and Fourier transform infrared reflection spectroscopy of the surfaces of A/W glass-ceramic soaked in simulated body fluid, the surface apatite layer consists of a carbonate-containing hydroxyapatite with small crystallite and/or defective structure.[13] The compositional and structural characteristics of this apatite are similar to those of the apatite in the natural bone. It is expected that a bone-producing cell, osteoblast, would proliferate preferentially over fibroblasts, on the surface of the apatite layer. Consequently, fibrous tissue, which usually forms around foreign material, is not formed around A/W glass-ceramic, and the surrounding bone can grow directly on the surface apatite layer. When this occurs, a tight chemical bond forms between the surface apatite and the bone apatite, in order to reduce the interfacial energy. This is confirmed by the observation that a pair of A/W glass-ceramic samples which were implanted, soaked in the simulated body fluid[14] or implanted subcutaneously into rats[15] were so tightly bonded together through a mutual apatite layer that they could not be separated. This suggests that the bone-like apatite layer which is formed on the surface of A/W glass-ceramic in the body plays an essential role in forming the chemical bond of the glass-ceramic to the bone. The same type of Ca,P-rich layer or apatite layer has been observed also for Bioglass®-type glasses,[16] Ceravital®-type glass-ceramic,[17] and sintered hydroxyapatite,[18] but not for non-bioactive glasses[19] and glass-ceramics.[20] The bone-like apatite layer plays an essential role in forming the chemical bond of all bioactive materials which bond to bone. The higher bioactivity of A/W glass-ceramic than that of sintered hydroxyapatite might be attributed to the higher rate of formation of the apatite layer on A/W glass-ceramic.[18]

The human body fluid is already supersaturated with respect to apatite under normal condition.[21] Therefore, once apatite nuclei are formed, they can spontaneously grow. A/W glass-ceramic releases appreciable amounts of calcium and silicate ions into the simulated body fluid.[4] It is probable that these ions promote apatite nucleation on the surface of A/W glass-ceramic. In order to study this mechanism in more detail, the compositional dependence of apatite formation on the surface of glasses in the simple ternary system CaO-SiO_2-P_2O_5 was investigated in simulated body fluid. It was found that only CaO,SiO_2-based glasses formed the apatite layer on their surfaces within 30 days, whereas the CaO,P_2O_5-based glasses did not. This behavior is in contrast to the conventional expectation, at least within the glass-forming compositional region, as shown in Fig. 10.[22] Even P_2O_5-free $CaO \cdot SiO_2$ binary glasses formed the apatite layer *in vitro* as well as *in vivo*.[23] The CaO,SiO_2-based glasses release an appreciable amount of calcium, whereas the CaO,P_2O_5-based glasses release phosphates. Both types of ions increase the ionic activity product of the apatite in the surrounding fluid. The magnitude of the increase is almost equal between the CaO,SiO_2-based glasses and the CaO,P_2O_5-based glasses, as shown in Fig. 11.[24] In spite of it, only the CaO,SiO_2-based glasses formed the apatite layer which decreased the ionic activity product of the fluid.

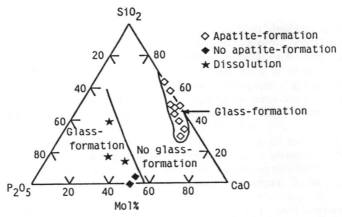

Fig. 10. Compositional dependence of apatite formation on glasses in the system CaO-SiO$_2$-P$_2$O$_5$ in simulated body fluid (soaking time: 30 days).

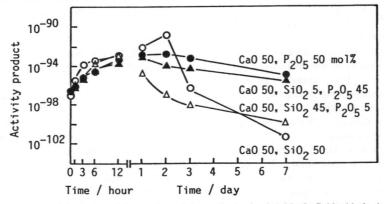

Fig. 11. Variation of the ionic activity product of the apatite in simulated body fluid with the immersion of CaO-SiO$_2$-P$_2$O$_5$ glasses.

This behavior is due to a peculiar surface structure of the CaO,SiO$_2$-based glasses which provides favorable sites for apatite nucleation. The CaO, SiO$_2$-based glasses form a silica hydrogel layer prior to the formation of the apatite layer.[23,24] It is probable that this hydrated silica induces apatite nucleation. This is confirmed by the observation that a pure silica gel, prepared by the sol-gel method, formed bone-like apatite on it when soaked in simulated body fluid at pH 7.4, as shown in Fig. 12.[25] Such apatite formation was observed neither on the surface of silica glass nor on that of a crystalline quartz silica. This suggests that a certain kind of silanol group, abundant on the surface of the silica gel, is responsible for the apatite nucleation.

Fig. 12. Apatite formation on a pure silica gel in simulated body fluid of pH 7.4.

SUMMARY

The mechanism of apatite formation on the surfaces of $CaO \cdot SiO_2$-based glasses and glass-ceramics, including A/W glass-ceramic, in the body can be interpreted as follows. The calcium ion dissolved from the glasses and glass-ceramics increases the ion activity product of the apatite in the surrounding body fluid, and the hydrated silica on the surfaces of the glasses and glass-ceramics provides favorable sites for apatite nucleation, as shown in Fig. 13. Consequently, the apatite nuclei are rapidly formed on their surfaces. Once the apatite nuclei are formed, they spontaneously grow by consuming calcium and phosphate ions from the surrounding body fluid. In the case of A/W glass-ceramic, although the presence of the silica gel layer on its surface could not be detected even under the high resolution transmission electron micrographs (see Figs. 8 and 9), the dissolution of an appreciable amount of the silicate ion from the glass-ceramic into the simulated body fluid indicates the formation of a large number of silanol groups at the surface of the glass-ceramic in the body.

If the explanation of the mechanism of the apatite formation described above is valid, it is expected that the bone-like apatite layer could be formed on the surfaces of various kinds of materials including metals, ceramics and organic polymers by the following biomimetic method at ordinary temperature and pressure.[26] When a material as a substrate is placed on or in granular particles of a CaO, SiO_2-based glass soaked in the simulated body fluid for a certain period, as shown in Fig. 14, a large number of apatite nuclei can be formed on the surface of the substrate as well as on the surfaces of the glass particles by the calcium and silicate ions dissolved from the glass particles. When the substrate is then soaked in another solution supersaturated with respect to the apatite, the apatite nuclei grow spontaneously *in situ* on the substrate by consuming the calcium and phosphate ions from the surrounding solution to form the apatite layer.[27] A dense and uniform layer of bone-like apatite was formed on various kinds of materials

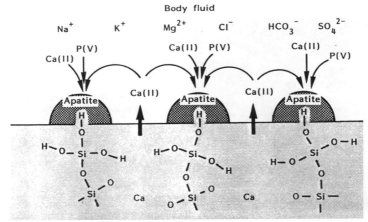

Fig. 13. Schematic representation of the mechanism of apatite formation on the surfaces of CaO,SiO$_2$-based glasses and glass-ceramics in the body.

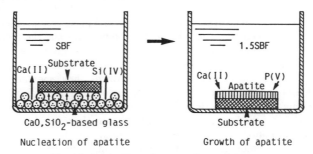

Fig. 14. Apatite formation on various substrates.

including stainless steel, titanium metal, platinum, gold, silicon, carbon, alumina, zirconia, polymethylmethacrylate, polyethylene, polyethyleneterephthalate and polyethersulfone, by this method, as shown in Fig. 15.[26,27] The thickness of the apatite layer continued to increase with increasing soaking time in the second solution, as shown in Fig. 16. The rate of growth of the apatite layer increased with increasing temperature (Fig. 16) and the degree of the supersaturation of the second solution.[28] The rate of the growth in the solution with ion concentrations 1.5 times of the simulated body fluid at 60°C was 7 μm/day. The adhesive force of the apatite layer thus formed to the substrate varied with the time and roughness of the substrate. Poly-ethyleneterephthalate and polyethersulfone showed remarkably high adhesive strength to the apatite layer among the examined polymers. Figure 17 shows the bone-like apatite layer coated on fine fibers of a cloth of polythethyleneterephthalate. This apatite-polymer

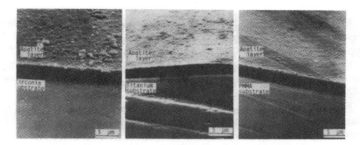

Fig. 15. Apatite layer formed on various substrates.

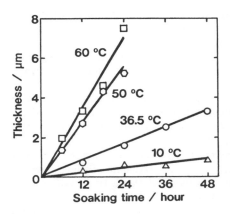

Fig. 16. The thickness of the apatite layer as a function of soaking time in a solution with ion concentrations 1.5 times those of simulated body, fluid at various temperatures.

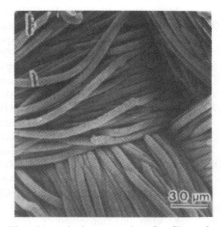

Fig. 17. Apatite layer coated on fine fibers of a cloth of polyethyleneteraphthalate.

composite can be bent sharply without peeling off the apatite layer. This type of composite may be useful for fabricating highly bioactive artificial bone with mechanical properties close to those of the natural bone, not only fracture strength and fracture toughness, but also elastic modulus. In addition, this type of organic polymer coated with apatite is expected to show high compatibility even with soft tissues.

The fundamental understanding of the mechanism of the apatite formation on A/W glass-ceramic in the body described above also provides the way for developing other kinds of high performance bioactive materials, such as self-setting bioactive materials useful as bone fillers and drug delivery system[29,30] and ferrimagnetic bioactive materials useful as thermoseeds for hyperthermia treatment of cancer.[31,32]

REFERENCES

1. T. Kokubo, M. Shigematsu, Y. Nagashima, M. Tashiro, T. Yamamuro and S. Higashi, "Apatite- and Wollastonite-Containing Glass-Ceramics for Prosthetic Application," *Bull. Inst. Chem. Res.*, Kyoto Univ., **60** (1982) 260-268.

2. T. Kokubo, S. Ito, S. Sakka and T. Yamamuro, "Formation of a High-Strength Bioactive Glass-Ceramic in the System $MgO-CaO-SiO_2-P_2O_5$," *J. Mater. Sci.* **21** (1986) 536-540.

3. T. Kokubo, Y. Nagashima and M. Tashiro, "Preparation of Apatite Containing Glass-Ceramics by Sintering and Crystallization of Glass Powders," *Yogyo-Kyokai-Shi* **90** (1982) 151.

4. T. Kokubo, H. Kushitani, C. Ohtsuki, S. Sakka, and T. Yamamuro, "Chemical Reaction of Bioactive Glass and Glass-Ceramics with a Simulated Body Fluid," *Materials in Medicine* **3** (1992) 79-83.

5. T. Kokubo, S. Ito, M. Shigematsu, S. Sakka and T. Yamamuro, "Mechanical Properties of a New Type of Apatite-Containing Glass-Ceramic for Prosthetic Application," *J. Mater. Sci.* **20** (1985) 2001-2004.

6. T. Kokubo, S. Ito, M. Shigematsu, S. Sakka and T. Yamamuro, "Fatigue and Life Time of Bioactive Glass-Ceramic A-W Containing Apatite and Wollastonite," *J. Mater. Sci.* **22** (1987) 4067-4070.

7. G. H. Takaoka, R. Tanaka, H. Usui, S. K. Kcoh, M.I. Current, I. Yamada, S. Akamatsu and T. Kokubo, "Surface Modification of Bioactive Ceramic (Artificial Bone) by Ion Implantation," in *Tissue-Inducing Biomaterials*, ed. L. G. Cima and E. S. Ron (Materials Research Society, Pittsburgh, 1992) pp. 23-28.

8. T. Kitsugi, T. Yamamuro, T. Nakamura, T. Kakutani, Y. Hayashi, S. Ito, T. Kokubo and T. Shibuya, "Aging Test and Dynamic Fatigue Test of Apatite-Wollastonite Containing Glass-Ceramics and Dense Hydroxyapatite," *J. Biomed. Mater. Res.* **21** (1987) 467-484.

9. K. Ono, T. Yamamuro, T. Nakamura and T. Kokubo, "Quantitative Study on Osteoconduction of Apatite-Wollastonite Containing Glass-Ceramic Granules, Hydroxyapatite Granules, and Alumina Granules," *Biomaterials* **11** (1990) 265-271.

10. T. Kitsugi, T. Nakamura, T. Yamamuro, T. Kokubo T. Shibuya and M. Takagi, "SEM-EPMA Observation of Three Types of Apatite-Containing Glass-Ceramics Implanted in Bone: The Variance of a Ca-P-rich Layer," *J. Biomed. Mater. Res.* **21** (1987) 1255-1271.

11. T. Kokubo, C. Ohtsuki, S. Kotani, T. Kitsugi and T. Yamamuro, "Surface Structure of Bioactive Glass-Ceramic A-W Implanted into Sheep and Human Vertebra," in *Bioceramics*, Vol. 2, ed. G. Heimke (German Ceramic Society, Cologne, 1990) pp. 105-112

12. M. Neo, S. Kotani, Y. Fujita, T. Nakamura, T. Yamamuro, Y. Bando, C. Ohtsuki and T. Kokubo, "Difference of Bone-Bonding Mechanism Between Surface-Active Ceramics and Reasonable Ceramics: A Study by Scanning and Transmission Electron Microscopy," *J. Biomed. Mater. Res.* **26** (1992) 255-267.

13. T. Kokubo, S. Ito, Z. Huang, T. Hayashi, S. Sakka, T. Kitsugi and T. Yamamuro, "Ca,P-rich Layer Formed on High-Strength Bioactive Glass-Ceramic A-W," *J. Biomed. Mater. Res.* **24** (1987) 331-343.

14. T. Kokubo, T. Hayashi, S. Sakka, T. Kitsugi and T. Yamamuro, "Bonding Between Bioactive Glasses, Glass-Ceramics or Ceramics in a Simulated Body Fluid," *Yogyo-Kyokai-Shi* **95** (1987) 785-791.

15. T. Kitsugi, T. Yamamuro, T. Nakamura, T. Kopkubo, M. Takagi, T. Shibuya, H. Takeuchi and M. Ono, "Bonding Behavior Between Two Bioactive Ceramics *In Vivo*," *J. Biomed. Mater. Res.* **21** (1987) 1109-1123.

16. L. L. Hench and A. E. Clark, "Adhesion to Bone," in *Fundamental Aspects of Biocompatibility,* Vol. 1, ed. D. F. Williams (CRC Press Boca Raton, FL. U.S.A, 1981) pp. 67-85.

17. C. Otsuki, H. Kushitani, T. Kokubo, S. Kotani, T. Yamamuro and T. Nakamura, "Apatite Formation on the Surface of Ceravital-Type Glass-Ceramic in the Body," *J. Biomed. Mater. Res.* **25** (1991) 1363-1370.

18. T. Kokubo, H. Kushitani, Y. Ebisawa, T. Kitsugi, S. Kotani, K. Oura and T. Yamamuro, "Apatite Formation on Bioactive Ceramics in Body Environment,"in *Bioceramics*, Vol. 1, ed. H. Oonishi, H. Aoki and K. Sawai, (Ishiyaku EuroAmerica, Tokyo, 1989) pp. 157-162.

19. T. Kitsugi, T. Yamamuro, T. Nakamura and T. Kokubo, "Bone Bonding Behavior of $MgO-CaO-SiO_2-P_2O_5-CaF_2$ Glass (Mother Glass of A-W Glass-Ceramics)," *J. Biomed. Mater. Res.* **23** (1989) 631-648.

20. T. Kitsugi, T. Yamamuro, T. Nakamura and T. Kokubo, "The bonding of Glass Ceramics to Bone," *International Orthopaedics (SICOT)* **13** (1989) 199-206.

21. W. Neuman and M. Neuman, in *The Chemical Dynamics of Bone Mineral* (University of Chicago, Chicago, 1958) p. 175.

22. C. Ohtsuki, T. Kokubo, K. Takatsuka and T. Yamamuro, "Composition Dependence of Bioactivity of Glasses in the System $CaO-SiO_2-P_2O_5$: Its *In Vitro* Evaluation," *J. Ceram. Soc.*, Japan, **99** (1991) 1-6.

23. K. Ohura, T. Nakamura, T. Yamamuro, T. Kokubo, Y. Ebisawa, Y. Kotoura and M. Oka, "Bone-Bonding Ability of P_2O_5-Free $CaO \cdot SiO_2$ Glasses," *J. Biomed. Mater. Res.* **25** (1991) 357-365.

24. C. Ohtsuki, T. Kokubo and T. Yamamuro, "Mechanism of Apatite Formation on $CaO-SiO_2-P_2O_5$ Glasses in a Simulated Body Fluid," *J. Non-Crystal. Solids* **143** (1992) 84-92.

25. P. Li, C. Ohtsuki, T. Kokubo, K. Nakanishi, N. Soga, T. Nakamura and T. Yamamuro, "Apatite Formation Induced by Silica Gel in a Simulated Body Fluid," *J. Am. Ceram. Soc.*, in press.

26. Y. Abe, T. Kokubo and T. Yamamuro, "Apatite Coating on Ceramics, Metals and Polymers Utilizing a Biological Process," *Materials in Medicine* **1** (1990) 233-238.

27. T. Kokubo, K. Hata, T, Nakamura and T. Yamamuro, "Apatite Formation on Ceramics, Metals and Polymers Induced by a CaO, SiO_2-Based Glass in a

Simulated Body Fluid," in *Bioceramics*, Vol. 4, ed. W. Bonfield, G. W. Hastings and K. E. Tunner (Butterworth-Heinemann, Guildford, 1991) pp. 113-120.

28. T. Kokubo, K. Hata, T. Nakamura and T. Yamamuro, "Growth of Bone-Like Apatite Layer on Various Materials in Simulated Body Fluids," *Transactions of Forth World Biomaterials Congress*, 1992, Berlin, p. 67.

29. T. Kokubo, S. Yoshihara, N. Nishimura, T. Yamamuro and T. Nakamura, "Bioactive Bone Cement Based on $CaO-SiO_2-P_2O_5$ Glass," *J. Am. Ceram. Soc.* **74** (1991) 1739-1741.

30. N. Nishimura, T. Yamamuro, Y. Taguchi, M. Ikenaga, T. Nakamura, T. Kokubo and S. Yoshihara, "A New Bioactive Bone Cement: Its Histological and Mechanical Characterization," *J. Applied Biomater.* **2** (1991) 219-229.

31. T. Kokubo, Y. Ebisawa, Y. Sugimoto, M. Kiyama, K. Ohura, T. Yamamuro, M. Hiraoka and M. Abe, "Preparation of Bioactive and Ferromagnetic Glass-Ceramic for Hyperthermia," in *Bioceramics*, Vol 3, ed. J. E. Hulbert and S. F. Hulbert (Rose-Hulman Institute of Technology, Terre Haute, 1992).

32. M. Ikenaga, K. Ohura, T. Nakamura, Y. Kotoura, T. Yamamuro, M. Oka, Y. Ebisawa and T. Kokubo, "Hyperthermic Treatment of Experimental Bone Tumor with a Bioactive Ferromagnetic Glass-Ceramic," in *Bioceramics*, Vol. 4, ed. W. Bonfield, G. W. Hastings and K. E. Tanner (Butterworth-Heinemann, Guildford, 1991) pp. 255-262.

Chapter 6

A/W GLASS-CERAMIC: CLINICAL APPLICATIONS

Takao Yamamuro

Department of Ortopedics - Kyoto University, Japan

INTRODUCTION

There are many kinds of ceramics currently available for the repair of extensive lesions or defects in bones and joints. Among them, alumina ceramic and synthetic hydroxyapatite (HA) have recently become widely used to fabricate bone and joint prostheses or as bone fillers. Alumina ceramic exhibits characteristics such as high resistance to elution and corrosion, high mechanical strength, good biocompatibility, and low friction. Alumina ceramic appears adequate for prolonged clinical use as a bearing surface of joint prostheses. (Chapter 2.) When it was used as a substitute for the pelvis or vertebra, loosening of the prosthesis frequently occurred in a relatively short period of time, presumably due to insufficient mechanical binding between the material and bone. It is well known that alumina ceramic is very compatible with bone, but does not chemically bond to the bone tissue. Microscopic observation of the interface between the alumina ceramic prosthesis and bone always reveals an intervening layer of connective tissue. Therefore, the use of alumina ceramic as a bone substitute under unfavorable mechanical conditions will result in proliferation of the connective tissue as the interface, leading eventually to loosening of the prosthesis.

On the other hand, Bioglass®, Ceravital®, and synthetic hydroxyapatite[1] have a strong chemical bond with bone *in vivo*. Mechanical strengths exceeding that of the human cortical bone were not reached, however, except for dense A/W glass-ceramic. As described earlier A/W glass-ceramic is capable of binding strongly to living bone in a few weeks and has a mechanical strength significantly higher than that of human cortical bone (Chapter 5). Since 1983, we have used A/W glass-ceramic in spine and hip surgery of patients with extensive lesions or bone defects, and the results were quite satisfactory in more than 200 cases.

Bonding of A/W Glass-Ceramic to Bone

The bonding ability of A/W glass-ceramic to bone was evaluated using rabbit tibial bones, and the load to failure was measured and compared with that of other bioactive ceramics. In the first experiment,[2] each bioactive ceramic was shaped into a rectangular plate (2mm X 10mm X 15mm), surfaces were polished by No. 1500 SiC paper, and they were placed in an ethanol-filled ultrasonic cleaner for 20 min. The tibial proximal metaphyses were exposed subperiosteally, a 15 mm long hole penetrating the medial and lateral cortex was made using a dental burr parallel to the longitudinal axis of the tibia, and a ceramic plate was implanted into each hole. Each experimental group comprised of 4-8 rabbits, which were killed 2, 4, 8, and 25 weeks postimplantation. A segment of the tibia containing a ceramic implant was excised and the bone dissected as shown in Fig. 1a. After dissection, the bone on either side of the implant was not directly connected with the other but was joined only through the intervening implant, Fig. 1b. Each segment was held with a hook connected to an Instron-type testing machine. The implant was placed horizontally and pulled in opposite direction at a cross-head speed of 3.5cm/min. (Fig. 1c) the load at which the implant, the bone, or the interface was broken was designated as failure load. The failure load 8 and 25 weeks after implantation is shown in Table 1. A/W glass-ceramic showed tight bonding to bone comparable with synthetic dense HA, and in 25 weeks its load was 70% of that of bone. The breaking of the bond occurred mostly in the bone in A/W glass-ceramic groups, while it occurred mostly in the ceramic in HA groups, but not at the bonded interface in either group. The bonded interface was observed by contact microradiography and SEM after being ground to about 80 μm thick. Histological examination revealed that A/W glass-ceramic bonded directly to the bone tissue (Fig. 2) as did synthetic HA by forming a Ca, P-rich layer at the interface.[3]

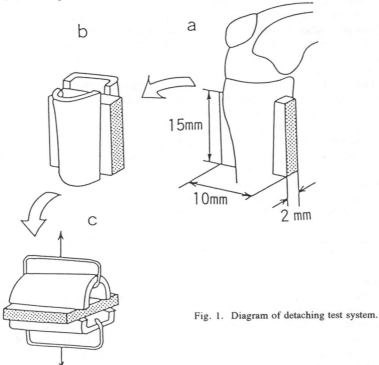

Fig. 1. Diagram of detaching test system.

Table 1. Failure Loads Obtained by Detaching Test.

Ceramics	Mean ± S.D. (kg)	
	8 Weeks	24 Weeks
Alumina	0.18 ± 0.018	
Bioglass®	2.75 ± 1.8	
Ceravital®	5.51 ± 1.48	4.35 ± 1.45
Dense Hydroxyapatite	6.57 ± 1.36	7.77 ± 1.91
AW Glass-Ceramic	7.44 ± 1.91	8.19 ± 3.6

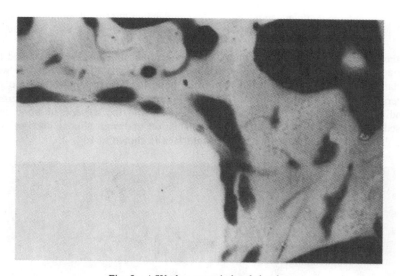

Fig. 2. A/W glass-ceramic bonded to bone.

In the second experiment,[4] to the bonding of A/W glass-ceramic to the surface of the cortical, the proximal metaphyseal cortex of rabbit tibia was exposed subperiosteally and an implant made of A/W glass-ceramic was fixed on the surface of cortex with a metal screw (Fig. 3). The rabbits were killed 2, 4, 8, and 25 weeks after the implantation. Each experimental group comprised of 5-6 animals. After removing the metal screw, the bone segment was excised and the bone covering the margin of the implant was completely removed. The implant and the piece of bone were held by hooks which were connected to an Instron-type testing machine. The implant was pulled perpendicularly to the base of the implant in contact with bone cortex, at a cross-head speed of 3.5cm/min. and the load required to detach the implant from the bone or to

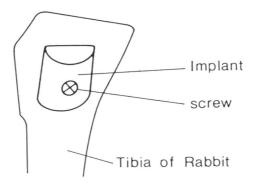

Fig. 3. Fixation of implant.

break the bone was measured. All the A/W glass-ceramic implants were detached from the surface of bone cortex leaving bone attached to the implants after 4, 8, and 25 weeks implantation. The fracture occurred within the bone in most instances, and the A/W glass-ceramic implants did not break (Fig. 4). Figure 5 shows that the tensile strength of bonding between the A/W glass-ceramic implant and the surface of bone cortex increased remarkably 4 weeks after the implantation and was almost maximal by 8 weeks. The histological appearance at the interface is shown in Fig. 2.

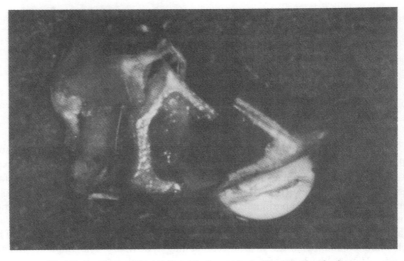

Fig. 4. Implant and bone after detachment test. The bone has broken.

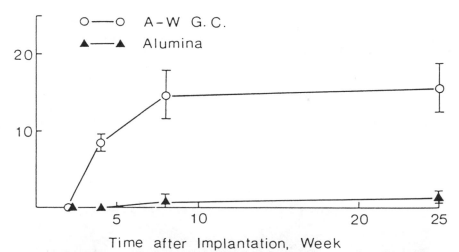

Fig. 5. Tensile strength in Kg/cm^2 (vertical axis) against implant time in weeks (horizontal axis).

It has been confirmed mechanically and histologically that implants made of A/W glass-ceramic bind firmly to living bone in a few weeks and the implants do not deteriorate in vivo.[5,6] Further-more, *in vitro* studies suggest that A/W glass-ceramic binds to bone in a shorter time than synthetic HA, presumably because it contains a glassy phase which releases more Ca ions in the early postimplantation stage than does hydroxyapatite and releases silicate ions which may initiate crystallization of apatites working as their nuclei on the surface of the implant[7] (Chapter 5).

Intercalary Replacement of a Segment of the Long Bone with an A/W Glass-Ceramic Implant in Animals

Intercalary replacement of the shafts of rabbit tibiae with A/W glass-ceramic implants was performed under weight bearing conditions to determine whether this glass-ceramic was useful as a material for load-bearing prostheses.[8] A 16mm segment of the middle of the shaft of each rabbit tibia was resected and the defect was replaced with a hollow cylindrical implant. The implant was 9mm in diameter and 15mm long, and had a central hole with a diameter of 3.05mm. It was fixed to the tibia by intramedullary nailing using a 3mm Kirschner wire. A cast was applied from thigh to toe for 6 weeks after operation, and the animals afterwards allowed free movement. There were four groups each comprising 8 rabbits. Two groups, one group with A/W glass-ceramic implants and the other with alumina ceramic implants, were killed 12 weeks after implantation. Two other similar groups were killed 25 weeks after implantation. The Kirschner wire was removed, the segment of the tibia that contained the implant was excised, and the proximal and distal bones were pulled apart with a tension-tester at a cross-head speed of 3.5cm/min.

The load to failure of specimens that contained the A/W glass-ceramic implant increased with time, 19.8 ± 7.06 Newtons after 12 weeks of implantation to 126.4 ± 32.54 Newtons after 25 weeks of implantation, while the failure loads with alumina ceramic implants at the same implantation period were 0 and 19.6 ± 13.92 Newtons, respectively. No glass-ceramic implants broke during the tension-testing. Histological observation of the interface between the implant and bone revealed that the gaps which had been observed at the interface immediately after operation were filled with woven bone tissue in the 12 week specimens with an A/W glass-ceramic implant. On the other hand, in specimens containing an alumina ceramic implant, gaps were always filled with a thin layer of fibrous tissue. On the basis of mechanical strength and the performance of the bone-implant interface, prostheses fabricated from A/W glass-ceramic should be usable under load-bearing conditions (Fig. 6).

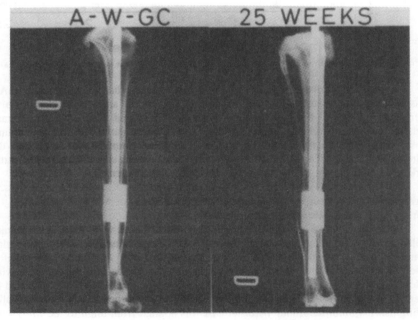

Fig. 6. Radiographs of segmental replacements in rabbit tibia.

Replacement of the Vertebrae of Sheep
with an A/W Glass-Ceramic Prosthesis

Ten castrated male sheep weighing from 40 to 60kg underwent replacement of the 3rd and 4th lumbar vertebrae with vertebral prostheses made of A/W glass-ceramic, without bone graft.[9] Under general anesthesia, the lumbar vertebrae were approached retroperitoneally. The intervertebral disc between the two lumbar vertebrae was removed with about half of the vertebral bodies above and below. The bone defect thus produced was then filled with a vertebral prosthesis. The prosthesis was cylindrical, 15mm in length and 10mm in diameter in 2 animals, and 30mm X 12mm in 8 animals. In the former group, the prostheses were implanted without any fixation. In the latter group, the prostheses were securely fixed with Zielke's instrumentation. The animals were immobilized for two days postoperatively with a specially made body brace, and they were allowed free movement thereafter. They were killed at various times from 3 to 27 months postoperatively. After removing the Zielke's instrumentation, the vertebrae containing A/W glass-ceramic prostheses were prepared for microscopic, contact micrographic, SEM, and EPMA observations.

In the two cases in which 15mm long prostheses were implanted and animals killed 24 months and 27 months after operation respectively, bone bonding between the prostheses and the two vertebrae was observed by radiography and contact microradiography. The prostheses were bonded directly to the trabeculae of the cancellous bone, and even at the sites where trabeculae were lacking, irregularly shaped new bone of about 12μm in thickness was observed covering the prosthesis surface. In other animals, which received a 30mm long implant, direct bonding of the prosthesis to both vertebrae was observed in two cases and bonding to one vertebra only in one case. SEM-EPMA and X-ray microdiffraction analysis of the interface whether bonded or not, always showed a Ca-P rich layer on the surface of the prosthesis. This layer was confirmed as apatite by crystallographic analysis. From this study together with previous studies using rabbit tibia,[2,3,4,8] we presumed that failure of bonding of the prostheses to the vertebrae might be due to a failure of fixation rather than chemical inactivity of the prosthetic surface, and this was later confirmed with clinical experience where a very firm fixation method was employed.

Replacement of the Vertebrae with A/W Glass-Ceramic Vertebral Prostheses in Clinical Cases

In the past, when the vertebral column was extensively damaged by tumors or trauma, its reconstruction had been attempted by the use of autogenous bone and allograft in combination with metals, PMMA bone cement, or alumina ceramic. However, autogenous bone and allograft have certain limits in their availability, and the long-term durability of non bone-bonding implants was not always satisfactory due to loosening and dislocation. To make a strong, bone-bonding vertebral anaplerosis, we prepared vertebral prostheses made of A/W glass-ceramic for clinical use. The prosthesis is in many different sizes so that the surgeon is able to choose the appropriate one in the operating theater (Fig. 7).

In 1983, the first replacement operation of thoracic vertebra using an A/W glass-ceramic prosthesis was performed at Kyoto University Hospital in a 50 year old female patient who developed a breast cancer metastasis in the 10th thoracic vertebra, associated with mild paraplegia.[10] At the operation, through a thoracotomy, total excision of the 10th thoracic vertebra and partial excision of the vertebrae above and below were

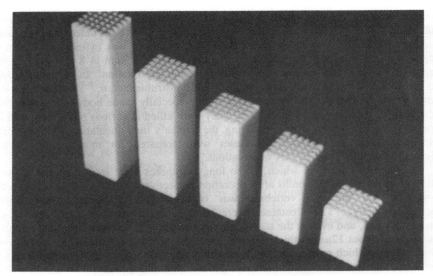

Fig. 7. Selection of A/W vertebral prosthetic devices.

performed, and a vertebral prosthesis was securely implanted in the bone defect combined with some autogenous bone graft. The patient has survived for 8 years without recurrence of the tumor in the operated site, although she developed another metastatic lesion in a lumbar vertebra 6 years postoperatively and it has also been replaced with another prosthesis (Fig. 8).

The A/W glass-ceramic vertebral prosthesis had been used in 70 clinical cases during the period from 1983 to 1990 at Kyoto University and Hokkaido University. The age distribution of the patients ranged from 22 to 79 years with an average of 51-5 years. There were 39 males and 31 females. Among them, 19 cases were metastatic tumors, 30 were burst fractures, 15 were compression fractures, and others were lumbar instabilities due to spondylolysis, spondylolisthesis or intervertebral disc hernia. Among 19 cases of metastatic tumors which underwent a vertebral prosthetic replacement, there were 12 different malignant tumors. Some patients, such as those with malignant melanoma, osteosarcoma and lung cancer died within a few months postoperatively. Many others survived for longer than 3 years, particularly, those with breast cancer, renal cancer, thyroid cancer, and prostate cancer who tended to survive for longer than 5 years. Nevertheless, it is important to be able to provide patients with a pain free and non-paralysed life to make their remaining, limited life time valuable. This prosthetic replacement operation seems particularly indicated for patients with vertebral metastases of slow growing malignant tumors.

In cases of burst fracture of the thoracic and lumbar vertebrae with paraplegia, the vertebral prosthetic replacement after decompression of the spinal cord always gives

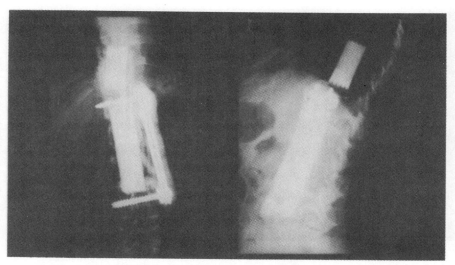

Fig. 8. Radiograph of vertebral prosthesis, 6 years and 6 months post operation.

satisfactory results, although more-or-less paralysis may remain, depending on the preoperative severity and duration of spinal cord compression. The patient illustrated in Fig. 9 sustained burst fracture of the 3rd lumbar vertebra with mild paraplegia after falling from a height. Anterior decompression of the spinal cord and reconstruction of the spinal column were performed shortly after the injury by the use of a 4cm long vertebral prosthesis which was firmly fixed with a Kaneda device. The patient has recovered completely from paraplegia. In cases of compression fracture of the thoracic and lumbar vertebra with paraplegia, the vertebral prosthetic replacement of the fractured vertebral body always gives good results, although posterior decompression and spinal instrumentation with Luque's rods may be indicated in some cases before anterior vertebral replacement. In any case, late collapse of the replaced spinal segments, which can occur when autogenous or allogenic bone is used was prevented by the use of an A/W glass-ceramic vertebral prosthesis (Fig. 10).

An A/W glass-ceramic vertebral spacer was used in 6 cases of lumbar instability for the purpose of posterior lumber interbody fusion fixed with Steffee's device (Fig. 11). In all cases, solid fixation and stability of the lumbar vertebrae had been achieved by laminectomy and posterior interbody fusion with vertebral spacers (Fig. 12). Based on our 9 years of clinical experiences with the A/W glass-ceramic vertebral prosthesis,[11] we conclude that this vertebral prosthesis is useful for the reconstruction of the spinal column which had been severely destroyed by tumors, trauma, or degenerative diseases, provided a very firm fixation of the prosthesis to the adjacent bone is accomplished by the use of various spinal instrumentations in addition to autogenous bone grafting.

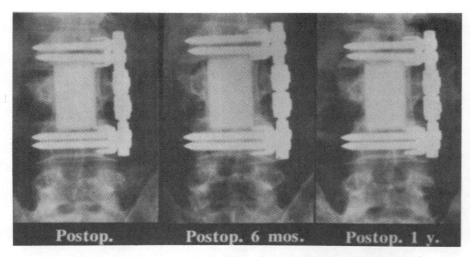

Fig. 9. Radiographs of patient with burst fracture of lumbar vertebra (L3).

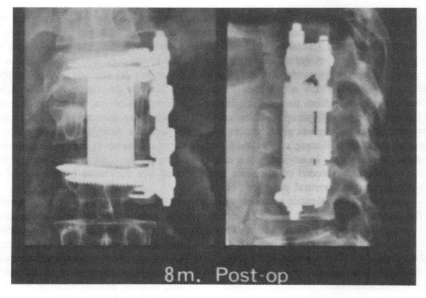

Fig. 10. Radiograph of patient with fracture of lumbar vertebra (L1).

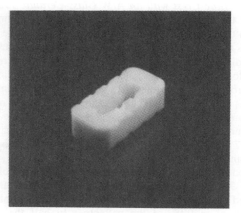

Fig. 11. A/W glass-ceramic vertebral spacer.

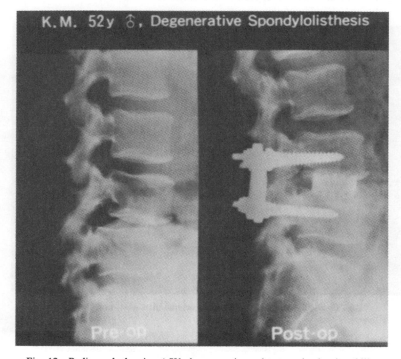

Fig. 12. Radiograph showing A/W glass-ceramic used to treat lumbar instability.

Reconstruction of the Iliac Crest with an A/W Glass-Ceramic Prosthesis

In surgery of the spine, skeletal tumors, and skeletal trauma, large autogenous bone grafts are often taken from the iliac crest. Taking such bone grafts from the iliac crest not only distorts the shape of the remaining ilium, but produces other complications such as tenderness, discomfort, pain on walking, fracture of the rest of the iliac crest, paresthesia, and abdominal herniations. To prevent such complications and fill bone defects, iliac crest prostheses of various sizes were fabricated from A/W glass-ceramic and used in 113 cases between 1989 and 1990 (Fig. 13). The age of the patients ranged from 14 to 75 years (average, 42.5 years), and the length of the bone defects ranged from 15 to 70cm (average, 43cm). Results obtained one to two years after the operation were excellent in 97% of the patients, there was no spontaneous pain in 96%, no tenderness in 91%, and no foreign body feeling in 96%. Overall, patients' satisfaction was excellent or good in 100% of the cases. Early fixation and ultimate stability were excellent in 94 and 96%, respectively. New bone formation around the iliac crest prostheses progressed steadily, and one year following the operation, good new bone formation was observed in 90% of the patients. In half the cases, there was no radiological clear zone on either side of the prosthesis (Fig. 14).[12]

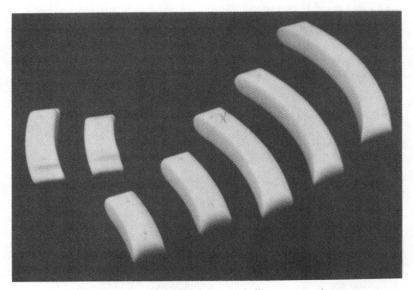

Fig. 13. A range of A/W glass-ceramic iliac crest prostheses.

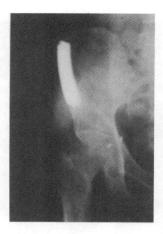

Fig. 14.   Radiograph of an iliac crest prosthesis after 1 year.

A/W Glass-Ceramic Granules as a Bone Defect Filler

Bone defects remaining after the excision of bone tumors are usually filled with either autogenous bone or allograft. However, sufficient amounts of autogenous bone and allograft are not necessarily available. In cases of aseptic loosening of hip prostheses combined with large bone loss, a large amount of autogenous or allogenic bone is required. In such cases, we have used A/W glass-ceramic granules in combination with autogenous cancellous bone and fibrin glue as bone substitute. In our animal experiments,[13] A/W glass-ceramic-fibrin mixture showed significantly better osteoconduction and acceleration of the bone repairing process than A/W glass-ceramic granules. A/W glass-ceramic granules mixed with autogenous cancellous bone and fibrin glue are recommended to fill this type of bone defect. The case illustrated in Fig. 15 was revised due to aseptic loosening of a Charnley hip prosthesis which had been inserted 15 years previously. At the revision operation, 45g of A/W glass-ceramic granules was used, mixed with autogenous bone and fibrin glue. Loosening of the stem of hip prostheses has been revised by the use of a long stem in combination with A/W glass-ceramic granules mixed with autogenous bone and fibrin glue.

Since 1989, we have used A/W glass-ceramic granules in 32 cases of revision surgery of hip prostheses. The age of the patients at the time of revision surgery ranged from 30 to 76 years (average, 60 years). Retrieved prostheses varied in type. In seven cases, only the socket of the prosthesis was retrieved. The amount of A/W glass-ceramic granules used at the operation ranged from 10 to 75g with an average of 31g. The results of the revision operations using A/W glass-ceramic granules have been satisfactory to date.

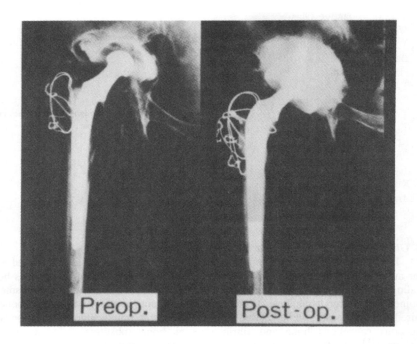

Fig. 15. Radiographs of a hip prosthesis revision in a 55 year old female patient.

REFERENCES

1.   M. Jarcho, J.F Kay, K.I. Gumaer, R.H. Doremus, H.P. Drobeck, "Tissue, Cellular and Subcellular Events at a Bone-ceramic Hydroxyapatite Interface," *J. Bioeng.* **1** (1977) 79-91.
2.   T. Nakamura, T. Yamamuro, S. Higashi, T. Kokubo, S. Itoo, "A New Glass-ceramic for Bone Replacement: Evaluation of its Bonding to Bone Tissue," *J. Biomed. Mater. Res.* **19** (1985) 685-698.
3.   T. Kitsugi, T. Yamamuro, T. Nakamura, T. Kokubo, "The Bonding of Glass-ceramics to Bone," *Int. Orthop.* **13** (1989) 199-206.
4.   S. Yoshii, Y. Kakutani, T. Yamamuro, T. Nakamura, T. Kitsugi, M. Oka, T. Kokubo, M. Takagi, "Strength of Bonding Between A/W Glass-ceramic and the Surface of Bone Cortex," *J. Biomed. Mater. Res.* **22:A3** (1988) 327-338.
5.   T. Kitsugi, T. Yamamuro, T. Nakamura, Y. Kakutani, T. Hayashi, S. Ito, T. Kokubo, M. Takagi, T. Shibuya, "Aging Test and Dynamic Fatigue Test of Apatite-Wollastonite-Containing Glass-ceramics and Dense Hydroxyapatite," *J. Biomed. Mater. Res.* **21** (1987) 467-484.

6. T. Kokubo, S. Ito, M. Shigematsu, S. Sakka, T. Yamamuro, "Fatigue and Life-time of Bioactive Glass-ceramic A/W Containing Apatite and Wollastonite," *J. Mater. Sci.* **22** (1987) 4067-4070.

7. Y. Ebisawa, T. Kokubo, K. Ohura, T. Yamamuro, "Bioactivity of $CaO.SiO_2$ Based Glasses: In Vitro Evaluation," *J. Mater. Sci., Materials in Medicine* **1** (1990) 239-244.

8. T. Kitsugi, T. Yamamuro, T. Kokubo, "Bonding Behavior of a Glass-Ceramic Containing Apatite and Wollastonite in Segmental Replacement of the Rabbit Tibia Under Load-bearing Conditions," *J. Bone Joint Surg.* **71-A** (1988) 264-272.

9. T. Yamamuro, J. Shikata, H. Okumura, T. Kitsugi, Y. Kakutani, T. Matsui, T. Kokubo, "Replacement of the Lumbar Vertebrae of Sheep With Ceramic Prostheses," *J. Bone Joint Surg.* **72-B** (1990) 889-893.

10. T. Yamamuro, T. Nakamura, S. Higashi, R. Kasai, Y. Kakutani, T. Kitsugi, T. Kokubo, "Artificial Bone For Use As A Bone Prosthesis," in *Progress in Artificial Organs-1983*, ed. K. Atsumi, M. Maekawa, and K. Ota, (Cleveland: ISAO Press, 1984) pp. 810-814.

11. T. Yamamuro, "Replacement of the Spine with Bioactive Glass-Ceramic prostheses," in *Handbook of Bioactive Ceramics, Vol 1*, eds. T. Yamamuro, L. L. Hench, J. Wilson (CRC Press, Boca Raton, 1990) pp. 345-352.

12. T. Yamamuro, "Reconstruction of the Iliac Crest With Bioactive Glass-Ceramic Prostheses" in *Handbook of Bioactive Ceramics, Vol 1*, eds. T. Yamamuro, L. L. Hench, J. Wilson, J. (CRC Press, Boca Raton, 1990) pp. 335-342.

13. K. Ono, T. Yamamuro, T. Nakamura, Y. Kakutani, T. Kitsugi, K. Hyakuna, T. Kokubo, M. Oka, Y. Kotoura, "Apatite-Wollastonite Containing Glass-Ceramic-Fibrin Mixture as a Bone Defect Filler," *J. Biomed. Mater. Res.* **22** (1988) 869-885.

Chapter 7

CERAVITAL® BIOACTIVE GLASS-CERAMICS

Ulrich M. Gross, Christian Müller-Mai* and Christian Voigt*

Institute of Pathology and Department of Traumatology and Reconstructive Surgery*
University Hospital Steglitz, Free University of Berlin,
Hindenburgdamm 30, D 1000 Berlin 45

INTRODUCTION

Shortly after L. L. Hench published his paper on "Bonding mechanisms at the interface of ceramic prosthetic materials"[1] in 1971 by which a new era of considerations, developments and products in the field of biomaterials was induced, E. Pfeil and H. Brömer came together to use this information for the design of some new glasses and glass-ceramic compositions.[2] They coined the term "Ceravital" which means a number of different compositions of glasses and glass-ceramics and not only one product. In the first enthusiasm these new materials were optimistically considered to be applicable as solids in load bearing conditions for the replacement of bone and teeth.[3] It turned out, however, that the mechanical properties were not compatible with this aim. Furthermore it could be shown that the surface reactivity which is the leading force for the bone bonding mechanism is operative at areas of the implant surface where soft tissue is interfacing and providing a milieu for dissolution of the material and activation of mononuclear and multinuclear resorbing cells. It could be expected from the experimental data that the long-term stability of the material was endangered by this process. Therefore some other compositions were designed in order to decrease the solubility of the glasses and glass-ceramics.[4-6] *In vitro* experiments showed that this approach was very successful and that the solubility of the material could be adjusted to various conditions, from high to very low solubilities, by addition of metal oxides to the melt. In contrast, *in vivo* investigations demonstrated that these new materials containing metal oxides were more or less inhibitory for the development, maturation and especially for the mineralization of the bone matrix adjacent to the implant interface.

In order to better understand the leading mechanisms for this inhibition of mineralization of bone-bonding and non-bonding glasses and glass-ceramics studies of the mineralization process were performed and the role of matrix vesicles in the promotion of mineral deposition analyzed. A role for matrix vesicles was shown in these experiments.[7,8]

There were several approaches to improve the biomechanical properties of the glassy materials. In some trials metal surfaces were coated with enamel and the enamel dotted with particles of bone-bonding glass-ceramic. This experiment was only partially successful. The bone-bonding glass ceramic particles were indeed covered and anchored with bone. The surfaces between these particles however exposed enamel which contained various metal oxides, which produced inhibition of mineralization of bone matrix.[9] Therefore, this approach was abandoned.

Another series of experiments was done using composites of glass-ceramic particles and metal. The metal used was titanium and various percentages of glass-ceramic and titanium composites were tested *in vivo*. The tests gave some very promising results.[10] The clinical application of this principle still awaits realization.

The only field in which glass-ceramic "Ceravital" implants are clinically applied is the replacement of the ossicular chain in the middle ear where the loads are minimal and the mechanical properties of the material are sufficient.[11]

MATERIALS AND METHODS

There have been quite a number of different compositions of glasses and glass-ceramics investigated *in vitro* for their behaviour regarding solubility in water and in physiological saline solution. Only a selection of these compositions was investigated *in vivo* over the years and the results compared.[6]

The *in vivo* tests used different animal models including mouse, rat, rabbit, pig and mini-pig. In principle the results of these animal experiments were comparable with clinical results gained with follow-up of patients and histology of surgically removed specimens. To explain these experiments and to introduce our current interest in model bioactive surfaces we describe a typical experiment with bone-bonding and non-bonding glass-ceramics.

Rectangular blocks (1.1 x 1.1 x 4mm) of bone-bonding glass-ceramics KG Cera, Mina 13 and the non bonding glass-ceramics KG y 213 and M8/1 (compositions in weight percent see Table 1) were implanted midshaft into the femur of adult male Sprague-Dawley rats weighing 350-400g. Prior to implantation the surface roughness was 4 μm maximally. Implants were sterilized by autoclaving and radio-frequency glow discharge treatment (GD) in an argon atmosphere for 2 minutes after repeated ultrasonication prior to implantation.[13] GD-sterilized implants were stored in boiled, i.e. gas-free, triple-distilled water for 2 weeks to preserve the high surface energy states of 100 dynes/cm (GD-treated implants) and 35 dynes/cm (autoclaved implants).

Implants were examined histologically using light microscopy (LM), scanning electron microscopy (SEM), and transmission electron microscopy (TEM). For LM, implant-containing femur segments were fixed in formaldehyde (Lillie), subjected to graded ethanols and infiltrated with methylmethacrylate.[10] The specimens were sawn perpendicularly to the longitudinal axis of the implant and subjected to a Giemsa surface layer-staining[14] or to a von Kossa/Fuchsin reaction and were embedded with Corbit. Quantitative analysis was performed by using the LM-slices and histomorphometry.[10] SEM and TEM of the same implant materials was performed. Details of specimen preparation are in reference 15.

Table 1. Composition of Bone-Bonding Glass-Ceramic KG Cera, Mina 13 and Non-Bonding Glass-Ceramic KGy213, M8/1 in Weight Percent.

	KG Cera	**Mina 13**	**KGy213**	**M8/1**
SiO_2	46.2	46	38	50
$Ca(PO_2)_2$	25.5	16	13.5	7.1
CaO	20.2	33	31	-
Na_2O	4.8	0	4	5
MgO	2.9	5	-	-
K_2O	0.4	-	-	-
Al_2O_3	-	-	7	1.5
Ta_2O_5	-	-	5.5	-
TiO_2	-	-	1	-
B_2O_3	-	-	-	4
$Al(PO_3)_3$	-	-	-	2.4
SrO	-	-	-	20
La_2O_3	-	-	-	6
Gd_2O_3	-	-	-	4

THE HOST RESPONSE

The process of tissue or wound healing comprises at least four overlapping phases. The duration of each period is influenced by the implant properties.

Formation of the Blood-Clot

The formation of the blood-clot starts immediately after implant insertion. At the LM magnification-level glasses or glass-ceramic implants are surrounded by the cells of the blood-clot comprising white blood cells, erythrocytes, and platelets. Between cells there are mainly dark-blue-strained fibers, probably fibrin, arranged in a three-dimensional network (Fig. 1). Some focal bone-contacts exist between implant corners and the old compact bone of the femur. This situation is comparable for different kinds of glasses or glass-ceramics. There is no obvious difference in the histological response at the LM level and at this stage to implants sterilized by different procedures. This phase lasts for several days until the cells and cellular products of the early phase are replaced by other cells and extracellular matrix.

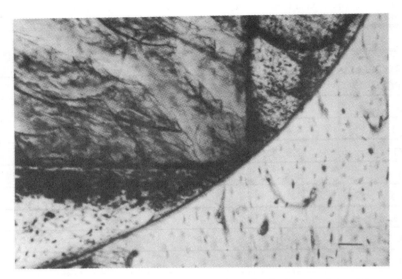

Fig. 1. Corner of a GD-sterilized KG Cera implant 3 days after implantation into rat femur. Fibrous material of the blood clot and old mineralized compact bone of the femur (at right). Dense zone at the implant surface due to accumulation of cells. (LM, bar 100 μm.)

At higher magnification, in the SEM and TEM, it was demonstrated that at this stage there were, as well as the blood-derived round cells and fibrous material condensed directly on the implant surface (Fig. 2), already some elongated fibroblastic or osteoblastic cells settled on the implant surface. Some cells contained a lot of rough endoplasmic reticulum, indicating a productive stage. Between them there were single, large macrophage-like cells with lobed nuclei. Such cells have been shown to stain positively for acid phosphatase.[12] Some macrophages contained phagocytosed particles of implant origin.[5,15] Macrophages were shown to be able to dissolve incorporated particles *in vitro*.[16] Whether these cells are able directly to resorb the bulk material is not yet clear. If implants were GD-sterilized instead of autoclaved and were stored in water to preserve the high surface energy, large osteoclast-like cells of 50 μm and more in length were observed on the implant surface only at bone-bonding KG Cera. Due to leaching in the water the KG Cera developed a rugosity of the surface due to attack of the crystal edges and grain boundaries.[13] Rough implant surfaces have been shown, to attract osteoclasts and macrophages.[17] Additionally, some Ca/P-phases attractive for such cells might have been produced by the storage in water (Fig. 3). A completely different morphology was observed on autoclaved KG Cera implants which showed no signs of leaching. The surface was covered with a thin layer of apparently mineralized material which protruded focally into the tissue (Fig. 4).

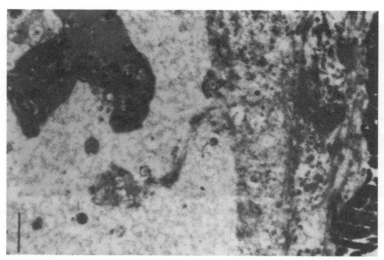

Fig. 2. GD-treated KG y 213 implant surface 3 days after implantation, covered with condensed fibrous material of the blood-clot and remnants of cells. Some vesicular stuctures in the fibrous material close to the interface. (TEM, bar 1 μm.)

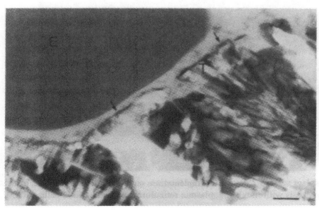

Fig. 3. GD-sterilized KG Cera implant surface 3 days after implantation with an approximately 0.1 μm wide layer showing liberated tips of the ceramic moiety (T) at the implant surface and partially liberated ceramic particles in the bulk (grey), probably due to leaching in water. Some ceramic particles were lost in cutting. Thin, electron-dense, probably adsorbed layer covering the ceramic tips (arrows). Part of an erythrocyte (E) close to the implant surface. (TEM, bar 0.2 μm.)

Formation of Organization Tissue

This phase of the wound healing starts 2 to 3 days after implantation with fibroblasts, osteoblasts, and capillaries invading the blood-clot and filling up the space to form new tissue. This process lasts for several days, depending on the volume of material to be resorbed and replaced (Fig. 5). It seems that the replacement of blood derived cells is slower with metal oxide substituted implants, such as KG y 213 or M8/1,

Fig. 4. Autoclaved surface of a KG Cera implant (3 days) after storage for 2 weeks in water covered with a mineralized appearing seam (S) of about 0.1 μm in width. Some mineralized spots already present in the fiber-rich extracellular matrix, probably fibrin of the blood-clot. Ceramic moiety totally covered by the glass moiety (black). (TEM, bar 0.2 μm.)

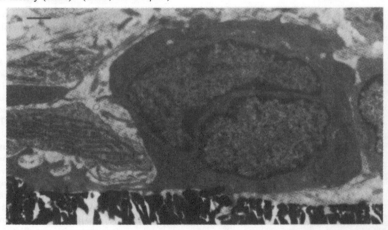

Fig. 5. GD-treated KG Cera implant surface 7 days after implantation, covered with extracellular matrix (top, left), processes of productive cells with rough endoplasmic reticulum, and white blood-cells. (TEM, bar 1 μm.)

which still have polymorphonuclear leukocytes and macrophages on their surfaces (Fig. 6). Beside these cells, the blood-clot is replaced by many capillaries and macrophages, especially in the implant vicinity where loose implant particles were detectable. Some groups of cells are already arranged in trabecular structures and extend from the edge of the drill hole to the implant surface. These structures are not yet mineralized and attach only focally to the implant surface.

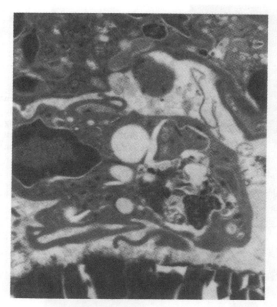

Fig. 6. Autoclaved surface of an implant (7 days) covered with parts of phagocytosing cells, probably polymorphonuclear leukocytes. There is no evidence for the production of extracellular matrix in this area. (TEM, bar 1 μm.)

In the TEM, there is already a prominent collagen-rich ECM containing a huge amount of extracellular matrix vesicles, visible, and many foci are already calcified. Such foci have been described as calcospheritic globules and calcospheritic structures.[18]

In the SEM, after making a fracture between implant and surrounding tissue, nearly the whole surface of glass-ceramic implants is covered with tissue. Some cells are arranged in whirl-like or hump-like structures. During this phase the amount of organic material still adhering to the implant surface after fracturing the interface is higher for bone-bonding materials than for non-bonding implants. During this phase still, only on GD-sterilized surfaces were osteoclast-like cells present. Some of them contained loose implant particles.[13]

Regeneration of Organ-Typic Cells and Tissues

At approximately 6 days this phase starts, due to the first mineralized foci in the extracellular matrix which are observable in the LM. Young trabeculae can be observed which continue to mineralize and grow wider. In the case of KG Cera some trabeculae attach to parts of the implant surface and additionally a thin bony lamella forms on the implant surface between the attaching trabeculae (Fig. 7). Histomorphometry yields already approximately 10% bone-contact to KG Cera implants at 7 days after implantation (Table 2). In the TEM the bonding zone is an electron-dense seam which is known to develop on surface-reactive bone-bonding glass-ceramics. Such a layer is approximately 0.5 μm wide and contains no, or only a few, collagen fibers (Fig. 8).

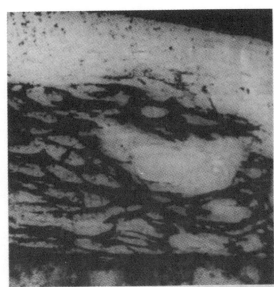

Fig. 7. GD-sterilized KG Cera implant surface 7 days after implantation into rat femur. Mineralization is beginning in the drill hole and on the implant surface. Old bone (black). (LM, von Kossa stain, bar 50 μm.)

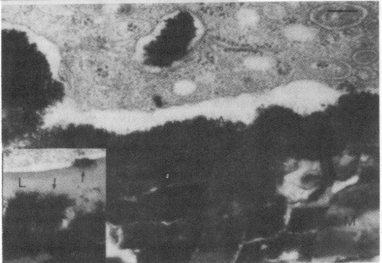

Fig. 8. Autoclaved KG Cera implant surface 7 days after implantation. Thin amorphous appearing seam (A) approximately 0.3 μm wide on the implant surface is the beginning of bone-bonding, probably due to the development of a Ca/P-rich layer on the implant surface. Lighter appearing ceramic moiety (M). Inset: Other area of the interface of the same implant representing an earlier stage with an electron lucent layer (L) containing some already mineralized spots (arrows). (TEM, bar 0.2 μm.)

Table 2. Bone (B), Osteoid (Os), Chondroid (Ch), and Soft Tissues (ST) ± Standard Error of the Mean (SEM) of GD-treated Implants and Autoclaved Controls. d = Days After Implantation, Cited From.[10,13]

Material	d	B	Os	Ch	ST
KG Cera GD	3	0.4 ±	0.8 ±	0.0 ± 0	98.8 ±
	7	<1	<1	0.8 ±	<1
	14	12.5 ± 3	4.5 ± 3	<1	82.2 ± 4
	28	57.8 ± 12	9.2 ± 4	2.1 ± 2	30.9 ± 7
		92.9 ±	2.7 ±	0.5 ±	3.9 ±
		<1	<1	<1	<1
KG Cera	3	0.1 ±	3.5 ± 1	0.0 ± 0	96.4 ± 1
	7	<1	6.2 ± 2	0.4 ±	87.0 ± 2
	14	6.4 ± 1	2.9 ±	<1	45.6 ± 3
	28	51.5 ± 3	<1	0.0 ± 0	15.2 ± 6
		74.2 ± 7	10.4 ± 2	0.2 ±	
				<1	
KGy213 GD	3	0.4 ±	1.6 ± 1	0.0 ± 0	98.8 ± 1
	7	<1	4.4 ± 1	0.0 ± 0	95.1 ± 2
	14	0.6 ±	32.2 ± 11	5.8 ± 3	58.5 ± 11
	28	<1	30.5 ± 2	8.1 ± 5	50.6 ± 4
		3.5 ± 1			
		10.8 ± 5			
KGy213	3	0.3 ±	3.4 ±	0.0 ± 0	96.3 ±
	7	<1	<1	0.0 ± 0	<1
	14	1.0 ±	5.9 ± 1	3.1 ±	93.1 ± 1
	28	<1	16.5 ± 3	<1	77.1 ± 5
		3.3 ±	34.4 ± 5	5.4 ± 3	52.3 ± 8
		<1			
		7.9 ± 1			
KG Cera	84	84.0 ± 3	1.0 ± 1	0.0 ± 0	15.0 ± 3
Ti/Cera	84	85.0 ± 12	0.0 ± 0	0.0 ± 0	15.0 ± 12

More collagen is incorporated in the outer part of the bonding zone at later stages and may represent the beginning of primary bone. In the case of glasses and glass-ceramics this bonding layer consists of Si-rich and Ca/P-rich parts which provide a chemical bonding of the implant to bone.[1,19-23] Such interfacial layers also develop *in vitro*[23] and after intraperitoneal implantation[24,25] and are able to produce chemical bonding even between two glass-ceramic implants.[26] In the case of hydroxyapatite such seams were also observed and consisted of carbonated hydroxyapatite.[27-29]

In the case of non-bonding implants the situation at the interface is completely different. On the whole implant surface there is no evidence for mineralization, i.e. there is a seam of non-mineralized extracellular matrix on the implant surface in the LM (Fig. 9) and there are no attaching layers of collagen-fibers showing different stages of mineralization in the TEM (Fig. 10). Instead of mineralization, large multinucleated cells of about 50 μm in length and more were found attaching to the surface (Fig. 11). This indicates that the difference of the healing around non-bonding and bone-bonding glass-ceramic implants occurred between the second phase (formation of granulation tissue) and third phase (formation of bone tissue) of wound healing (Fig. 12).

Some bone-bonding and non-bonding specimens were prepared for SEM evaluation with NaOCl solution to remove the non calcified tissues. These specimens showed trabecular structures fixing the implants within the marrow cavity. The amount of trabeculae is higher in the case of the bone-bonding material KG Cera.[7,18] Former TEM studies demonstrated that the insertion of bone-bonding and non-bonding glass-ceramic implants affected the mineralization process differently. The production of matrix vesicles was changed in an implant-dependant, and time-dependent manner, which might alter the calcification of osteoid differently. Total MV numbers around KG Cera and KG y 213 showed significant higher values (p < 0.05) around the bone-bonding material KG Cera as compared to the non-bonding material KG y 213. This clearly demonstrates that more MV/μm^2 were produced around the bonding material. Similar results were gained comparing Mina 13 and M8/1.[32] The significant higher value of MV per area of non-calcified ECM points to the possibility that the capacity of osteoblasts to produce MV is reduced around the metal oxide containing non-bonding implants. These results led to the suggestion that mineralization around non-bonding implants leads to an inhibited tissue maturation, whereas bone-bonding implants seem to promote primary calcification via an increased number of matrix vesicles.

Additionally, the matrix vesicle distance to the calcifying front was lower in the case of the non-bonding material KG y 213 at 3 days and significantly statistically lower at 6 days after implantation as compared to the bone-bonding material. This might suggest that the extracellular matrix is not as mature in the KG y 213 material as in KG Cera. These results point to an affected osteoblastic function in the case of the non-bonding material where less extracellular matrix was produced. The decreased rate of extracellular matrix in the non-bonding material is supported by *in vitro* ultrastructural results using 45S5 bone-bonding glass and a non-bonding quartz glass. It was demonstrated in this study that the bonding material exhibited multi-layers of osteoblasts and abundant extracellular matrix, whereas the non-bonding quartz glass did not show significant production of extracellular matrix.[33]

From several *in vitro* experiments it is well known, that aluminum delays the formation and growth of hydroxyapatite in a dose-dependent manner.[34,35] Similar results were gained in using Ti^{4+} and V^{5+} ions *in vitro*. Titanium decreased the hydroxyapatite formation in a dose-dependent manner without delaying the onset of hydroxyapatite formation, and vanadium decreased the hydroxyapatite formation.[36] These data lead to

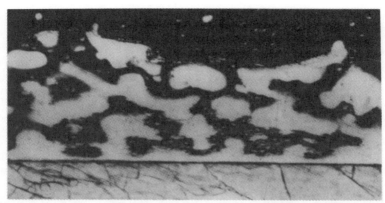

Fig. 9. Surface of an autoclaved KG y 213 implant 7 days after implantation into rat femur with young, not fully mineralized trabeculae. Small unmineralized seam at the implant surface. (LM, von Kossa stain, bar 200 μm.)

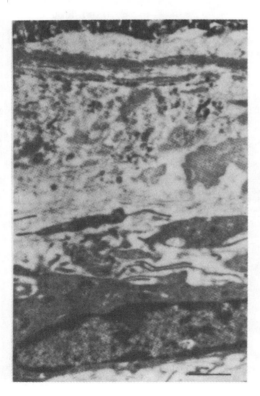

Fig. 10. GD-sterilized KG y 213 implant surface with unchanged surface morphology 7 days after implantation. Non-productive cells in the implant vicinity, single collagen fibers and still condensed fibrous material on the implant surface. (TEM, bar 1 μm.)

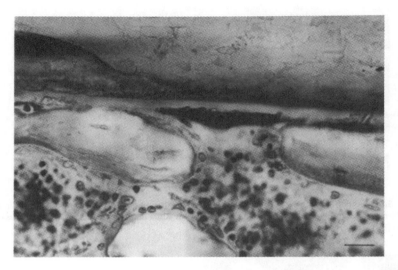

Fig. 11. Surface of an autoclaved KG y 213 implant 28 days after implantation into rat femur, with focal bone contact and a multinucleated giant cell on the implant surface. (LM, Giemsa stain, bar 200 μm.)

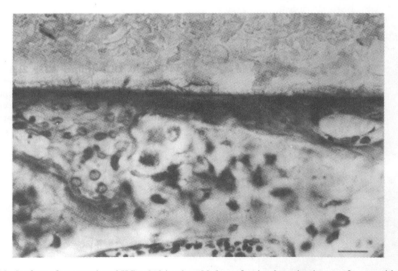

Fig. 12. Surface of an autoclaved KG y 213 implant 28 days after implantation into rat femur, with young, fully mineralized trabeculae in the former drill hole. Disturbed mineralization at the implant surface with osteoid. (LM, Giemsa stain, bar 200 μm.)

the conclusion that the negative effect of non-bonding implants might be on the growth of calcifying fronts.   To achieve bone-bonding, the surface solubility is of utmost importance.   Up to a certain level a higher solubility results in increased bioactivity.   In a previous study, it was shown that up to 1.6% Al_2O_3 can be tolerated without diminishing bone-bonding behaviour.   The percentage of metal ions can be slightly higher if the surface solubility is increased up to a defined maximum.[37]

The amount of bone in the interface does not only depend on the chemical composition, it can also be influenced by the sterilization procedure.   Higher amounts of osteoid and chondroid and more bone at the implant surface of bone-bonding and non-bonding implants were measured histomorphometrically if implants were GD-sterilized as compared to autoclaving.[13,30]   Some slices of the KG Cera yielded 100% bone contact when GD-sterilized, at 28 days after implantation.[13]   Even non-bonding implants show slightly higher amounts of bone in the interface.   In the TEM such areas with bone in the interface of non-bonding material shows a morphology similar to bone-bonding implants.   There is a seam of amorphous material between the "true" bone which consists mainly of mineralized collagen fibers (Fig. 13).   The reason for the development of such zones with bone-contact or bone-bonding in non-bonding implants is not yet clear.   This might depend on an inhomogenous composition of the bulk material, i.e. there might be zones with low metal ion content.

A possible explanation for higher amounts of bone in contact to GD-sterilized implant surfaces could be a chemical difference between GD-sterilized and autoclaved surfaces which could be produced by storage in water prior to implantation. TEM-observations demonstrated morphologic differences between such surfaces in the case of bone-bonding implants (Figs. 3,4) leading to the assumption that the rate of leached implant ions is higher in the GD-group since the protective layer of organic and

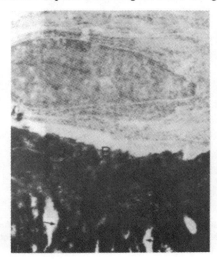

Fig. 13. GD-treated KG y 213 implant surface (bottom) 28 days after implantation. Bone-bonding via an afibrillar appearing, probably Ca/P-rich layer (arrows) and mineralized bone (B) in the vicinity of a cell containing rough endoplasmatic reticulum TEM, bar 1 μm.

inorganic contaminants usually occurring on the implant surface, was removed.[13] This hypothesis is supported by *in vitro* experiments showing higher dissolution rates of GD-sterilized calcium oxide and even dense hydroxyapatite implants after GD-sterilization as compared to autoclaved controls.[31] These results point to the solubility of the implant surface which might differ in GD-treated autoclaved glass-ceramic implants.

REMODELING

This phase starts at approximately 14 days when tissues typical of the organ, i.e. mineralized bone or bone marrow, are produced and last for a life-time. This phase comprises the replacement of trabecular bone by lamellar bone or bone marrow. The remodeled tissue will be adapted to the local needs, e.g. mechanical load.

The Material Response

The surfaces of glass-ceramic implants show degradation phenomena which are dependent on the chemical composition of the bulk material as well as on the tissue in contact to the surface. If there is bone-bonding the surface is protected and stabilized against further degradative processes. Other tissues, e.g. soft tissue, allow ongoing degradation by various processes. Most obvious changes occur in bonding glass-ceramics, which show changes in surface morphology in the SEM and TEM. The ceramic phase of the implant is liberated, partially, as early as 3 days after implantation (Fig. 14). The process continues until the ceramic phase is almost totally liberated

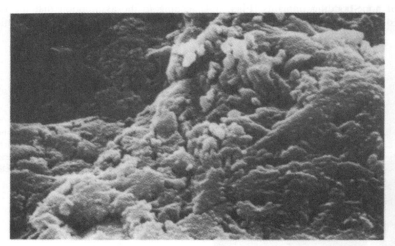

Fig. 14. GD-sterilized KG Cera implant surface after storage in water for 2 weeks, 3 days after implantation, with liberated tips of the ceramic moiety on the implant surface. This stage represents leaching which is observed on autoclaved implants at 14 days or later. (SEM, bar 10 μm.)

approximately 28 days after implantation (Figs. 15,16). Leaching phenomena start at the phase transition between glass and ceramic phases.[12] Such processes are enhanced on GD-sterilized KG Cera implants which show changes at 3 days which were seen on autoclaved implants as early as 14 and more days after implantation[13] (Figs. 3,14) with zones displaying partially or totally freed ceramic particles. Leaching leads additionally to the liberation of the ceramic phase to loss of small implant particles in the range of 1 μm in diameter which then are incorporated by cells, such as macrophages and osteoblast-like cells.[15] A direct bioresorption of bulk materials seems to be also possible. Recently it was shown that osteoclast-like cells were able to phagocytose dense hydroxyapatite implants.[15] Such cells were also observed on the KG Cera glass-ceramic with loose implant particles in the cytoplasm.[13] Ruffled borders were not observed. But it seems possible that these cells are able to increase the degration just by lowering the pH underneath their cellular membrane which in turn increases the implant solubility.

On the other hand, non-bonding implants with reduced solubility due to the addition of metal oxides to the bulk material do not show evidence of leaching phenomena up to 14 days after implantation. Sawing marks are still detectable. Later, in higher magnifications, there are still small elevations of the implant's surface with sawing marks with only few circumscribed areas showing discrete signs of leaching, evidenced by an accentuation of elevations and loosening of small implant particles indicating beginning corrosion. At this stage macrophages can be observed with incorporated particles of implant origin. Similar observations were made when comparing surfaces of autoclaved and GD-sterilized implants.

CONCLUSIONS

• Some glasses and glass-ceramics were designed and characterized *in vitro* studies to display a range of different surface solubility and surface reactivity, and termed "Ceravital".

• *In vivo* studies with different species showed there was a very close resemblance of the cellular reaction and the behaviour of the extracellular matrix.

• This reaction was different for glasses and glass-ceramics depending on the amount of metal oxides being released and influencing negatively the cellular function and the development and maturation of the extracellular matrix.

• There were at least two major groups, namely bone-bonding and non-bonding materials.

• The mechanical stability of the glassy material could be improved by metal-ceramic composites.

• The surface reactivity could be improved by special surface treatment. An example was given with the use of glow discharge treatment.

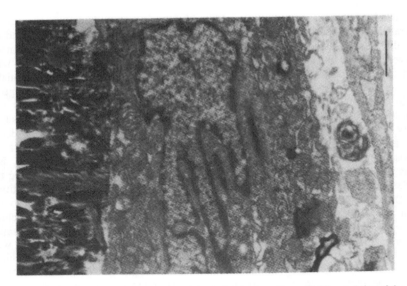

Fig. 15. Autoclaved KG Cera implant surface 7 days after implantation, with liberated tips of the ceramic moiety.   Star like greyish structures within the bulk material represent cross-cut ceramic particles, longitudinal cut ceramic particle (L).   (TEM, bar 1 μm.)

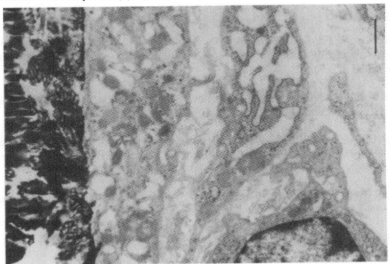

Fig. 16. GD-treated KG Cera implant surface (black, LHS) 14 days after implantation, with almost totally liberated ceramic moiety in a non-bonding area. Productive cells in the interface. (TEM, bar 1 μm.)

REFERENCES

1. L. L. Hench, R. J. Splinter, W. C. Allen and T. K. Greenlee, "Bonding Mechanisms at the Interface of ceramic Prosthetic Materials," *J. Biomed. Mater. Res. Symp.* **2** (1971) 117-141.

2. H. Brömer, E. Pfeil and H. H. Käs German Patent 2,326,100 (1973).

3. B. A. Blencke, H. Brömer, E. Pfeil and H. Käs "Implantate aus Glaskeramik in der Knochenchirurgie (Tierexperimentelle Untersuchungen)," *Langenbecks Arch. Klin. Chir. Suppl. Chir. Forum* (1973) 116-119.

4. U. M. Gross and V. Strunz, "The Anchoring of Glass Ceramics of Different Solubility in the Femur of the Rat," *J. Biomed. Mater. Res.* **14** (1980) 607-618.

5. U. Gross, J. Brandes, V. Strunz, I. Bab and J. Sela, "The Ultrastructure of the Interface Between a Glass Ceramic and Bone," *J. Biomed. Mater. Res.* **15** (1981) 291-305.

6. U. M. Gross and V. Strunz, "Interface of Various Glasses and Glass Ceramics in a Bony Implantation Bed," *J. Biomed. Mater. Res.* **19** (1985) 251-271.

7. C. Müller-Mai, D. Amir, H. Wendland, Z. Schwartz, J. Sela and U. Gross "The Effect of Glass-Ceramic Implants on Matrix Vesicle Calcification after Two Weeks of Rat Tibial Bone Healing," *J. Biomed. Mater. Res.* **24** (1990) 1571-1584.

8. C. M. Müller-Mai, D. Amir, Z. Schwartz, J. Sela, B. D. Boyan, H. Wendland and U. M. Gross, "Ultrastructural Histomorphometry of Extracellular Matrix Vesicles in Primary Calcification Around Bone-Bonding and Non-Bonding Glass-Ceramics," *Cells & Materials* **1** (1991) 341-352.

9. G. Zeiler, V. Strunz and U. Gross, "Bone Implant Surfaces with Coatings Containing Surface-Reactive Particles," in *Material Sciences and Implant Orthopedic Surgery*, eds. R. Kossowsky and N. Kossowsky (NATO ASI Series, Martinus Nijhoff Publ., Dordrecht, 1986) pp 249-259.

10. C. Müller-Mai, H. J. Schmitz, V. Strunz, G. Fuhrmann, T. Fritz and U. M. Gross, "Tissues at the Surface of the New Composite Material Titanium/Glass-Ceramic for Replacement of Bone and Teeth," *J. Biomed. Mat. Res.* **23** (1989) 1149-1168.

11. R. Reck, "Bioactive Glass-Ceramics in Ear Surgery: A Animal Studies and Clinical Results," *Laryngoscope* **94(2)** (1984) Suppl.33, 1.

12. C. Müller-Mai, C. Voigt, W. Knarse, J. Sela and U. M. Gross, "The Early Host and Material Response of Bone-Bonding and Non-Bonding Glass-Ceramic Implants as Revealed by Scanning Electron Microscopy and Histochemistry," *Biomaterials* **12** (1991) 865-871.

13. C. M. Müller-Mai, C.Voigt, R. E. Baier and U. M. Gross, "The Incorporation of Glass-Ceramic Implants in Bone After Surface Conditioning Glow-Discharge Treatment," *Cells & Materials* (in press).

14. U. M. Gross and V. Strunz, "Surface Staining of Sawed Sections of Undecalcified Bone Containing Alloplastic Implants," *Stain Technol.* **52** (1977) 217-219.

15.   C. M. Müller-Mai, C. Voigt and U. M. Gross, "Incorporation and Degradation of Hydroxyapatite Implants of Different Surface Roughness and Surface Structure in Bone," *Scanning Microsc.* **4** (1990) 613-624.

16.   C. H. Kwong, W. B. Burns and H. S. Cheung, "Solubilization of Hydroxyapatite Crystals by Murine Bone Cells, Macrophages and Fibroblasts," *Biomaterials* **10** (1989) 579-584.

17.   K. Gomi, B. Lowenberg, G. Shapiro and J. E. Davies, "Resorption of Sintered Synthetic Hydroxyapatite by Osteoclast-Like Cells *In Vitro*," *Transactions, 4th World Biomat. Congress* (1992) 531.

18.   D. Amir, C. Müller-Mai, H. Wendland, U. Gross and J. Sela, "Effect of Glass-Ceramic Implants on Primary Calcification in Rat Tibial Bone After Injury," *Biomaterials* **10** (1989) 585-589.

19.   L. L. Hench, C. G. Pantano, P. J. Buscemi and D. C. Greenspan, "Analysis of Bioglass Fixation of Hip Prostheses," *J. Biomed. Mater. Res.* **11** (1977) 267-282.

20.   T. Kitsugi, T. Nakamura, T. Yamamuro, T. Kokubo, T. Shibuya and M. Takagi "SEM-EPMA Observations of Three Types of Apatite-Containing Glass-Ceramics Implanted in Bone: The Variance of a Ca-rich Layer," *J. Biomed. Mater. Res.* **21** (1987) 1255-1271.

21.   C. Ohtsuki, H. Kushitani, T. Kokubo, S. Kotani and T. Yamamuro, "Apatite Formation on the Surface of Ceravital-Type Glass-Ceramic in the Body," *J. Biomed. Mat. Res.* **25** (1991) 1363-1370.

22.   L. L. Hench, "Surface Reaction Kinetics and Adsorption of Biological Moieties: A Mechanistic Approach to Tissue Attachment," in *The Bone-Biomaterial Interface*, ed. J. E. Davies (University of Toronto Press, Toronto, Canada, 1991) pp 33-48.

23.   L. L. Hench, "Bioactive Ceramics," *Ann. NY Acad. Sci.* **523** (1988) 54-71.

24.   Ö. H. Andersson and I. Kangasniemi, "Calcium Phosphate Formation at the Surface of Bioactive Glass *In Vitro*," *J. Biomed. Mat. Res.* **25** (1991) 1019-1030.

25.   F. Pernot, J. Zarzycki, P. Baldet, F. Bonnel and P. Rabischong, *In Vivo* Corrosion of Sodium Silicate Glasses," *J. Biomed. Mat. Res.* **19** (1985) 293-301.

26.   T. Kitsugi, T. Yamamuro, T. Nakamura, T. Kokubo, M. Takagi, T. Shibuya, H. Takeuchi and M. Ono, "Bonding Behavior Between Two Bioactive Ceramics *In Vivo*," *J. Biomed. Mater. Res.* **21** (1987) 1109-1123.

27.   M. Jarcho, "Calcium Phosphate Ceramics as Hard Tissue Prosthetics," *Clin. Orthop. & Rel. Res.* **157** (1981) 259-278.

28.   G. Daculsi, R. Z. LeGeros, M. Heughebaert and I. Barbieux, "Formation of Carbonate-Apatite Crystals After Implantation of Calcium Phosphate Ceramics," *Calcif. Tissue Int.* **46** (1990) 20-27.

29.   I. Orly, M. Gregoire, J. Menanteau, M. Heughebaert and B. Kerebel, "Chemical Changes in Hydroxyapatite Biomaterial Under *In Vivo* and *In Vitro* Biological Conditions," *Calcif. Tissue Int.* **45** (1989) 20-26.

30. C. M. Müller-Mai, C. Voigt, R. E. Baier and U. Gross, "Enhanced Bone Development After Implantation of Glow-Discharge Treated Surface Reactive Glass-Ceramics," *Transactions, 4th World Biomat. Congress* (1992) 237.

31. V. J. Sendax and R. E. Baier, "Improved Integration Potential for Calcium-Phosphate-Coated Implants After Glow-Discharge and Water Storage," *Dent. Clin. North Am.* **36** (1992) 221-224.

32. Z. Schwartz, D. Amir, B. D. Boyan, D. Cochavy, C. Müller-Mai, L. D. Swain, U. Gross and J. Sela, "Effect of Glass Ceramic and Titanium Implants on Primary Calcification During Rat Tibial Bone Healing," *Calcif. Tissue Int.* **49** (1991) 359-364.

33. T. Matsuda and J. E. Davies, "The *In Vitro* Response of Osteoblasts to Bioactive Glass," *Biomaterials* **8** (1987) 275-284.

34. N. C. Blumenthal and A. S. Posner, "*In vitro* Model of Aluminium-Induced Osteomalacia: Inhibition of Hydroxyapatite Formation and Growth," *Calcif. Tissue Int.* **36** (1984) 439-441.

35. R. Z. LeGeros, M. H. Taheri, G. B. Quirolgico and J. P. LeGeros, "Formation and Stability of Apatites: Effects of Some Cationic Substituents," in *Proc. 2nd Int. Congress on Phosphorous Compounds* (Boston, April 21-25, 1980) pp 89-103.

36. N. C. Blumenthal and V. Cosma, "Inhibition of Apatite Formation by Titanium and Vanadium Ions," *J. Biomed. Mater. Res.* **23** (1989) 13-22.

37. Ö Andersson, K. H. Karlsson, K. Kangasniemi and A. Yli-Urpo, "Models for Physical Properties and Bioactivity of Phosphate Opal Glasses," *Glastech. Ber.* **61** (1988) 300-305.

Chapter 8

MACHINEABLE AND PHOSPHATE GLASS-CERAMICS

W. Höland[*] and W. Vogel

Friedrich-Schiller-University Jena/Otto-Schott-Institute
Fraunhoferstrasse 6, D/O - 6900 Jena, Germany

PROCESSING

The development of new materials for bone implants and substitutes in man has gained major importance over the past twenty years. In addition to metals, alumina, and organic polymers, glasses and glass-ceramics have come especially to the fore. These new implant materials are biocompatible, and may have bioactive properties. They are not regarded as foreign bodies and encapsulated by fibrotic tissue; instead, direct bonding takes place. The special combinations of properties required for medical indications can be adjusted and varied in glasses and glass-ceramics.

Glass-ceramics consist of at least one glassy and at least one crystalline phase. The processing to develop a glass-ceramic is characterized by a formation of a base glass and an additional heat treatment of the glass. During this heat treatment process, nucleation and crystallization has to be controlled to form the crystals in the base glass (Chapter 1).

Biomaterials for bone substitution are called bioactive if a stable bond to the bone is formed. Hench[1] showed that Bioglass® can bond to bone in animals. The interface reactions between Bioglass® and bone, the formation of different Ca-, P-, and SiO_2-rich layers, the dependence on the composition of the implant, the environment, and reaction time were studied.[2] It was shown that the preferred bonding of bone and biomaterial (glass-ceramic or sintered ceramic) can be achieved, if the biomaterial contains apatite crystals in the basic material[3-5] or develops an apatite layer. Mica crystals will permit predictable mechanical machining properties.[6]

We believe that for a material to be machineable and bioactive, it should contain both mica and apatite crystals and we have developed glass-ceramics for medicine derived from different base glasses.[7-9]

BIOVERIT I is a mica-apatite glass-ceramic with a chemical composition from the SiO_2-(Al_2O_3)-MgO-Na_2O-K_2O-F-CaO-P_2O_5 base glass system. These glasses are from the silico-phosphate type. The key to the development of BIOVERIT I was to form a phase separated base glass consisting of three glassy phases and to control the nucleation and crystallization by heat treating the glass. BIOVERIT II glass-ceramic contains mica as the main crystal phase and secondary crystals, e.g. cordierite crystals.

---

[*]Present Address: Ivoclar AG, Bendererstrasse 2, 9494 Schaan, Principality of Liechtenstein

The base glass is derived from the SiO_2-Al_2O_3-MgO-Na_2O-K_2O-F system, called silicate glasses. The base glass is phase separated into two glassy phases and micas were formed during heat treatment of the glass. Because of the high mica content, BIOVERIT I and II are machineable glass-ceramics.

The chemical composition of BIOVERIT III glass-ceramic is characterized by glasses from the CaO-Al_2O_3-P_2O_5-Na_2O (ZrO_2-FeO/Fe_2O_3) system, so called "invert"-glasses of a phosphate type. In comparison to BIOVERIT I and II, the base glass of BIOVERIT III does not show phase separation. Apatite, $AlPO_4$-crystals and other phosphate crystals grow via a special process in the base glass during heat treatment.[10,11]

COMPOSITIONS

Machineable mica-based glass-ceramics were developed for technical applications by Beall.[12] The composition of the base glasses are characterized by the alkali-fluoroboro-silicate system. Grossman[13] formed special tetra-silicic-mica in alkali-fluorosilicate glasses. The glass-ceramic implants known as Dicor®, have been used for dental restorations. Similar glass-ceramics have been recently developed in Japan.

Typical chemical compositions of BIOVERIT-type glass-ceramics are shown in Tables 1-3. The range of possible chemical compositions are given and examples show materials which have been developed specially. Example 1 of BIOVERIT I (Table 1) is a mica-apatite glass-ceramic with special fluorophlogopite mica crystals (Na/K $Mg_3(AlSi_3O_{10})F_2$). Example 2 demonstrates a mica-apatite glass-ceramic with tetrasilicic mica crystals. The formation of these crystals was possible by heat treating the base glass compositions at temperatures between 610 and 1050°C. The controlled crystallization takes place via phase separation of the base glass.[7-9]

Table 1. Composition of BIOVERIT I.

	Weight Percent		
	Composition Range	Example 1	Example 2
SiO_2	29.5-50	30.5	38.7
MgO	6-28	14.8	27.7
CaO	13-28	14.4	10.4
Na_2O/K_2O	5.5-9.5	2.3/5.8	0/6.8
Al_2O_3	0-19.5	15.9	1.4
F	2.5-7	4.9	4.9
P_2O_5	8-18	11.4	8.2
TiO_2	additions	--	1.9

Table 2. Composition of BIOVERIT II.

	Weight Percent	
	Composition Range	Example 1
SiO_2	43-50	44.5
Al_2O_3	26-30	29.9
MgO	11-15	11.8
Na_2O/K_2O	7-10.5	4.4/4.9
F	3.3-4.8	4.2
Cl	0.01-0.6	0.1
CaO	0.1-3	0.2
P_2O_5	0.1-5	0.2

Table 3. Composition of BIOVERIT III.

	Weight Percent
	Composition Range
P_2O_5	45-55
Al_2O_3	6-18
CaO	13-19
Na_2	11-18
$MeO/Me_2O_5/MeO_2$ ($MnO,CoO,NiO,FeO,$ Fe_2O_3,Cr_2O_3,ZrO_2)	1.5-10

BIOVERIT II (Table 2) is a glass-ceramic consisting of a new type of curved mica of fluorophlogopite-type. A second crystal phase has also been precipitated in the glass. The basis for the nucleation and crystallization is also the phase separation of the glasses which has been caused by Na_2O-, K_2O- and F^--additions to the glass.

BIOVERIT III is a pure phosphate glass which does not contain silica. During the development of BIOVERIT III as a biomaterial for bone substitution in medicine, a new way for the control of the crystallization in phosphate glasses was found.[10] It starts from phosphate invert glasses of the P_2O_5-Al_2O_5-CaO-Na_2O system, the structure of which is formed only by mono- and diphosphate units. Doping with a suitable nucleation agent, e.g. iron oxide or ZrO_2, leads to a supersaturation of the glass with this component in a certain concentration range. This supersaturation is reduced by a subsequent thermal treatment, which results in the precipitation of a primary crystal phase in the glass. In this case this phase may be an Na-Ca-Fe-phosphate of the varulite-type. Crystal nuclei have developed which initiate the precipitation of the actual main crystal phase apatite, $AlPO_4$, in various modifications and specific complex phosphate structures.

The BIOVERIT III glass-ceramic, which is SiO_2-free and is bioactive (i.e. the material can bond to living bone) is a material containing the following phases: apatite, $AlPO_4$, (berlinite) and complex phosphate structures.[11]

PROPERTIES

Characteristic properties of BIOVERIT I, II and III are shown in Tables 4, 5, and 6. It is possible to control the properties by the chemical composition and the content of the crystals in the glass matrix.

BIOVERIT I and II are machineable glass-ceramics; these biomaterials are workable with standard metal tools and instruments. They can also be easily modified during surgical procedures (see middle ear implants of BIOVERIT II). The workability of the biomaterial depends on the mica content as well as the morphology of the glass-ceramic, e.g. a glass-ceramic with high mica content has excellent machinability, thus the workability of BIOVERIT II is better than that of BIOVERIT I.

The content of the crystals within the glass-ceramic is also responsible for the translucency of the material. A higher content of crystals gives a lower translucency than a lower crystal content.

The color can be controlled by adding special pigments, namely the oxides NiO, Cr_2O_3, MnO_2, FeO, Fe_2O_3 and others in small amounts to the base glass of the glass-ceramic.

Table 4. Properties of BIOVERIT I Glass Ceramic.

density	2.8	g/cm^3
linear thermal expansion coefficient (20-400°C)	$8\text{-}12 \cdot 10^{-6}$	m/mK
bending strength	140-180	MPa
fracture toughness, K_{IC}	1.2-2.1	$MPa \cdot m^{1/2}$
Young's modulus	70-88	GPa
compressive strength	500	MPa
Vickers hardness, HV 10	5000	MPa
hydrolytic class (DIN 12111)	2-3	

Table 5. Properties of BIOVERIT II Glass Ceramic.

density	2.5	g/cm^3
linear thermal expansion coefficient (20-400°C)	$7.5\text{-}12\cdot10^{-6}$	m/mK
bending strength	90-140	MPa
fracture toughness, K_{IC}	1.2-1.8	$MPa\cdot m^{1/2}$
Young's modulus	70	GPa
compressive strength	450	MPa
Vickers hardness, HV 10	up to 8000	MPa
hydrolytic class (DIN 12111)	1-2	
roughness (after polishing)	0.1	μm

Table 6. Properties of BIOVERIT III Glass Ceramic.

density	2.7-2.9	g/cm^3
linear thermal expansion coefficient (20-400°C)	$14\text{-}18\cdot10^{-6}$	m/mK
bending strength	60-90	MPa
fracture toughness, K_{IC}	0.6	$MPa\cdot m^{1/2}$
Young's modulus	45	GPa
hydrolytic class (DIN 12111)	2-3	

The translucency and color of the materials are important in dental applications. Mechanical properties, such as bending strength and fracture toughness, of BIOVERIT I and II allow them to be used as biomaterials for bone substitution (Tables 4, 5).

The mechanical properties of BIOVERIT III (primarily bending strength) are lower than those of mica glass-ceramics. But their thermal properties make them suitable for the preparation of composites with certain metals, especially the Co-Cr-Mo alloys widely used in implantology. All these glass-ceramics (BIOVERIT I, II, III) have good chemical properties (hydrolytic stability).

SURFACE CHEMISTRY

Electronic microscopic investigations show the surface of the BIOVERIT glass-ceramics at a high magnification. BIOVERIT I (Fig. 1) contains apatite crystals with a diameter of about 1 - 2 μm and large mica crystals embedded in the glassy matrix.

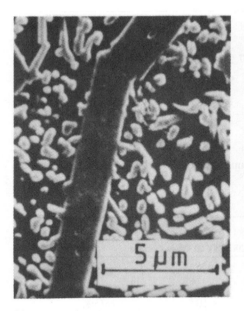 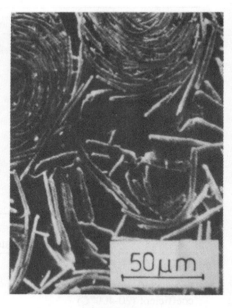

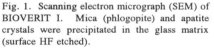

Fig. 1. Scanning electron micrograph (SEM) of BIOVERIT I. Mica (phlogopite) and apatite crystals were precipitated in the glass matrix (surface HF etched).

Fig. 2. SEM of BIOVERIT II. Curved mica crystals of phlogopite-type and cordierite crystals grow in the glass matrix (HF etched).

BIOVERIT II (Fig. 2) shows a special type of curved mica crystals. This type of crystal does not exist in nature, it is only formed in glasses. The mica crystals are spherical in shape and the cordierite crystals are precipitated between the mica platelets.

The crystals of apatite and $AlPO_4$-type of BIOVERIT III are shown in Fig. 3. It was also possible to precipitate berlinite-type $AlPO_4$-crystals, which show piezoelectric properties. Other complex phosphate crystals can grow in the glass-ceramics.

Reactions on the surface of the biomaterials are studied from *in vitro*. Glass-ceramic BIOVERIT I was kept for one week in Ringer's solution. Figure 4b compared with Fig. 4a shows more prominent crystals due to solubility of the glassy phase. Solubility of alkali ions and magnesium ions from BIOVERIT I and II was found in Tris-buffer-solution (Fig. 5). This indicates an ion exchange between the glass-ceramics and simulated body fluid. New, high resolution measurements by SIMS show a tendency for phosphate group enrichment at the surface of BIOVERIT I glass-ceramics, in addition to the phosphate contained in the original apatite crystals.

Fig. 3. BIOVERIT III contains apatite, AlPO$_3$-crystals and other "complex" phosphates (TEM replica, HCl-etched).

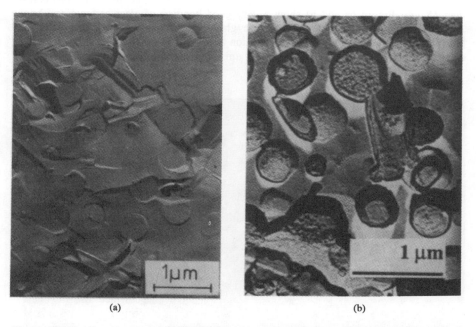

(a)   (b)

Fig. 4. TEM/replica micrograph of BIOVERIT I: (a) untreated specimen (b) specimen after one week in Ringer's solution.

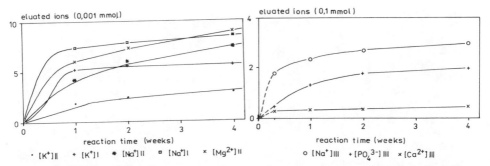

Fig. 5. Eluation of ions from BIOVERIT I, II and III (T = 37°C, pH = 7.4, sample 3 g, grain size: 160-315 μm 100 ml Tris-buffer solution).

TISSUE RESPONSE

Results of *in vivo* tests of BIOVERIT I show that one year after operation, the reaction interface between bone (tibia of guinea pig) and glass-ceramic implants is less than 15 μm. Figure 6a and b) show a very good bonding of glass-ceramic and bone. Apatite is the main crystal phase of the glass-ceramic; in addition, the Ca^{2+} and phosphate ion content is higher in the reaction zone than in the glass-ceramic. Leaching of alkali ions occurs to a depth of about 5 μm. Optical microscopic investigations

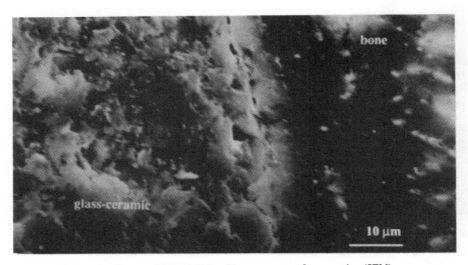

Fig. 6a. The interface between BIOVERIT I and bone one year after operation (SEM).

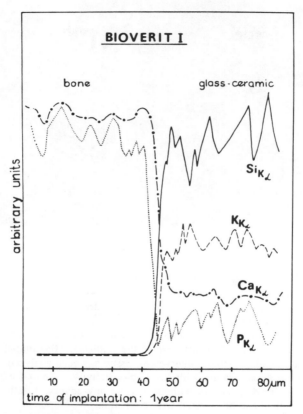

Fig. 6b. The interface between BIOVERIT I and bone one year after operation (electron microprobe).

(histology) and the calculated bone connection (Fig. 6) show a bioreactive behavior of BIOVERIT I glass-ceramic in comparison to alumina implants. These results of bone connection of BIOVERIT I and Al_2O_3 can be compared directly because of the experimental model. The shearing strength of the implant-bone boundary has been determined by measuring the mechanical force necessary to push out the implant of BIOVERIT I. The value of about 2.3 MPa found for glass-ceramic implants was on average eight times greater than for alumina implants, measured for comparison in the same model. The surface of the implants after push out tests was determined by SEM. A typical surface is demonstrated in Fig. 7, in comparison to Fig. 4a, which shows the surface of an untreated BIOVERIT I glass-ceramic. These investigations show the very small solubility of the glassy phase of the glass-ceramic, of less than 0.5 μm. The

Fig. 7. BIOVERIT I after pull out test, 8 weeks after implantation in tibia of guinea pig (TEM/replica).

apatite and mica crystals are directly deposited at the surface of the biomaterial. The black parts of the photograph, where the crystals are not clearly visible, are bone which has remained adherent to the surface of the glass-ceramic. The process of bioactivity is complex. It includes the reaction of the apatite, shown by dense hydroxyapatite implants, new apatite formation, ion transport and surface activity, the role of SiO_2, as described by Hench[1,2] and biomechanical factors. Animal experiments with BIOVERIT III implants showed good bioactive behavior of the implants. Biocompatibility has been tested and a good bonding without intervening tissue has been analyzed. BIOVERIT II is a biocompatible glass-ceramic with lower bioreactivity. Animal experiments have demonstrated that bonding occurs without causing any adverse reactions and that the biocompatible implant can be covered with epithelium.

CLINICAL APPLICATIONS

BIOVERIT I and II can be used as biomaterials for bone substitution. More than 850 implants have been successfully used in clinical cases in the following fields:

- orthopaedic surgery (especially different types of spacers)
- head and neck surgery (especially middle ear implants)
- stomatology (especially tooth root (Pinkert[16]) and veneer laminates)

Orthopaedic Surgery:[14]
- reconstruction of the root of the acetabulum in a dislocated hip at the dysplasic stage (pericapsular iliumosteotomy according to Pemperton)

- ligament fixation in capsule-ligament plastic surgery of the knee joint
- osteotomy of the tibial head and augmentation of the tibial plateau
- partial replacement of vertebrae in the dorsal part of the spine
- ventral spondylodesis in the cervical vertebrae, according to Robinson
- distraction osteotomy for maintaining of space
- plastic surgery of the shoulder joint, according to Eden-Hybinette
- substitution of vertebrae

Head and Neck Surgery:
- collumellization in tympanoplasty
- augmentation of the stapes
- reconstruction of the posterior wall of the auditory canal
- reconstruction of the skull base
- maintenance of the base of the orbit
- construction of the anterior wall of the frontal sinus
- rhinoplasty

Figure 8 shows spacers of BIOVERIT I for treating recurrent dislocations of the shoulder by the Eden-Hybinette operation and Fig. 9 shows middle ear implants of BIOVERIT II.

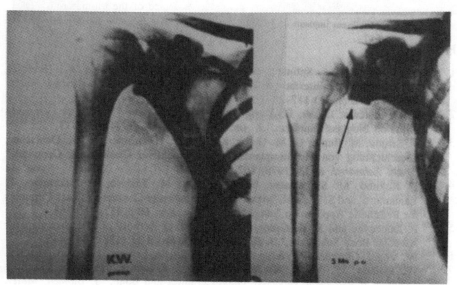

Fig. 8. BIOVERIT I spacers for treating recurrent dislocation of the shoulder by an Eden-Hybinette operation.

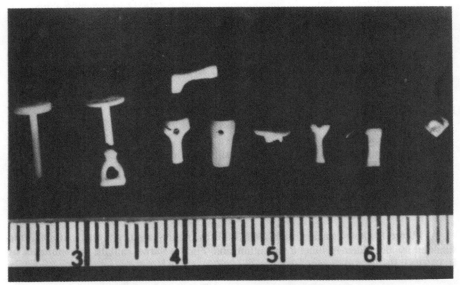

Fig. 9. Middle-ear implants of BIOVERIT II that have been shaped by the surgeon.

REFERENCES

1. L. L. Hench, R. J. Splinter, W. C. Allen and T. K. Greenlee, Jr., "Bonding Mechanisms at the Interface of Ceramic Prosthetic Materials," *J. Biomed. Mater. Res. Symp.* **2** (1971) 117-141.

2. C. G. Pantano, Jr., A. E. Clark, Jr. and L. L. Hench, "Multilayer Corrosion Films on Bioglass® Surface," *J. Am. Ceram. Soc.* **57** (1974) 412-413.

3. V. Strunz, M. Bunte, U. M. Gross, K. Männer, H. Brömer, and K. Deutscher, "Beschichtung von Metallimplantanten mit Bioaktiver Glaskeramik Ceravital," *Dtsch. Zahnärztl. z.* **33** (1978) 862-865.

4. T. Kukubo, M. Shigematsu, Y. Nagashima, M. Tashiro, T. Nakamura, T. Yamamuro, and S. Higashi, "Apatite and Wollastonite-Containing Glass-Ceramics for Prosthetic Application," *Bull. Inst. Chem. Res.* **60** (1982) 260-268.

5. P. Ducheyne and K. de Groot, "*In Vivo* Surface Activity of a Hydroxylapatite Alveolar Bone Substitute," *J. Biomed. Mater. Res.*d **15** (1981) 441-445.

6. W. Vogel, *Glaschemie* (Verlag Grundstoffindustrie, Leipzig, 1979) pp. 310-322.

7. W. Höland, W. Vogel, K. Naumann and J. Gummel, "Interface Reactions Between Machineable Bioactive Glass-Ceramics and Bone," *J. Biomed. Mater. Res.* **19** (1985) 303-312.

8. W. Vogel and W. Höland, "Development of Glass-Ceramics for Medical Application," *Angew. Chem. Int., Ed. Engl.* **26** (1987) 527-544.

9. W. Höland, P. Wange, K. Naumann, J. Vogel, G. Carl, C. Jana, and W. Götz, "Control of Phase Formation Process in Glass-Ceramics for Medicine and Technology," *J. Non-Crystalline Solids* **129** (1991) 152-162.

10. W. Vogel, J. Vogel, W. Höland, and P. Wange, "Zur Entwicklung Bioaktiver Kieselsäurefreier Phosphatglas Keraminken für die Medizin Wissenschaftliche Zeitschrift FSK Jena," *Natur wiss. R.* **36** (1987) 841-854.

11. P. Wange, J. Vogel, L. Horn, W. Höland, and W. Vogel, "The Morphology of Phase Formations in Phosphate Glass Ceramics," *Silic. Incl.* **7-8** (1990) 231-236.

12. G. H. Beall, M. R. Montierth and P. Smith, "Workable Glass-Ceramics," *Glas-Email-Keramo-Technik* **22** (1991) 409-415.

13. D. G. Grossman, "Stain-Resistant Mica Compositions and Articles Thereof in Particular Dental Constructions," Patent EP 0 083 828 (1983).

14. T. Schubert, W. Purath, P. Liebscher and K.-J. Schulze, "Klinische Indikationen für die Anwendung der Janaer Bioaktiven Maschinell Bearbeitbaren Glaskeramik in der Orthopädie und Traumatologie," *Beiträge Orth. Traumatol.* **35** (1988) 7-16.

15. E. Beleites, H. Gudziol, and W. Höland, "Maschinell Bearbeitbare Glaskeramik für die Kopf-Hals-Chirurgie," *HNO-Praxis* **13** (198) 121-125.

16. R. Pinkert, K. Naumann and W. Vogel, "Individuelle Herstellung Enossaler Zahnimplantate aus der Maschinellen Barbeitbaren Bioaktiven Glaskeramik Jena," *Philip Journal* **6** (1987) 339-342.

Chapter 9

DENSE HYDROXYAPATITE

Racquel Z. LeGeros and John P. LeGeros
New York University College of Dentistry
345 East 24th Street, New York, New York 10010

INTRODUCTION

Successful repair of a bony defect with a calcium phosphate reagent, described as 'triple calcium phosphate compound' was first reported by Albee in 1920.[1] A half a century later, Levitt et al. in 1969[2] and Monroe et al. in 1971[3] described a method of preparing a ceramic apatite from mineral fluorapatite, $Ca_{10}(PO_4)_6F_2$ and suggested the possible use of this apatite ceramic in dental and medical applications. Clark et al.[4] and Hubbard[5] described methods of preparing calcium phosphate ceramics from commercially available calcium phosphate reagents. However, these ceramics were not appropriately characterized so that what was described as 'tricalcium phosphate' or TCP ceramic in the first clinical application reported by Nery et al.[6] was actually a mixture of beta-tricalcium phosphate (b-TCP, $Ca_3(PO_4)_6$), and hydroxyapatite (HA, $Ca_{10}(PO_4)_6(OH)_2$).[7]

In the mid-seventies, three groups: Jarcho et al. in USA;[8-10] deGroot et al.[11,12] and Denissen[13] in Europe and Aoki et al. in Japan[14,15] simultaneously but independently worked towards the development and commercialization of hydroxyapatite, (HA), more accurately, calcium hydroxyapatite, as a biomaterial for bone repair, augmentation and substitution. This was based on the rationale that the bone mineral has been usually described as 'hydroxyapatite'[16] and followed the observation reported in the early seventies by Hench[17] of a 'chemical bonding' of bone with a bioactive glass ceramic through a calcium phosphate-rich layer.

HA can be prepared as dense or as macroporous forms, with pores as large as 500 μ. Dense HA which is the subject of this chapter, is described as having a maximum microporosity of 5% by volume with the micropores measuring about 1 μm in diameter[11,13,18] and consisting of crystals with size exceeding 2000Å. This chapter will attempt to provide basic information on the general properties of pure HA and biological apatites and then discuss the processing, composition, properties, surface chemistry, tissue response and clinical applications of dense HA.

GENERAL STRUCTURE AND PROPERTIES OF CALCIUM HYDROXYAPATITE

The term APATITE describes a family of compounds having similar structures but not necessarily having identical compositions. Hence, apatite is a description and not a composition. Hydroxyapatite (HA), specifically, calcium hydroxyapatite, is a compound of a definite composition, $Ca_{10}(PO_4)_6(OH)_2$, and a definite crystallographic structure. The structure of calcium hydroxyapatite (HA), showing the exact atomic positions in the crystal was determined by Beevers and McIntyre from a mineral[19] and later refined by Kay et al. with a synthetic HA.[20] Calcium hydroxyapatite belongs to the hexagonal system, with a space group, $P6_3/m$. This space group is characterized by a six-fold c-axis perpendicular to three equivalent a-axes (a_1, a_2, a_3) at angles 120° to each other. The smallest building unit, known as the unit cell, contains a complete representation of the apatite crystal, consisting of Ca, PO_4, and OH groups closely packed together in an arrangement shown in Fig. 1.[21]

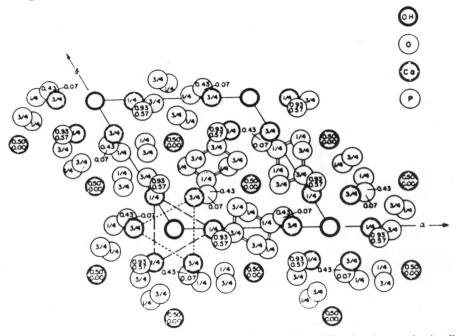

Fig. 1. The atomic arrangement of calcium hydroxyapatite, $Ca_{10}(PO_4)_6(OH)_2$, in a hexagonal unit cell. The OH ions located in the corners of the unit-cell are surrounded by two groups of Ca(II) atoms arranged in a triangle positions at z = 0.25 and at 0.75; by two groups of PO_4 tetrahedra also arranged in triangle positions; and by a hexagonal array of Ca (I) atoms at the outermost distance (courtesy, Prof. R. Young[21]).

The ten calcium atoms belong to either Ca(I) or Ca(II) subsets depending on their environment. Four calcium atoms occupy the Ca(I) positions: two at levels z = 0 and two at z = 0.5. Six calcium atoms occupy the Ca(II) positions: one group of three calcium atoms describing a triangle located at z = 0.25, the other group of three at z = 0.75, surrounding the OH groups located at the corners of the unit cell at z = 0.25 and z = 0.75, respectively (Fig. 2). The six phosphate (PO_4) tetrahedra are in a helical

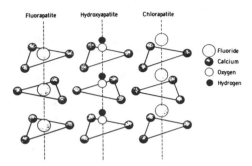

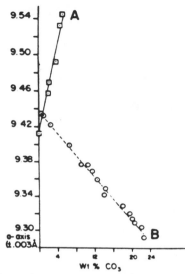

Fig. 2. The comparative positions of OH, F and Cl atoms at the center of the Ca(II) triangles in HA, $Ca_{10}(PO_4)_6(OH)_2$; FA, $Ca_{10}(PO)_6F_2$; and ClA, $Ca_{10}(PO_4)_6Cl_2$. The respective positions for the OH groups, O is at z = 0.20 or 0.30, and at 0.70 or 0.80; H, at 0.06 or 0.44 and at z = 0.56 or 0.94; for F, at z = 0.25 and at 0.75; for Cl, at z = 0 and at 0.44 (courtesy, Prof. R. Young[21]).

Fig. 3. The effect of two types of carbonate substitution on the a-axis dimensions of synthetic apatites. A, CO_3-for-OH substitution, $Ca_{10}(PO_4)_6CO_3$, observed in synthetic apatites prepared at 1000°C[25,26] causes an expansion in the a- and contraction in the c-axis dimensions. B, CO_3-for-PO_4 coupled with Na-for-Ca substitution, $(Ca,Na)_{10}(PO_4,CO_3)_6(OH)_2$, observed in synthetic apatites prepared from aqueous systems either by precipitation or hydrolysis methods[28,32,33,36,41] causes an expansion in the a- and contraction in the c-axis dimensions compared to the a- and c-axis dimensions of unsubstituted HA.

arrangement from levels z = 0.25 to z = 0.75. The network of PO_4 groups provide the skeletal framework which gives the apatite structure its stability. The oxygens of the phosphate groups are described as one O_I, one O_{II}, and two O_{III}. The atomic arrangements of F-apatite, $Ca_{10}(PO_4)_6F_2$, and of Cl-apatite, $Ca_{10}(PO_4)_6Cl_2$, in which fluoride (F) and chloride (Cl), respectively, substituted for the OH groups in the apatite structure, are similar. The F or Cl atoms substituted for OH differ in the respective position of the OH for which they substitute. The O-H, F and Cl atoms lie along the c-axis at the center of the Ca (II) triangles (Fig. 2) as described by Young and Elliott.[21]

The apatite structure is very hospitable one allowing the substitutions of many other ions. Substitutions in the apatite structure for (Ca), (PO_4) or (OH) groups result in changes in properties: e.g., lattice parameters (Table 1), morphology, solubility without significantly changing the hexagonal symmetry. However, Elliott and Young[21] showed that Cl substitution causes the loss of hexagonal symmetry and exhibits monoclinic symmetry because of the alternating positions of the Cl atoms and an enlargement of the cell in the b direction. For example, substitution of F for OH, cause a contraction in the a-axis dimensions without changing the c-axis (Table 1), is usually associated with an increase in the crystallinity, reflecting increase in crystal size (and/or decrease in crystal strain); and imparts greater stability to the structure (Fig. 2). Increased stability is

Table 1. Lattice Parameters of Mineral, Synthetic and Biological Apatites.

Apatite	Major Substituent*	Lattice Parameters	(+0.003A)
Mineral			
OH Apatite (Holly Springs)	-	9.422	6.880
F-apatite (Durango, Mex)	F	9.375	6.880
Dahllite (Wyoming)	CO_3	9.380	6.885
Staffelite (Staffel, Germany)	CO_3, F	9.345	6.880
Marine phosphorite (w.USA)	CO_3, F	9.322	6.882
Synthetic (non-aqueous)[a]			
OH-apatite	-	9.441	6.882
F-apatite	F	9.375	6.880
Cl-apatite	Cl	9.646	6.771
CO_3 apatite	CO_3	9.544	6.859
Synthetic (aqueous)[b]			
OH-apatite (Ca-deficient)	HPO_4**	9.438	6.882
F-apatite	F	9.382	6.880
(Cl,OH)-apatite	Cl**	9.515	6.858
CO_3-OH-apatite	CO_3**	9.298	6.924
CO_3-F-apatite	CO_3**, F	9.268	6.924
Sr-apatite	Sr	9.739	6.913
Pb-apatite	Pb	9.894	7.422
Ba-apatite	Ba	10.162	7.722
Biological			
CO_3-OH-apatite (human enamel)	HPO_4, Cl, CO_3, Mg	9.441	6.882
F-apatite (Shark enameloid)	F, CO_3, HPO_4	9.382	6.880

[a]Prepared at high temperature (1000°C) by solid state diffusion.[25-27]
[b]Prepared at 100°C either by precipitation or by hydrolysis methods.[22,23,28-40]
Biological apatites.[22,28,28,31,41]
**Maximum incorporation in case of Cl⁻ is less than one mole; in the case of CO_3 is 3 moles; in cases of HPO_4 is unknown.[42] In these cases the substitutions are: F- or Cl-for-OH; CO_3-for OH (Type 1); CO_3-for PO_4 coupled with Na-for-Ca (Type B); HPO_4-for PO_4; Sr, Pb or Ba-for-Ca.

reflected in the observation that F-substituted apatites are less soluble than F-free synthetic and biological apatites.[22-24]

Carbonate, CO_3, can substitute either for the hydroxyl (OH) or the phosphate (PO_4) groups; designated as Type A or Type B substitutions, respectively . These two types of substitution have opposite effects on the lattice parameters, a-axis and c-axis dimensions (Fig. 3 and Table 1). In the case of Type A, the substitution of larger planar

CO_3 group for smaller linear OH group, cause an expansion in the $\underline{a}$-axis and contraction in the $\underline{c}$-axis dimensions;[25,26] while for Type B, the substitution of smaller planar CO_3 group for a larger tetrahedral PO_4 group, cause a contraction in the $\underline{a}$-axis and expansion in the $\underline{c}$-axis dimensions compared to the CO_3-free apatites.[28,32,33,36,41] Differences in infrared spectral properties were also observed.[26,28,35] LeGeros and co-workers also demonstrated that the coupled CO_3-for-PO_4 and Na-for-Ca substitution cause changes in the size and shape of the apatite crystal: from acicular crystals to rods to equi-axed crystals with increasing carbonate content (Fig. 4); and in dissolution properties: the CO_3 substituted apatite being more soluble than CO_3-free synthetic apatites.[43]

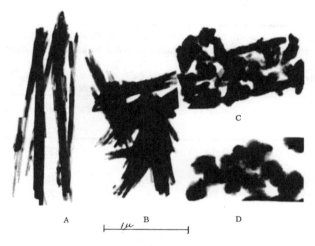

Fig. 4. Electron micrographs of CO_3-apatite crystals obtained by the conversion of $CaHPO_4$ in solution containing increasing concentration of carbonate ions. Carbonate incorporation causes changes in crystal sizes and shape. Carbonate content: (A) 2.5 wt%; (B) 4.5 wt%; (C) 15.2 wt%; (D) 17.25 wt%. Magnification, 40,000X. Apatite crystals change in morphology from acicular to rods to equi-axed crystals with increasing carbonate incorporation.[28,32,36,41]

Other ions to which the apatite structure may play host include: strontium (Sr), magnesium (Mg), barium (Ba), lead (Pb), etc. for calcium;[28,30,44-46] vanadates, borates, manganates, etc., substituting for phosphate.[22,37,47] Differences in lattice parameters between substituted and unsubstituted HA reflect the size and the amount of the substituting ions (Tables 1 and 2). Various substitutions in the apatite besides those of F- or Cl-for-OH or CO_3-for-OH or CO_3-for-PO_4 mentioned above, also affect properties, e.g., crystallinity (Table 2), thermal stability, and dissolution properties or solubility of the apatite crystals. Sr-for-Ca or Mg-for Ca substitution cause an increase in the extent of dissolution of the apatite.[33,30,39,46] When simultaneously present, the substituents in the apatite structure can have synergistic or antagonistic effects on the properties of the apatite. For example, magnesium and carbonate have synergistic effects on the crystallinity (Table 2) and dissolution properties of synthetic apatites; magnesium and fluoride or carbonate and fluoride have antagonistic effects, the fluoride effect being the more dominant one.[22,30,43]

A comprehensive understanding of both the individual effects of substituents in the structure on the properties of apatite is important to the development of substituted HA as new biomaterials and to an improved understanding of the interaction between bone mineral and the HA materials presently used for bone repair, augmentation, substitution and coatings of metal dental and orthopedic implants.

BIOLOGICAL APATITES

Biological apatites, which comprise the mineral phases of calcified tissues (enamel, dentin, bone) and of some pathological calcifications (e.g., human dental calculi, salivary and urinary stones) are usually referred to as calcium hydroxyapatite,

Table 2. Qualitative Effects of Some Substituents for Ca^{2+}, PO_4^{3+} or OH in $-Ca_{10}(PO_4)_6(OH)_2$ on the Lattice Parameters and Crystallinity of Apatites.

Substituent	Ionic Rad(A)*	Lattice Parameters (+003A)		Crystallinity
		a-axis	c-axis	
for Calcium, Ca^{2+}	0.99	9.438	6.882	
Strontium, Sr^{2+}	1.12	(+)	(+)	(nc)
Barium, Ba^{2+}	1.34	(+)	(+)	(-)
Lead, Pb^{2+}	1.20	(+)	(+)	(-)
Potassium, K^+	1.33	(nc)	(nc)	(nc)
Sodium, Na^+	0.97	(nc)	(nc)	(nc)
Lithium, Li^+	0.68	(nc)	(nc)	(nc)
Magnesium, Mg^{2+}	0.66	(-)**	(-)**	(-)**
Cadmium, Cd^{2+}	0.97	(-)	(-)	(-)
Manganese, Mn^{2+}	0.80	(-)	(-)	(-)
Zinc, Zn^{2+}	0.74	(+)**	(+)**	(-)**
Aluminum, Al^{3+}	0.51	(+)	(+)	(-)
for OH				
Fluoride, F^-	1.36	(-)	(nc)	(+)
Chloride, Cl^-	1.81	(+)	(-)	(nc)
for PO_{43-}				
Carbonate, CO_3^{2-}		(-)	(+)	(-)
HPO_4^{2-}		(+)	(nc)	(nc)

*Handbook of Chemistry of Physics, 68th Edition, ed. R.C. Weast (CRC Press, Boca Raton, FL, 1988). (+)increase, (-)decrease, (nc)no change, **TCP formed in addition to AP. If prepared at temperatures below 60°C, CO_3, P_2O_7, Mg, Al, when present at a critical concentration in solution cause the formation of amorphous calcium phosphate.

HA, meaning, $Ca_{10}(PO_4)_6(OH)_2$. Actually, the biological apatites differ from pure HA in stoichiometry, composition, and crystallinity and in other physical and mechanical properties (Table 3). Biological apatites are usually calcium-deficient and are always carbonate substituted. It is therefore more appropriate that biological apatites be referred to as carbonate apatite and not as hydroxyapatite or HA.[28,41,48,49] The carbonate in

Table 3. Comparative Composition, Crystallographic and Mechanical Properties of Human Enamel, Bone and Hydroxyapatite (HA) Ceramic.

	Enamel	Bone	HA
Constituents (wt %):			
Calcium, Ca^{2+}	36.0	24.5	39.6
Phosphorus, P	17.7	11.5	18.5
(Ca/P) molar	1.62	1.65	1.67
Sodium, Na^+	0.5	0.7	tr
Potassium, K^+	0.08	0.03	tr
Magnesium, Mg^{2+}	0.44	0.55	tr
Carbonate CO_{32+}	3.2	5.8	-
Fluoride, F^-	0.01	0.02	-
Chloride, Cl^-	0.30	0.10	-
Ash (total inorganic)	97.0	65.0	100
Total organic	1.0	25.0	-
Absorbed H_2O^*	1.5	9.7	-
Trace elements: Sr^{2+}, Pb^{2+}, Ba^{2+}, Fe^{3+}, Zn^{2+}, Cu^{2+}, etc.			
Crystallographic properties			
Lattice parameters (+/- 0.003A)			
a-axis	9.441	9.419	9.422
c-axis	6.882	6.880	6.880
Crystallinity index(**)	70-75	33-37	100
Crystallite size, A	1300 x 300A	250 x 25-50	
Products after sintering (950°C)	HA + TCP	HA + CaO	HA
Mechanical properties			
Elastic modulus (10^6 MPa)	0.014	0.020*	0.01
Tensile strength (MPa)	70	150*	100

*Values for cortical bone.[52] Mechanical properties for human enamel.[18]

biological apatites substitutes primarily for the phosphate groups in a coupled manner, i.e., Ca-for-Na, CO_3-for-PO_4, referred to as a Type B substitution.[22,28,32-35,41] The coupled substitution is necessary to balance charges for the substitution of a CO_3(divalent) for PO_4 (trivalent). Biological apatites of enameloids of some species of fish or of shark enameloid are substituted with F and CO_3.[22,31] Other minor elements, e.g., sodium (Na^+), magnesium (Mg^{2+}), potassium (K^+), acid phosphate ($HPO_4)_{2-}$, chloride (Cl⁻) and fluoride (F⁻) and some trace elements (e.g., Sr^{2+}, Pb^{2+}, Ba^{2+}, etc.) are associated with biological apatites and may be substituents in the apatite structure, as described below:

$$(Ca,M)_{10}(PO_4,CO_3,Y)_6(OH,F,Cl)_2 \qquad (1)$$

where M represents other minor (e.g., Mg^+, Na^+, K^+, etc.) and trace elements (e.g., strontium, lead, barium); Y represents acid phosphate, HPO_4^{2-}, sulfates, borates, vanadates, etc. Some of these minor and trace elements may be surface- rather than lattice-bound.[40,50,51] Lattice bound elements will contribute to changes in lattice parameters; surface elements will not,but contribute to changes in crystal properties.

The biological apatites of enamel differ from those of dentin or bone in crystallinity (Figs. 5, 6 and 7) and in the concentration of the minor elements, principally, CO_3, and Mg^{2+} (Table 3). Enamel apatites contain the least amount of carbonate and magnesium and have the largest crystal size compared to either dentin or bone apatites (Table 3). In terms of dissolution properties, enamel apatite is less soluble than dentin or bone but much more soluble than dense HA which is prepared at high temperature as ceramic HA (Fig. 8). The difference in crystal size (Table 3) and in dissolution properties among enamel, dentin and bone may be attributed in part to the differences in the carbonate and magnesium concentrations. Magnesium and carbonate have been shown to cause a decrease in crystallinity (Table 2) and an increase in the extent of dissolution of synthetic apatites.[22,30,43]

Biological apatites of human enamel, dentin and bone also give different end products after sintering above 800°C (Table 3) reflecting the differences in their composition and calcium deficiencies.[28,52] Sintering enamel and dentin apatites above 800°C gives HA and small amounts of b-TCP (about 5 wt% in enamel and about 12 wt% in sintered dentin). The substituted b-TCP phase is magnesium-substituted as determined from lattice parameters and chemical analyses and is therefore sometimes referred to as b-TCMP.[28,41] Sintering human bone apatite above 800°C gives mainly HA and minor amounts of CaO (Table 3). Sintering enamel, bone or dentine apatite results in changes in crystal morphology, crystallinity (Fig. 9) and composition (Table 3). Morphologically, the enamel apatite crystals changed in shape from small acicular to large hexagonal or rhombohedral crystals (Fig. 9). Compositional changes in biological apatites include the loss of CO_3 and the formation of Mg-substituted b-TCP phase.[22] The formation of the b-TCP after heat treatment of enamel and dentin and calcium-deficient synthetic apatites is attributed to the HPO_4 content of the apatite before sintering.[22,28,41]

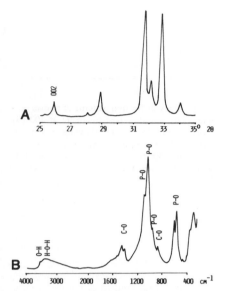

Fig. 5. (A) X-ray diffraction pattern and (B) IR absorption spectrum of human enamel mineral.

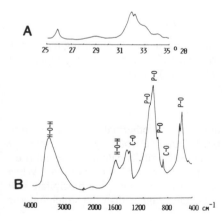

Fig. 7. (A) X-ray diffraction pattern and (B) IR absorption spectrum of human bone mineral.

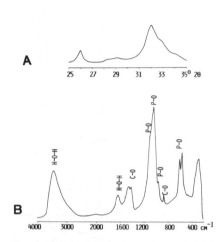

Fig. 6. (A) X-ray diffraction pattern and (B) IR absorption spectrum of human dentin mineral.

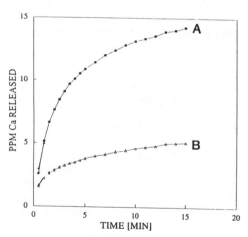

Fig. 8. Comparative extent of dissolution of powdered human enamel apatite (A) and ceramic HA (B) in acetate buffer (0.1M KAc, pH 6, 25°C). [Kijkowska/LeGeros, 1992, unpublished]

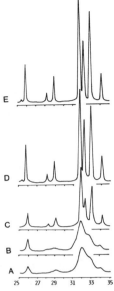

Fig. 9. X-ray diffraction patterns of human bone before (A) and after heat treatment at 400 (B), 700 (C), and 950°C (D) showing increase in crystal size as a result of heat treatment. Heat treatment at 400°C resulted in the removal of the organic phase which constitutes about 35 wt% of the bone.

Sintered animal bones result in the formation of b-TCP (Mg-substituted) or CaO depending on the age and specie of the animals.[22]

PREPARATION OF DENSE HYDROXYAPATITE

Preparation of dense HA consists of the following steps: (1) preparing the apatite powder or using commercially available apatite reagents; (2) compacting or compressing into a desired size and shape under high pressure; and (3) sintering.

The different steps are described below.

Preparation of HA Powder

Pure HA can be obtained from reactions in hydrothermal systems or from solid-state reactions . However, when prepared from aqueous systems either by precipitation or hydrolysis methods, the apatite obtained is usually calcium deficient (i.e., the Ca/P molar ratio is lower than the stoichiometric value of 1.67 for pure HA). When the precipitation reaction is carried out under very basic conditions, the precipitate will contain carbonate which will make the Ca/P molar ratio higher than the stoichiometric value.

Precipitation methods.   The apatite preparation methods commonly used in commercial preparations were based on the method of Rathje[53] and of Hayek and Newesely.[54] Rathje's method consisted of dropwise addition of phosphoric acid, H_3PO_4, to a stirring suspension of calcium hydroxide, $Ca(OH)_2$ in water (reaction 1):

$$10\ Ca(OH)_2 + 3\ H_3(PO_4)_2 \ \text{---} \ \text{"}Ca_{10}(PO_4)_6(OH)_2\text{"} \ \text{(Apatite or AP)} \qquad (2)$$

This method is modified by the addition of ammonium hydroxide, NH_4OH, to keep the Ph of the reaction very alkaline to insure the formation of HA after sintering the apatite (AP) precipitate.[14,47] Hayek and Newesely's method[54] consisted of reaction between calcium nitrate, $Ca(NO_3)_2$, and ammonium phosphate, $(NH4)_2HPO_4$, with added NH4OH (reaction 2):

$$10\ Ca(NO_3)_2 + 6\ (NH_4)_2HPO_4 + 2\ NH_4OH \ \text{---} \ \text{"Apatite"} \ \text{(AP)} \qquad (3)$$

This method is sensitive to the concentrations of each of the reactants and the Ph of the reaction for the formation of HA upon sintering of the apatite precipitate.[55,56] Calcium acetate, $Ca(CH_3COO)_2$ instead of calcium chloride or nitrate is recommended as the calcium source in precipitation   reactions because the acetate ions will not BE incorporated into the apatite[22,28,30,32,33,376] unlike nitrate or chloride[29] ions which may. The temperature of precipitation ranges from room temperature (about 24°C) to boiling (95-100°C). The concentration of calcium can be adjusted if substitution for calcium (e.g. strontium, magnesium, manganese, etc.) in the apatite is desired. Similarly, the phosphate concentration can be adjusted and replaced with carbonate if carbonate apatite is desired; other ions, e.g., vanadate, borate, manganate, etc., ions can be similarly added replacing part of the phosphate component.[22,37,39,41,46] Fluoride- or chloride-substituted apatite can be prepared by the addition of F- or Cl- ions in the reaction.[23,28,29,31,41]

The precipitation method, like the hydrolysis method, usually results in calcium-deficient apatite which causes the formation of b-TCP with HA upon sintering. If the reaction, either precipitation or hydrolysis, is carried out under very basic pH, the Ca/P approaches the stoichiometric value or exceeds it, depending on the formation of carbonate-apatite before sintering. The designation, "Apatite" indicates that the product is not pure hydroxyapatite.

Hydrolysis method. Apatite (AP) can also be prepared by the hydrolysis of acid calcium phosphates, e.g., dicalcium phosphate dihydrate, DCPD, $CaHPO_4 \cdot 2H_2O$; octacalcium phosphate, OCP, $Ca_8H_2(PO_4)_6 \cdot 5H_2O$; or monetite, dicalcium phosphate anhydrous, DCP, $CaHPO_4$[22,29-30,37,38] in ammonium, sodium or potassium hydroxide, carbonate, fluoride or chloride solutions depending on the desired composition of the apatite. Calcium carbonate, $CaCO_3$, can also be hydrolyzed to apatite in ammonium or sodium phosphate solutions; or to F-apatite in fluoride solutions.[37] a- or b-TCP and tetracalcium phosphate (TTCP) and amorphous calcium phosphate (ACP) of special composition can also easily hydrolyze to calcium-deficient apatite.

Calcium phosphate reagents. Commercially available calcium phosphate reagents (labeled or mislabeled as 'calcium phosphate, tribasic', 'calcium hydroxyapatite') is used as the apatite powder, with or without the addition of appropriate amounts of $CaCO_3$ or $Ca(OH)_2$ or CaO[11-13,18] to make the apatite or apatitic reagent less calcium-deficient and thus minimize the formation of b-TCP phases upon sintering.

Apatites prepared from aqueous systems by precipitation or hydrolysis are usually calcium deficient and HPO_4-enriched as shown by the expanded a-axis dimensions compared to mineral or ceramic HA (Table 1) and the formation of a b-TCP phase upon heat treatment above 800°C.[38,40,41,57] Commercially available calcium phosphate reagents labeled as 'calcium phosphate tribasic' are sometimes mixed phases of apatitic calcium phosphate and monetite (Fig. 10); reagents labeled as 'spheroidal hydroxyapatite' were shown by x-ray diffraction analysis to consist mostly of b-TCP mixed with small amounts of HA[22,57] and not HA (Fig. 11).

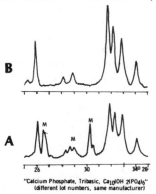

"Calcium Phosphate, Tribasic, $Ca_{10}OH_2(PO_4)_6$"
(different lot numbers, same manufacturer)

Fig. 10. X-ray diffraction patterns of commercially available 'apatite' reagents from different manufacturers labeled 'calcium phosphate phosphate tribasic', 'calcium hydroxyapatite' showing apatitic calcium phosphate with poor crystallinity, some mixed with another calcium phosphate phase, $CaHPO_4$, dicalcium phosphate anhydrous or monetite (M).

Fig. 11. X-ray diffraction patterns of commercial reagents labeled as 'spheroidal hydroxyapatite" from the same manufacturer but different lot numbers showing that the reagents are no hydroxyapatite but beta-tricalcium phosphate, b-TCP in one lot (A) and b-TCP mixed with small amounts of HA in the other (B).

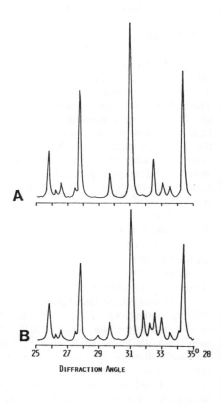

DIFFRACTION ANGLE

All preparations, commercial or non-commercial, must be characterized using x-ray diffraction, infrared spectroscopy and chemical analyses (for calcium and phosphate concentrations) before use in the preparation of dense HA. This is important since apatite preparations after sintering above 800°C can produce materials consisting of intimate mixtures of b-TCP and HA of varying b-TCP/HA ratios.[7] Some dense or macroporous biphasic calcium phosphates, BCP, are intentionally prepared with the desired b-TCP/HA ratio.[58,59]

<u>Solid-state reactions.</u> HA can also be prepared by solid state reactions as follows:

$$6\ CaHPO_4 + 4\ Ca(OH)_2 \longrightarrow Ca_{10}(PO_4)_6(OH)_2 + 6\ H_2O \qquad (4)$$
  monetite   HA

$$3\ Ca_3(PO_4)_2 + Ca(OH)_2 \longrightarrow Ca_{10}(PO_4)_6(OH)_2 + H_2O \qquad (5)$$
  b-TCP   HA

The mixed calcium compound, is compressed and sintered above 950°C.[28] Substituted apatites (e.g., Sr for Ca; F or Cl for OH) can also be prepared by adding appropriate compounds.

<u>Hydrothermal reactions.</u> The above reactions can also be carried out hydrothermally at 275°C, under steam pressure of 12,000 psi.[22,28] In addition, b-TCP, $Ca_3(PO_4)_2$, and tetracalcium phosphate, TTCP, $Ca_4P_2O_9$ or $Ca_4(PO_4)_2O$, can be easily converted to HA hydrothermally under these conditions.

Calcium carbonate, $CaCO_3$, in the presence of the appropriate amounts of $CaHPO_4$ or $(NH_4)_2HPO_4$ can be transformed to HA as follows:

$$4\ CaCO_3 + 6\ CaHPO_4 \longrightarrow Ca_{10}(PO_4)_6(OH)_2 + 6\ H_2O + 4\ CO_2 \qquad (6)$$
  calcite or aragonite monetite   HA

$$10\ CaCO_3 + 6\ (NH_4)_2HPO_4 \longrightarrow Ca_{10}(PO_4)_6(OH)_2 + H_2O + CO_2 \qquad (7)$$

Compacting and Sintering

The apatite powder prepared according to any of the methods described in the previous section can be made into either of two forms: dense or macroporous. As mentioned earlier, dense HA is described as having porosity of less than 5 % by volume. Dense HA may also be described as microporous. The microporosity is unintentionally introduced and is dependent on the temperature and duration of sintering. For dense HA, the maximum pore size is less than 1 μm in diameter.[11,13,18]

Macroporosity, on the other hand, can be deliberately introduced by mixing the powder with a volatile component, e.g., hydrogen peroxide or naphthalene, then evaporating the volatile component at low temperature, about 80°C, before sintering.[11,60] Macroporosity can also be a property of the original material such as bone or coral used in the fabrication of apatite. Macro-porous HA from coral (coralline HA) is discussed in chapter 10.

Apatite powder is compressed or compacted into a mold at a pressure of 60 to 80 MPa. The powder may be mixed with a binder, e.g., 1 w% cornstarch and water,[47] or stearic acid in alcohol[13,18] or low molecular weight hydrocarbons[61] applied to the die as a lubricant. The compressed body can be sintered in air (conventional method) at the desired temperature, usually 950 to 1300°C, heating at the rate of about 100°C per hour and holding at the maximum temperature for several hours before cooling at the same rate as the heating rate.

Hot-pressing techniques are also used in which heat and pressure are simultaneously and continuously applied. This procedure allows densification to take place at a much lower temperatures than in the conventional sintering process (e.g., 900 vs 1300°C). The lower temperature of densification prevents the formation of other calcium phosphate phases, e.g., a- and b-TCP, TTCP, which usually form when HA is sintered at temperatures above 900°C.[62] The disadvantage of this technique is the expensive equipment required and the limited geometry of the end product. Another method of compressing is by hot-isostatic pressing (HIP) where materials are compressed by gaseous pressures at high temperatures. This process results in greater density and higher compressive strength than the conventional method (uniaxial pressing) of compacting.

Using the above methods, dense HA is prepared in tooth forms (Fig. 12) or in blocks. Particulates in irregular or spherical shapes (Fig. 13) are obtained by milling and rolling the compacted and sintered blocks. The block form is carved to prepare middle ear implants.[63,64] Dense HA as tooth forms has been used as immediate tooth root

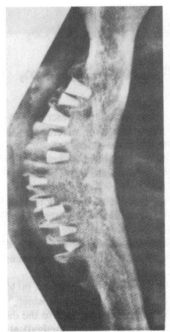

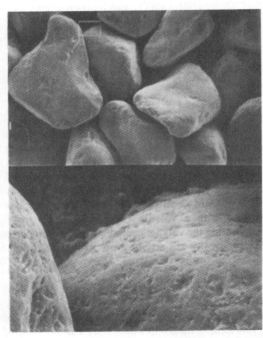

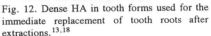

Fig. 12. Dense HA in tooth forms used for the immediate replacement of tooth roots after extractions.[13,18]

Fig. 13. SEM of dense HA particles showing the presence of micropores (SEM, in collaboration with Prof. B. Penugonda).[22]

replacements to minimize alveolar ridge resorption which follows tooth loss, and to maintain ridge width and height.[13,16,65] Dense HA in block and particulate forms is used in the augmentation of alveolar ridge for better denture fit[11,66,67] or in orthopedic surgery.[68] Blocks of dense HA have also been used as target materials for ion-sputtered coatings. Dense HA in particulate form is used as a filler in bony defects in dental and orthopedic surgery;[59,68-72] as a filler in association with placing of metal implants or for repair of failing metal dental implants;[74,74] as a filler in composites[75,76] or cements;[77] and for plasma-sprayed coating on metal implants.[78-81]

COMPOSITION OF DENSE HA

Pure HA, $Ca_{10}(PO_4)_6(OH)_2$, has the theoretical composition of: 39.68 wt% Ca; 18.45 wt% P; Ca/P wt ratio, 2.151; Ca/P molar ratio of 1.667. Dense HA materials, commercial or non-commercial, vary in Ca/P ratios reflecting the b-TCP/HA ratios in the sintered material which in turn reflect the purity (whether consisting of only the apatite phase or mixed with other Ca-P phases); and/or composition or calcium deficiency of the apatite preparation before sintering. If the Ca/P is 1.67, only HA will be observed in the x-ray diffraction (Fig. 14a) and infrared spectrum (Fig. 14b); if the Ca/P is lower than 1.67, b-TCP, and other phases such as tetracalcium phosphate, TTCP, $Ca_4P_2O_9$, or $Ca_4(PO_4)_2O$ will be present with the HA phase in the sintered material, depending on the temperature and condition of sintering. If Ca/P is higher than 1.67, CaO will be present with the HA phase. In addition, there will be minor and trace elements from in the original reagents used to prepare the apatite powder. Some of the commercial and non-commercial dense HA materials contain up to 10 wt% b-TCP mixed with HA.[13,18] According to ASTM designation: F 1185-88, 1990 Annual Book of

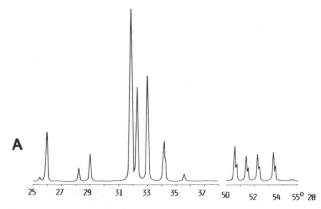

Fig. 14a. X-ray diffraction pattern of powdered dense HA prepared by precipitation and subsequent sintering showing the presence of only the HA phase. The crystallinity, reflecting crystal size is comparable to that of mineral HA (Holly Springs, GA).

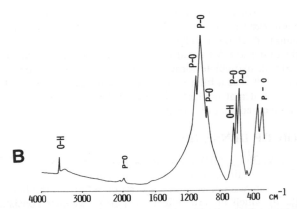

Fig. 14b. IR absorption spectrum of the same material as in (A) showing the characteristic absorption bands of O-H and P-O reflecting the vibrations of the OH and PO_4 groups in the calcium hydroxyapatite, $CA_{10}(PO_4)_6(OH)_2$.

ASTM Standards, Section 13,[82] the acceptable composition for commercial HA is a minimum of 95% HA, as established by x-ray diffraction analyses, and the acceptable concentration of trace elements is limited (maximum ppm) as follows: As = 3; Cd = 5; Hg = 5; Pb = 30; total heavy metals (as lead) = 50. The HA is to be associated with less than 5 wt% b-TCP.

The purity, composition and the particle size of the apatite preparation before sintering, the sintering temperature and conditions (e.g., with or without water pressure present) also affect the type and amount of other calcium phosphate phases and/or other calcium compounds which will be present with the HA phase. Many of the inter-relationships of these factors are descried by the phase diagrams of deGroot et al.[62] Reported sintering temperatures for commercial and non-commercial HA range from 950 to 1500°C. In this temperature range, the following calcium phosphates can form with or without the additional calcium oxide phase: b-TCP, a-TCP (resulting from the transformation of b-TCP at temperatures above 1300°C), TTCP, and oxyapatite according to the reaction outlined below:

$$\begin{array}{lll}
\text{"Apatite" preparation (AP)} \text{------} & Ca_{10}(PO_4)_6(OH)_2 + Ca_3(PO_4)_2 & \\
\text{Ca-deficient AP} \qquad >900°C & \text{HA} \qquad\qquad \text{b-TCP} & (8)
\end{array}$$

$$\begin{array}{ll}
\text{b-TCP ------ a-TCP} & \\
\qquad >1100°C & (9)
\end{array}$$

$$\begin{array}{lll}
Ca_{10}(PO_4)_6(OH)_2 \text{--------} & 2\,Ca_3(PO_4)_2 + Ca_4(PO_4)_2O & \\
\text{HA} \qquad\qquad >1300°C & \text{a-TCP} \qquad\quad \text{TTCP} & (10)
\end{array}$$

$$\begin{array}{lll}
Ca_3(PO_4)_2 + CaO \text{-------} & Ca_4(PO_4)_2O & \\
\text{b-TCP} \qquad\qquad >1400°C & \text{TTCP} & (11)
\end{array}$$

Apatite prepared from highly alkaline solution in the presence of air may often contain CO_3 and can form CaO and HA upon sintering above 900°C. TTCP can also result from the reaction between b-TCP and CaO.

Another calcium phosphate compound reported to form with HA upon sintering the apatite preparation above 900°C is dicalcium phosphate anhydrous, DCP, $CaHPO_4$. However, DCP is not stable at these temperatures, transforming instead to b-TCP and calcium pyrophosphate, $Ca_2P_2O_7$. If sintering is with water vapor pressure of about 500mm Hg, the formation of the other Ca-P phases (b- and a-TCP, TTCP) will be minimized and HA will be the more stable phase.[11,62] Thus the Ca/P molar ratio of the apatite preparation, the sintering temperature and conditions determine the final composition of the dense HA. Comparisons of Ca/P molar ratios of several commercial HA ceramics reported values ranging from 1.57 to 1.70.[83]

The composition of dense HA is appropriately determined by using x-ray diffraction (XRD), infrared spectroscopy (IR) and chemical analyses.

XRD determines: the purity (whether single or multiphasic); crystallinity of the HA phase; approximate ratios of the other phases (e.g. a- and b-TCP or TTCP or CaO) with the HA phase); and the lattice parameters of the HA and other phases.

Infrared spectroscopy gives information to support XRD data to detect additional phases, indicate relative crystallinities, and provide evidence of substituents (e.g., CO_3, F).

The combined XRD, IR and chemical analyses will provide information on the presence of substituents in the HA and b-TCP structures. The XRD patterns and IR spectra of some of the relevant calcium phosphate materials are shown in Figs. 15, 16, 17.

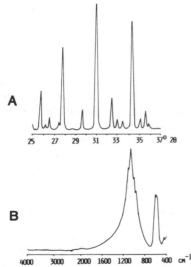

Fig. 15.   Xray diffraction (A) and IR absorption spectrum (B) of beta-tricalcium phosphate, b-TCP, $Ca_3(PO_4)_2$.

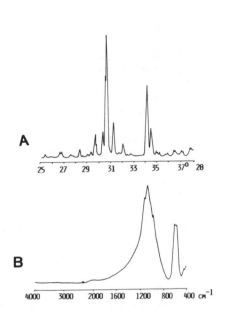

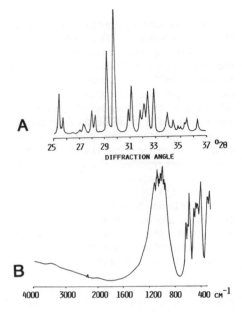

Fig. 16. X-ray diffraction (A) and IR absorption spectrum (B) of alpha-tricalcium phosphate, a-TCP, prepared by sintering b-TCP above 1300°C.

Fig. 17. X-ray diffraction (A) and IR absorption spectrum (B) of tetracalcium phosphate, TTCP, $Ca_4(PO_4)_2O$.

PROPERTIES

Crystallographic Properties

Powdered HA (from dense or macroporous forms) gives an XRD pattern characterized by diffraction peaks with small line broadening, $B_{1/2}$, and high intensities (Fig. 14a) indicating a high degree of crystallinity similar to mineral OH apatite (Holly Springs, Ga). The lattice parameters (a- and c-axis dimensions) are 9.422 and 6.881 + 0.003A, similar to mineral HA (Table 3 compared to Table 2). Other crystalline phases, e.g. b-TCP, b-TCP, TTCP, and CaO can be detected with x-ray diffraction when present above 1 wt %. IR absorption spectrum (Fig. 14b) shows the characteristic O-H and P-O absorption bands representing the vibrations of the OH and PO_4 groups, respectively, in the HA, $Ca_{10}(PO_4)_6(OH)_2$. Ceramic HA crystals are large and assume rhombic shapes (Fig. 18) compared to the much smaller acicular apatite crystals before sintering.

Daculsi et al.[84] reported for the first time the presence of hexagonal parallelapiped (void type) lattice defects in addition to other defect structure in crystals of ceramic HA sintered at 950°C (Fig. 19) but not in those prepared at 1250°C.[84] It may be logical to assume that the differences in the amount and types of lattice defects could cause

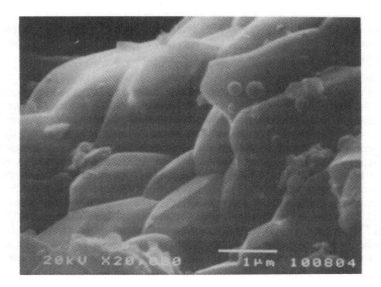

Fig. 18. SEM of dense HA particles showing the large rhombic crystals similar to those observed in human enamel sintered at 950°C (Fig. 9).

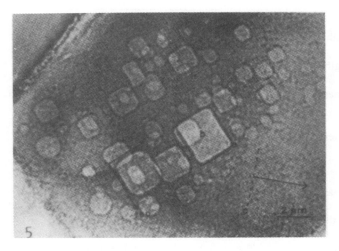

Fig. 19. TEM of HA crystals sintered at 950°C showing the presence of characteristic lattice defects (courtesy, Dr. G. Daculsi).[84]

differences in reactivity *in vivo*; the material with more lattice defects would be expected to be more reactive than the material with fewer lattice defects. This could explain the observations reported by Niwa et al.,[85] that materials sintered at lower temperatures were more reactive than those sintered at higher temperature.

Mechanical Properties

The properties of the apatite powder and the compression and sintering conditions influence the mechanical properties of the dense HA. Several mechanical properties (compressive strength, etc.) were shown to decrease with increasing amount of microporosity.[18] The density, grain size, compressive, flexural, torsional and dynamic torsional strengths, and moduli of elasticity in compression and bending increased with sintering temperature from 1150 to 1350°C.[47,61] The fracture toughness for HA ceramic sintered at from 1100 to 1150°C increased but no significant change was observed for HA ceramic sintered from 1150 to 1250°C. At sintering temperatures above 1250°C the fracture toughness drops down to a value lower than the value obtained for HA sintered at 1100°C. In addition, the presence of b-TCP also causes a decrease in fracture toughness.[47] The difference in values of mechanical properties has also been attributed to differences in the preparation of apatite powder. The difference in preparation methods causes difference in grain size (small grain size tends to give greater fracture toughness) and in composition.

The various mechanical properties of dense HA are several times greater than those of cortical bone, dentin or enamel (Table 4).

Table 4. Comparative Mechanical Properties of Dense HA and Human Enamel.

Properties	HA(1)	HA(2)	Enamel
Color	blue	white	
Compr strength MN/m^2	410 +75	430 + 95	270
Tensile strength MN/M^2	39 + 4	38 + 4	70
Vickers hardness MN/m^2	4500	4500	3400
Density	97%	99.9%	80%
Starting powder	Commercial reagent	Precipitated	
Preparation	Compressing and sintering	Compressing and sintering	
Modulus of elasticity MN/M^2	$1.1 - 1.3 \times 10^4$	$1.1 - 1.3 \times 10^4$	1.4×10^4
Impact strength MN/m^2	0.18	0.16	
Bending momentum	2.8 + 0.2	3.1 + 0.3	

Denissen et al.[18]

de Groot et al. reported that the flexural strength and fracture toughness of dense HA are much less in a dry than in a wet condition.[62] The Weibull factor, n, which describes the resistance of a material to fatigue failure is 50 for HA in a dry environment and 12 in a wet physiological implant bed. Implants with values of n = 10 to 20 are expected to fail in several months of clinical use.[62] This property makes dense HA ceramic an unsuitable material for load-bearing situations in spite of its good biocompatibility and osteoconductivity.[62]

Dissolution Properties

In vitro dissolution of HA depends on the type and concentration of the buffered or unbuffered solutions; pH of the solution; degree of saturation of the solution; solid/solution ratio; the length of suspension in the solutions; and the composition and crystallinity (reflecting crystal size and strain) of the HA.[24,43,86-89] In the case of ceramic HA, the degree of micro- and macro-porosities, defect structure and the amount and type of other phases present also have significant influence. Comparative dissolution of different calcium phosphates are shown in Fig. 20. The extent of dissolution of the HA ceramic was less in lactic acid buffer compared to that in acetic acid buffer.[90] For HA ceramic containing other calcium phosphate phases, the extent of dissolution will be affected by the type and amount of the non-HA phases. The extent of dissolution decreases in the following order:

$$TTCP >> a\text{-}TCP >> b\text{-}TCP >> HA \tag{12}$$

On the crystal level the dissolution of the ceramic HA crystals is non-site specific, i.e., dissolution is observed both on the surface and at the crystal core (Fig. 21). In comparison, dissolution of biological apatite crystals is site specific, showing preferential dissolution of the crystal core (Fig. 22).

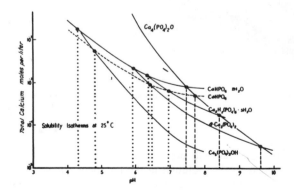

Fig. 20. Solubility diagrams of different calcium phosphate (Ca-P) compounds (courtesy, Dr. W. Brown, NIST). Some of these Ca-P compounds are associated with the preparation of HA biomaterial (e.g., b-TCP, TTCP) and others, with the possible dissolution, precipitation and transformation of the HA biomaterials *in vivo*.

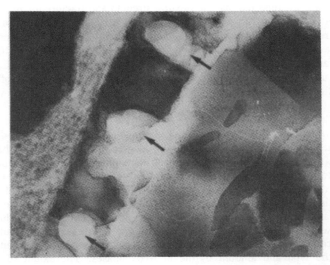

Fig. 21. TEM of ceramic HA crystals from implant retrieved after 3 months from non-bony sites in rabbit[111] showing non-site specific dissolution of the crystals, dissolving from the surface and from the crystal core.

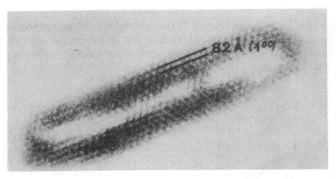

Fig 22. TEM of biological apatite (human enamel) after exposure in acid showing preferential dissolution at the crystal core.

SURFACE CHEMISTRY

The surface chemistry of HA ceramic will depend on the composition of the ceramic and on the composition and pH of the solution in the microenvironment. An acid environment will cause partial dissolution of the surface enriching the population of Ca^{2+}, $H_2PO_4^-$, HPO_4^{2-}, PO_4^{3-}, H^+, OH^-, and ion pairs such as $CaH_2PO_4^+$, and $CaOH^{+1}$[40] in a hydrated layer. Biological apatites have also been described as having a hydrated layer with ions reflecting the composition of the bone mineral and the biological fluid.[50,53,91] *In vivo*, electrolytes from the biological fluids will also be part of the surface chemistry and will contribute to the development of surface charges on the HA implant.[47,92-94] The surface charges will influence cellular interactions at the interface.[47,93,95] Ducheyne et al. reported that the absolute values of zeta potential measured were higher for the unsintered and Ca-deficient apatite compared with those of ceramic and stoichiometric HA[93] and suggested that these values affect the cellular activities involved in bone formation.

In addition, proteins will adsorb on modified HA surfaces,[96] a phenomenon well known in the use of apatite for column chromatography.[92,97] The relationship of specific functional groups of amino acids to the formation of biological apatites and eventual mineralization has been strongly suggested in studies on mineral-organic matrix interactions.[49,98,99]

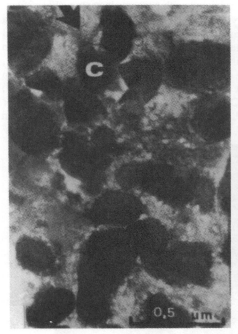

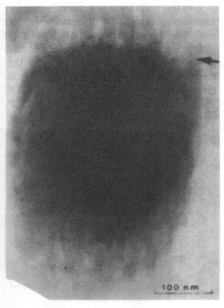

Fig. 23a. TEM of dense HA particles after 4-months implantation in human periodontal pockets. Undecalcified ultrathin section from the implant surface. Magnification, X 100,000 showing the presence of microcrystals (arrow) on the surfaces of the much larger HA crystals (C).[108]

Fig. 23b. TEM of dense HA particles after 2 months implantation in human periodontal pocket, showing epitaxial growth of the new crystals on the HA ceramic crystals. Magnification, X 1,350,000.[108]

Formation of Carbonate-Apatite Crystals on HA Surfaces: Dissolution/Precipitation Processes

Microcrystals observed on the surface of ceramic HA after implantation in bony sites[100-102] were identified also as apatite by selective electron diffraction with transmission electron microscopy (TEM).[103] Similar observations were made on the surfaces of crystals of ceramic HA, coralline HA, biphasic calcium phosphate (BCP), after suspension in cell culture or in serum[104-106] and after implantation in non-bony sites[67] and in bony sites.[58,108] These microcrystals (Fig. 23a) sometimes appear to exhibit epitaxial growth on the HA ceramic crystals (Fig. 23b) and have been identified as apatite, similar to bone apatite, by electron diffraction (Fig. 24) confirming earlier observations by Tracy et al.[103] In addition, of IR spectroscopy has helped to identify these crystals as carbonate-apatite (Fig. 25), which was intimately associated with an organic matrix (Figs. 25, 26).[107] Bone apatite crystals, like other biological apatites, are carbonate apatite and are also intimately associated with an organic matrix (Fig. 25).

The formation of these microcrystals of CO_3-apatite is believed to be a dissolution-precipitation processes [shown schematically in Fig. 27]. The partial dissolution of the calcium phosphate material, HA, is initiated by the acid condition which results from cellular activity[109] causing the release of Ca^{2+}, HPO_4^{2-} and PO_4^{3-}, and increasing the supersaturation of the micro-environment with respect to calcium phosphate phases which are stable at the pH in this environment. DCPD, $CaHPO_4.2H_2O$; octacalcium phosphate, OCP, $Ca_8H_2(PO_4)_6.5H_2O$, which can form under acid conditions; or Mg-substituted b-TCP, $(Ca,Mg)_3(PO_4)_2$, or whitlockite, which can form under either acid or basic conditions, can hydrolyze in the presence of CO_3^{2-} ions

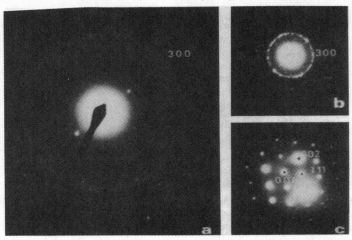

Fig. 24. Electron diffraction of large crystals of HA ceramic (A); of the microcrystals (B) similar to those shown in Fig. 23a; and of bone apatite crystals (C). The rings on (A) correspond to the 111 and 002 apatite lattice plane; those in (B) and (C) correspond to the (300) plane of apatite (in collaboration with Dr. G. Daculsi).[58]

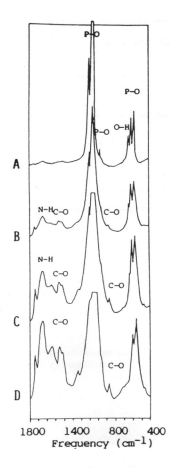

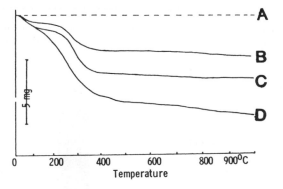

Fig. 25. Infrared (IR) absorption spectra of BCP (b-TCP/HA = 15/85) material before (A) and after implantation in surgically created dental defects (in collaboration with Drs. E. Nery and K Lynch). IR spectra of materials obtained from the core (B), the implant surface (C), and from an area in the bone furthest away from the implant (D), show increasing intensities of the N-H (related to the organic phase) and of the C-O (related to the CO_3 in the CO_3-apatite) absorption bands. In addition the intensity of the O-H (related to the OH groups in HA material before implantation) shown in (A) become unresolved in materials in (B) and (C) and become more similar to the bone apatite (D).[58]

Fig. 26. Thermogravimetric analyses (TGA) of dense HA before implantation (A); and of implants recovered after 60 (B), 180 (C), and 365 (D) days from non-osseous sites in hamster. The weight loss below 400°C is due to the loss of adsorbed water and organic phases; above 500°C due to loss of carbonate from the CO_3-apatite (lost as CO_2).[107]

in the biological fluid to CO_3-apatite.[58,110] Alternatively, or in addition, CO_3-apatite can form directly at physiological pH, using the calcium and phosphate ions released from partially dissolving ceramic HA and from the biological fluids which contain other electrolytes, notably, CO_3^{2-} and Mg^{2+}. These become incorporated in the new CO_3-apatite microcrystals forming on the surfaces of the much larger crystals of the ceramic HA (Fig. 23a). The CO_3-apatite can also form by precipitation, deduced from the observed uptake of calcium ions from serum[105] (shown in Fig. 28) or seeded growth on the ceramic HA crystals. However, cell-induced resorption or dissolution of the HA ceramic crystals is frequently observed *in vitro* and *in vivo*.[104,106,111]

The significance of the observed intimate association of the CO_3-apatite microcrystals with an organic matrix (Figs. 25 and 26) is not completely understood.

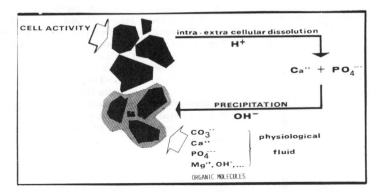

Fig. 27. Schematic representation of the dissolution/precipitation processes involved in the *in vivo* formation of CO_3-apatite on surfaces of Ca-P (e.g. HA) implant materials.

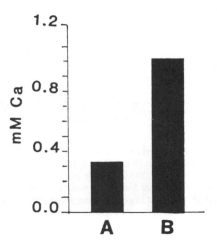

Fig. 28. Calcium uptake after suspension of dense HA (A) and coralline HA (B) in calf serum for 5 hours.[65]

The formation of CO_3-apatite has also been seen *in vitro* and *in vivo* on surfaces of bioactive glasses and glass ceramics.[17,112-114] (See chapters 3, 5 and 7.) In studies on biphasic calcium phosphate (BCP) materials consisting of different b-TCP/HA ratios, it was observed that the abundance of the CO_3-apatite microcrystals on the surfaces of the large BCP crystals was influenced by the b-TCP/HA ratio: the higher the ratio, the greater the abundance of the CO_3-apatite micro-crystals. This was thought to be due to the higher dissolution properties of the b-TCP component of the BCP causing an increase in the concentration of the calcium and phosphate ions in the microenvironment leading to   precipitation of the CO_3-apatite.[58]   Reactions on material surfaces, including the formation of CO_3-apatite may be important in establishing the strong 'bonding zone' at the bone-material interface unique to bioactive materials.

TISSUE RESPONSE

Cellular Interaction with HA

HA surfaces appear to be biocompatible with several cell types such as macrophages, fibroblasts, osteoclasts, osteoblasts, periodontal ligament cells.[59,65,85,104,105,113,115-118] The favorable response in terms of cell attachment and cell proliferation (Fig. 29a) of different types of cells to dense HA and other bioactive materials has been demonstrated by many studies such as those cited above. The cells cause the dissolution of the HA ceramic crystals intra-cellularly by phagocytosis (Fig. 29b); or extracellularly by producing an acid environment (10) which causes the partial dissolution of the HA ceramic crystals. The HA material allows the proliferation of fibroblasts;[106] osteoblasts;[116] and other bone cells.[104,117,118] The cells do not seem to distinguish between HA and bone surfaces, which indicates a significant similarity in the surface chemistry.

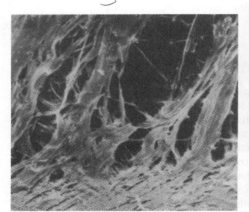
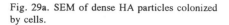
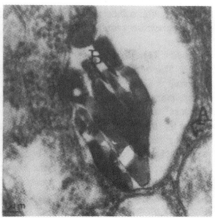

Fig. 29a. SEM of dense HA particles colonized by cells.

Fig. 29b. TEM showing phagocytosed HA crystals (A) and showing the dissolution of HA crystals (B).[110,111]

Osteoconductive Properties

Dense HA, like other calcium phosphate biomaterials is osteoconductive but not osteoinductive.[8,11,70,119,120] A discussion of osteoinduction is not within the scope of this chapter. An osteoconductive material allows the formation of bone on its surface by serving as a scaffold or a template (Fig. 30). In the case of bioactive glasses and glass ceramics, and calcium phosphate materials such as dense HA, their role is not a passive one but participatory, contributing to the formation of the CO_3-apatite on surfaces and

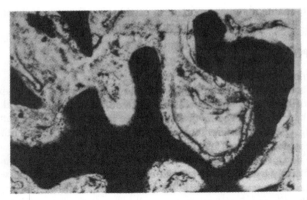

Fig. 30. Microradiography showing the osteoconductive properties of HA materials: the new bone is shown on the surface of the HA acting as a template.

promoting the adhesion of matrix-producing cells and organic molecules as a result of surface chemistry and surface charges.

Bone-HA Interface

The type of bonding at the material-bone interface depends on the nature of the material[121] (Table 5). The interfacial strength between bone and implant biomaterial, as determined from push-out tests, is much greater for the bioactive implant materials (e.g., Bioglass® and HA) compared to other materials, such as titanium, zirconia or alumina[122] (Table 6). In the case of bioactive materials, the fracture occurs either in the material or the bone but not at the interface. In the case of inert materials, the separation occurs at the interface. This phenomenon is attributed to the 'bonding osteogenesis' occurring at the interface between bone and the bioactive materials which does not occur with non-bioactive materials.[121] This interface has been referred to as the 'bonding zone', described as being electron-dense, consisting of mineralized organic meshwork. Greater interfacial strength was also observed with metal implants coated with plasma-sprayed HA than those without.[78-81]

The description of the bone-HA interface depends on the analytical methods used in studying the interface. Light microscope studies and low resolution scanning electron microscopy (SEM) shows intimate contact between the bone and the HA implant surface (Figs. 31a, 32a). Electron microprobe analyses showed that the relationship between the calcium and phosphorus concentrations was not significantly different when scanning from the implant to the interface, to the bone regions (Fig. 31b).

Histological examination showed that dense HA particles placed in periodontal bony defects[70,123] or dense HA tooth forms[65] become surrounded by fibrous tissue in the initial period of wound healing, followed by resorption in some areas and by osteoid deposition and bone formation in other parts of the implant.[70] Other reports indicated bone formation without intervening fibrous tissue (Fig. 33).[8,11,59,68,85,108]

Table 5. Effect of Type of Biomaterial on Type of Bonding at the Material-Bone Interface.

Type of Material	Biodynamics	Type of Bonding
1. Metals and Polymers	Biotolerant	Distance Osteogenesis ⎯⎯⎯ ⏐ ⏐ ⎯⎯⎯ Bone Implant
2. Ceramic oxides (alumina, zirconia) titanium	Bioinert	Contact Osteogenesis ⎯⎯⎯⎯⏐⎯⎯⎯ B I
3. Ca-P ceramics (HA, b-TCP) Bioglass®	Bioactive	Bonding Osteogenesis B⎯⎯⎯⎯ ⎯ ⎯ ⎯ ⎯ ⎯ ⎯ ⎯ ⎯⎯⎯⎯ I

Osborne and Newesely, 1980.[121]

Table 6. Comparative Properties of Biomaterials.

Materials	Fracture Strength	Bone Contact	Fracture
Bioglass®*	28.9 MPa	92.9%	Cohesive
Hydroxyapatite*	19.6 MPa	95.4%	Cohesive
Titanium**	1.9 MPa	59.5%	Interfacial
Zirconia**	1.3 MPa	33.3%	Interfacial

Fracture strengths per unit area after 24 weeks. Fracture surfaces indicated cohesive failure of the materials and/or the bone tissue occurred with bioactive materials (*); interfacial failure occurred with non-bioactive materials (**). Niki et al.[122]

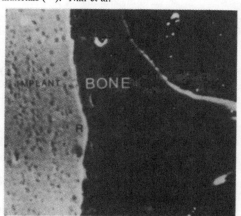

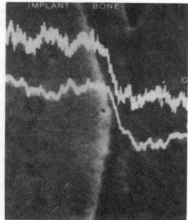

Fig. 31a. SEM of the interface between dense HA ceramic implant and bone (X300).

Fig. 31b. Electron microprobe analyses showing calcium (Ca) and phosphate (P) concentrations on the regions of the implant, interface and bone (X2100). (Courtesy of Dr. H. Denissen.)[13,18]

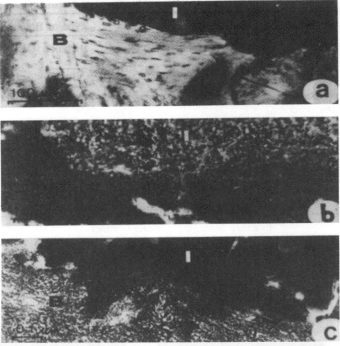

Fig. 32. Bone-BCP interface showing the formation of new bone (B) on the implant surface as seen by light microscopy (a) with resolution 0.2 to 2um (X250); by SEM (b) resolution 60 nm (X900) and TEM, (c) resolution 0.14 nm (X100,000).[58]

High resolution TEM of the HA-bone interface showed the presence of highly mineralized collagen in the vicinity of the large HA ceramic crystals which are associated with much smaller crystals of CO_3-apatites.

Events in the Formation of 'Bone-Bonded' Interface

The following events are proposed as responsible for the formation of the strong interface between bone and a bioactive material:

- acidification of the micro-environment due to the cellular action on the bioactive material;
- dissolution/precipitation processes resulting in the formation of CO_3-apatite intimately associated with an organic matrix similar to bone apatite;
- production of adhesive proteins and collagen fibrils containing extracellular matrix;

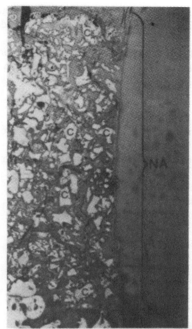

Fig. 33a. Histological observations showing a section of the implant site with BCP (85 HA/15b-TCP) ceramic with new attachement level (NA) and accelerated new bone formation around the ceramic particles (C); (b) is old bone indicating the original base of the pocket prior to implant procedure (H & E; original magnification X10).

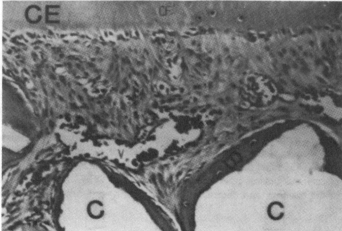

Fig. 33b. Higher magnification of Fig. 33a showing new cementum (CE) with insertin collagen fibers (CF), vascularity (V), and new bone formation (B) surrounding the ceramic particles (C). (H & E; original magnification X160, courtesy Dr. E. Nery.)[59]

- simultaneous mineralization of the collagen fibrils and incorporation of the CO_3-apatite crystals (originating from the material) in the remodelling new bone;
- interdigitation of the mineralized collagen between the host bone and the bioactive ceramic surfaces and within the pores provides the interfacial strength.

CLINICAL APPLICATIONS

The various clinical dental and medical applications of dense HA includes:

(1) repair of bony defects in dental and orthopedic applications;[6,59,66,69-72,85,91,101,123]
(2) immediate tooth root replacement;[13,18,65,124-126]
(3) augmentation of alveolar ridge;[66,67,73,127]
(4) adjuvant to the placement of metal implants;[73,74]
(5) pulp-capping materials;[81,128]
(6) enhancement of guided tissue regeneration;[129]
(7) maxillo-facial reconstruction;[8,21]
(8) percutaneous devices;[47,56]
(9) middle ear reconstruction;[63,64]
(10) for plasma-sprayed coatings for dental and orthopedic implants;[78-80,100] and
(11) as bioreactors.[118]

Coatings on Dental and Orthopedic Implants

Dense HA particles are plasma-sprayed on dental and orthopedic metal implants combine the strength of the metal and the bioactivity and osteoconductivity of the HA. These coated implants are described as 'HA'-coated. However, x-ray diffraction analysis of plasma-sprayed coupons or implants shows that the coating may differ significantly in crystallinity (i.e., per cent of crystalline phases vs. non-crystalline phases and crystallinity of the HA phase) and in composition from that of the dense HA ceramic used for plasma-spraying (Fig. 34). The coating may consist of: HA with lower crystallinity than the dense HA used in plasma spraying; a-TCP; b-TCP; TTCP and a large amount of amorphous calcium phosphate, ACP.[22,130] It was also observed that the coating composition is not homogeneous, the layer closer to the metal substrate tends to contain more ACP than the outermost layer of the coating (Fig. 35). Since different calcium phosphate phases have different solubilities (Fig. 20); it is possible that the stability of the coating, and therefore of the implant, will depend on the composition of the coating.

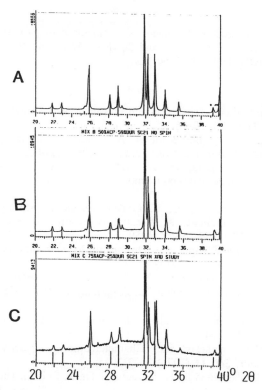

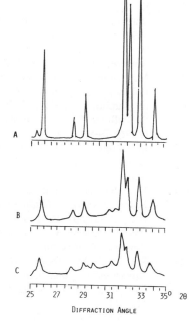

Fig. 34. X-ray diffraction of coatings on dental implant prepared by plasma-spraying dense HA particles on titanium substrate. The coating composition in the order of increasing relative concentration is: (1) HA (with crystallinity lower than starting material (A); (b) amorphous calcium phosphate; and (c) a- and b-TCP of poor crystallinity. Other coatings also contain small amounts of TTCP.[22,130]

Fig. 35. X-ray diffraction analyses of coating prepared as in Fig. 34, showing composition gradient from the layer closer to the metal substrate (B) to the coating surface (C). The dense HA material used for plasma spraying is shown in (A).

ACKNOWLEDGEMENTS

The work involving the authors cited in this paper was supported in part by research grants from NIH nos. DE 07223, DE 04123, and SO7RRO76226 and from HiTemco Medical Applications Inc. (Old Bethpage, New York).

REFERENCES

1.   F. H. Albee, "Studies in Bone Growth. Triple Calcium Phosphate as a Stimulus to Osteogenesis," *Ann. Surg.* **71** (1920) 32-36.

2.   G. E. Levitt, P. H. Crayton, E. A. Monroe and R. A. Condrate, "Forming Methods for Apatite Prosthesis," *J. Biomed. Mater. Res.* **3** (1969) 683-685.

3.   Z. A. Monroe, W. Votawa, D. B. Bass, and J. McMullen, "New Calcium Phosphate Ceramic Material for Bone and Tooth Implants," *J. Dent. Res.* **50** (1971) 860-862.

4.   W. J. Clark, T. D. Driskell, C. R. Hassler, V. J. Tennery and L. McCoy "Calcium Phosphate Resorbable Ceramics, A Potential Alternative to Bone Grafting," *IADR Prog. & Abst.* **52** (1973) Abstr. #259.

5.   W. Hubbard, *Physiological Calcium Phosphate as Orthopedic Implant Material #6* (Ph.D. Thesis, Marquette University, 1974).

6.   E. B. Nery, K. L. Lynch, W. M. Hirthe and K. H. Mueller, "Bioceramic Implants in Surgically Produced Infrabony Defects," *J. Periodont.* **46** (1975) 328-339.

7.   R. Z. LeGeros, "Calcium Phosphate Materials in Restorative Dentistry: A Review," *Adv. Dent. Res.* **2** (1988) 164-183.

8.   M. Jarcho, "Hydroxylapatite Synthesis and Characterization in Sense Polycrystalline Forms," *J. Mater. Sci.* **11** (1976) 2027-2035.

9.   M. Jarcho, J. F. Kay, H. P. Drobeck and R. H. Doremus, "Tissue Cellular and Subcellular Events at Bone-Ceramic Hydroxylapatite Interface," *J. Bioeng.* **1** (1976) 79-92.

10.   M. Jarcho, C. H. Bolen, M. B. Thomas, J. Bobick, J. F. Kay and R. H. Doremus, "Synthesis and Characterization in Dense Polycrystalline Form," *J. Mat. Sci.* **11** (1976) 2027.

11.   K. de Groot, "Ceramic of Calcium Phosphates: Preparation and Properties," in *Bioceramics of Calcium Phosphate*, ed. K. de Groot (CRC Press, Boca Raton, FL., 1983) pp. 100-114.

12.   J.G.C. Peelen, B. V. Rejda and K. de Groot, "Preparation and Properties of Sintered Hydroxyapatite," *Ceramurgia Int.* **4** (1980) 71-73.

13.   H. Denissen, *Dental Root Implants of Apatite Ceramics. Experimental Investigations and Clinical Use of Dental Root Implants Made of Apatite Ceramics* (Ph.D. Thesis, Vrije Universiteit te Amsterdam, 1979).

14.   H. Akao, J. Aoki and K. Kato, "Mechanical Properties of Sintered Hydroxyapatite for Prosthetic Application," *J. Mater. Sci.* **16** (1981) 809-812.

15. H. Aoki, K. Kato, M. Ogiso and T. Tabata, "Sintered Hydroxyapatite as a New Dental Implant Material," *J. Dent. Outlook* **49** (1977) 567-575.

16. W. F. DeJong, "La Substance Minerals das les os.," *Tec. Trav. Chim.* **45** (1926) 445-458.

17. L. L. Hench, "Bioceramics: From Concept to Clinic," *J. Am. Ceram. Soc.* **74** (1991) 1487-1510.

18. H. Denissen, C. Mangano and G. Cenini G, *Hydroxylapatite Implants* (India: Piccin Nuova Libraria, S.P.A., 1985)

19. C. A. Beevers and D. B. McIntyre, "The Atomic Structure of Fluorapatite and its Relation to that of Tooth and Bone Mineral," *Miner. Mag.* **27** (1956) 254-259.

20. J. F. Kay, "Calcium Phosphate Coatings for Dental Implants," *Dent. Clin. North. Amer.* **36** (1992) 1-18.

21. R. A. Young and J. C. Elliot, "Scale Bases for Several Properties of Apatites," *Archs. Oral. Biol.* **11** (1966) 699-707.

22. R. Z. LeGeros, *Calcium Phosphates in Oral Biology and Medicine*, Monographs in Oral Sciences. Vol. 15, ed. H. Myers (S. Karger, Basel, 1991).

23. R. Z. LeGeros, L. Singer, R. Ophaug and G. Quirolgico, "The Effect of Fluoride on the Stability of Synthetic and Biological (Bone Mineral) Apatites," in *Osteoporosis*, eds. J. Menczel, G. C. Robin, and M. Makin (J. Wiley and Sons, New York, (1982) pp. 327-341.

24. E. C. Moreno, M. Kresak and R. T. Zahradnik, "Physicochemical Aspects of Fluoride-Apatite Systems Relevant to the Study of Dental Caries," *Caries. Res.* (suppl 1) **11** (1977) 142-177.

25. G. Bonel, "Contribution a l'etude de la Carbonation des Apatites," *Ann. Chim.* **7** (1972) 65-88.

26. J. C. Elliot, *The Crystallographic Structure of Dental Enamel and Related Apatites* (PhD Thesis, University of London, 1964).

27. R. A. Yukna, E. T. Mayer and D. B. Brite, "Longitudinal Evaluation of Durapatite Ceramic as an Alloplastic Implant in Periodontal Osseous Defects After 3 Years," *J. Periodontol.* **55** (1984) 633-637.

28. R. Z. LeGeros, *Crystallographic Studies of the Carbonate Substitution in the Apatite Structure* (PhD Thesis, New York University, 1967).

29. R. Z. LeGeros, "The Unit-Cell Dimensions of Human Enamel Apatite: Effect of Chloride Incorporation," *Arch. Oral. Biol.* **20** (1974) 63-71.

30. R. Z. LeGeros, "Incorporation of Magnesium in Synthetic and in Biological Apatites," in *Tooth Enamel IV*, eds. R.W. Fearnhead and S. Suga (Elsevier Science Publishers, Amsterdam, 1984) pp.32-36.

31. R. Z. LeGeros and S. Suga, "Crystallographic Nature of Fluoride in Enameloids of Fish," *Calcif. Tiss. Int.* **32** (1980) 169-174.

32. R. Z. LeGeros, O. R. Trautz, J. P. LeGeros and W. P. Shirra, "Apatite Crystallites: Effect of Carbonate on Morphology," *Science* **155** (1967) 1409-1411.

33.   R. Z. LeGeros, O. R. Trautz, J. P. LeGeros and E. Klein, "Carbonate Substitution in the Apatite Structure," *Bull. Soc. Shim. (Fr)*, (1968) pp. 1712-1713.

34.   R. Z. LeGeros, J. P. LeGeros, O. R. Trautz and W. P. Shirra, "Comparative Crystallinity and Morphology of Enamel, Dentine Bone and Amorphous Concretions," EMSA 28th Annual Proc. (1970).

35.   R. Z. LeGeros, J. P. LeGeros, O. R. Trautz and E. Klein, "Spectral Properties of Carbonate in Carbonate-Containing Apatites," *Dev. Applied Spectroscopy* **7B** (1970) 3-12.

36.   R. Z. LeGeros, J. P. LeGeros, O. R. Trautz and W. P. Shirra, "Conversion of Monetite, $CaHPO_4$, to Apatites: Effect of Carbonate on the Crystallinity and the Morphology of the Apatite Crystallites," *Adv. in X-Ray Anal.* **14** (1971) 57-66.

37.   R. Z. LeGeros, M. H. Taheri, G. Quirolgico and J. P. LeGeros, "Formation and Stability of Apatites: Effects of Some Cationic Substituents," *Proc. 2nd International Congress on Phosphorous Compounds* (Boston, 1980) pp. 41-53; 89-103.

38.   R. Z. LeGeros, R. Kijkowska and J. P. LeGeros, "Formation and Transformation of Octacalcium Phosphate OCP: A Preliminary Report," *Scanning Electron Micro.* **4** (1984) 1771-1777.

39.   R. Z. LeGeros, R. Kijkowska, M. Tung and J. P. LeGeros, "Effect of Strontium on Some Properties of Apatites," in *Tooth Enamel V*, ed. R. W. Fearnhead, (Florence Publishers, Tokyo, 1989) pp. 393-402.

40.   P. Somasundran, "Zeta Potential of Apatite in Aqueous Solution and its Change During Equilibrium," *J. Colloid. Interface. Sci.* **27** (1968) 659-666.

41.   R. Z. LeGeros, "Apatites in Biological Systems," *Prog. Crystal. Growth. Charact.* **4** (1981) 1-45.

42.   R. Z. LeGeros, G. Bonel and R. Legros, "Types of H_2O in Human Enamel and in Precipitated Apatites," *Calc. Tiss. Res.* **26** (1978) 111-116.

43.   R. Z. LeGeros and M. S. Tung, "Chemical Stability of Carbonate and Fluoride-Containing Apatites," *Caries. Res.* **17** (1983) 419-429.

44.   R. L. Collins, "Strontium-Ccalcium Hhydroxyapatite: Solid-Solutions Preparations and Lattice Constants Measurements," *J. Am. Chem. Soc.* **82** (1960) 5067-5072.

45.   M. Okazaki and R. Z. LeGeros, "Crystallographic and Chemical Properties of Mg-Containing Apatites Before and After Suspension in Solutions," *Magnesium Res.* **5** (1992).

46.   A. Sillen and R. Z. LeGeros, "Solubility Profiles of Synthetic Apatites and of Modern and Fossil Bones," *J. Arch. Sci.* **18** (1991) 385-397.

47.   H. Aoki, *Science and Medical Applications of Hydroxyapatite* (Japan Association of Apatite Science (JAAS), Takayama Press System Center Co., Tokyo, 1991).

48.   W. E. Brown and L. Chow, "Chemical Properties of Bone Mineral," *Ann. Res. Mater. Sci.* **6** (1976) 213-226.

49. M. J. Glimcher, "Recent Studies of the Mineral Phase in Bone and its Possible Linkage to the Organic Matrix by Protein-Bound Phosphate Conds," *Phil. Trans. R Soc. Lond.* B **304** (1984) 479-508.

50. W. D. Armstrong and L. Singer L, "Composition and Constitution of the Mineral Phase of Bone," *Clin. Orthop. Relat. Res.* **196** (1965) 179-190.

51. W. F. Neuman and M. W. Neuman, *The Chemical Dynamics of Bone Mineral* (University of Chicago Press, Chicago, 1958).

52. R. Z. LeGeros, "Materials for Bone Repair, Augmentation and Implant Coatings," in *Proceedings of the International Seminar of Orthopedic Research*, Nagoya, 1990, ed. S. Niwa, (Springer-Verlag, 1992).

53. W. Rathje, " Zur Kentnis de Phosphate I. Uber Hydroxyapatite," *Bodenk Pflernah* **12** (1939) 121-128.

54. E. Hayek and H. Newesely, "Pentacalcium Monohydroxyorthophosphate," *Inorg. Syn.* **7** (1963) 63-65.

55. G. Bonel, J-C. Heughebaert, M. Heughebaert, J. L. Lacout and A. Lebugle, "Apatitic Calcium Orthophosphates and Related Compounds for Biomaterials Preparation," in *Bioceramics: Materials Characteristics Versus In Vivo Behavior*, eds. P. Ducheyne and J. E. Lemons (Ann. NY Acad. Sci., New York, 1988) Vol. 523, pp. 115-130.

56. J. A. Jansen, J.P.C.M. van der Waerden, H.B.M. van der Lubbe and K. deGroot, "Tissue Response to Percutaneous Implants in Rabbits," *J. Biomed. Mater. Res.* **24** (1990) 295-307.

57. R. Z. LeGeros, "Variability of b-TCP/HAP Ratios in Sintered Apatites," *J. Dent. Res.* **65** (1986) 292, Abstr. No. 110.

58. R. Z. LeGeros and G. Daculsi, "*In Vivo* Transformation of Biphasic Calcium Phosphate Ceramics: Ultrastructural and Physico-Chemical Characterizations," in *Handbook of Bioactive Ceramics Vol. II*, eds. T. Yamamuro, L. L. Hench, and J. Wilson (CRC Press, Boca Raton, Florida, (1990) pp. 17-28.

59. E. Nery, R.Z. LeGeros and K. L. Lynch, "Tissue Response to Biphasic Calcium Phosphate Ceramic with Different Ratios of Biphasic Calcium Phosphate Ceramic with Different Ratios of HA/TCP in Periodontal Osseous Defects," *J. Periodontol.* **63** (1992) 729-735.

60. W. I. Higuchi, N. A. Mir, R. R. Patel, J. W. Becker and J. J. Heffersen, "Mechanisms of Enamel Dissolution in Acid Buffers," *J. Dent. Res.* **44** (1969) 330-341.

61. S. Best, W. Bonfield and C. Doyle, "Optimization of Toughness in Dense Ceramics," in *Bioceramics Vol. 2*, ed. G. Heimke (German Ceramic Society, Cologne, 1990) pp. 57-64.

62. K. de Groot, C.P.A.T. Klein, J.G.C. Wolke and J.M.A. de Bliek-Hogervost, "Chemistry of Calcium Phosphate Bioceramics," in *Handbook of Bioactive Ceramics. Vol. II*, eds. T. Yamamuro, L. L. Hench and J. Wilson (CRC Press, Boca Raton, Florida, 1990) pp. 3-16.

63.   C. A. van Blitterswijk, *Calcium Phosphate Middle-Ear Implants* (Ph.D. Thesis, Rijksuniversitiet te Leiden, 1985).

64.   C. A. van Blitterswijk, S. C. Hesseling, J. J. Grote, H. K. Korerte and K. de Groot, "The Biocompatibility of Hydroxyapatite Ceramic: A Study of Retrieved Human Middle Ear Implants," *J. Biomed. Mater. Res.* **24** (1990) 433-453.

65.   J. H. Quinn and J. N. Kent, "Alveolar Ridge Maintenance with Solid Nonporous Hydroxylapatite Root Implants," *Oral. Surg.* **58** (1984) 511-516.

66.   A. N. Cranin, G. P. Tobin and J. Gelbman, "Applications of Hydroxyapatite in Oral and Maxillofacial Surgery, Part II: Ridge Augmentation and Repair of Major Oral Defects," *Compend. Contin. Educ. Dent.* **8** (1987) 334-345.

67.   M. I. Kay, R.A. Young and A. S. Posner, "Crystal Structure of Hydroxyapatite," *Nature* **204** (1964) 1050-1053.

68.   H. Oonishi, E. Tsuji, H. Ishimaru, M. Yamamoto and J. Delecrin, "Clinical Significance of Chemical Bonds Between Bioactive Ceramics and Bone in Orthopaedic Surgery," in *Bioceramics Vol. 2*, ed. G. Heimke (German Ceramic Society, Cologne, 1990) *pp. 286-293.

69.   S. J. Froum, J. Kushner, L. Scopp and S. S. Stahl, "Human Clinical and Histologic Responses to Durapatite Implants in Intraosseous Lesion, Case Reports," *J. Periodontol.* **53** (1986) 719-725.

70.   P. N. Galgut, I. M. Waite and S.M.B. Tinkler, "Histological Investigation of the Tissue Response to Hydroxyapatite Used as an Implant Material in Periodontal Treatment," *Clin. Mater.* **6** (1990) 105-121.

71.   J. Wilson and G. E. Merwin, "Biomaterials for Facial Bone Augmentation: Comparative Studies," *J. Biomed. Mater. Res. Appl. Biomat.* **22** (1988) 159-177.

72.   J. Wilson and S. B. Low, "Bioactive Ceramics for Periodontal Treatment: Comparative Studies in the Patus Monkey," *J. Appl. Biomat.* **3** (1992) 123-129.

73.   M. Jarcho, "Calcium Phosphate Ceramics as Hard Tissue Prosthetics," *Clin. Orthopaed.* **157** (1981) 259-278.

74.   L. Linkow, "Bone Transplants Using the Symphysis, the Iliac Crest and Synthetic Bone Materials," *J. Oral. Implantol.* **11** (1984) 211-217.

75.   W. Bonfield, "Hydroxyapatite-Reinforced Polyethylene as an Analogous Material for Bone Replacement," in *Bioceramics: Materials Characteristics Versus In Vivo Behavior*, eds. P. Ducheyne and J. E. Lemons (Ann. NY Acad. Sci., New York, 1988) Vol. 523, pp. 173-177.

76.   R. Z. LeGeros and B. Penugonda, "Potential Use of Calcium Phosphate as Fillers in Composite Restorative Biomaterials," Second World Congress on Biomaterials, Washington, D.C. (1984).

77.   H. Alexander, J. R. Parsons, J. L. Ricci, P. K. Bajpai, and A. B. Weiss, "Calcium Phosphate Ceramic Based Composite as Bone Graft Substitutes," in *Critical Reviews in Biocompatibility*, ed. C. William (CRC Press, Boca Raton, FL, 1987) pp. 43-77.

78. S. D. Cook, K. A. Thomas, R. J. Haddad, M. Jarcho and J. Kay, Hydroxylapatite-Coated Titanium for Orthopedic Impant Applications," *Clin. Orthop.* **232** (1988) 225-231.

79. S. D. Cook, K. A. Thomas, J. E. Dalton, R. K. Volkman, T. S. Whitecloud, and J. E. Kay, "Hydroxylapatite Coating of Porous Implants Improves Bone Ingrowth and Interface Attachment Strength," *J. Biomed. Mater. Res.* **26** (1992) 989-1001.

80. K. de Groot, "Hydroxylapatite Coatings for Implants in Surgery," in *High Tech Ceramics*, ed. P. Vincenzini (Elsever Science Publishers, B.V., Amsterdam, 1987) pp. 381-386.

81. A. Jean, B. Kerebel and R. Z. LeGeros, "Effects of Various Calcium Phosphatematerials on the Reparative Dentin Bridge," *J. Endo.* **14** (1988) 83-87.

82. ASTM, *Annual Book of ASTM Standards*, Section 13, F (1990) 1185-1188.

83. E. Fischer-Brandeis, E. Dietert and G. Bauer, Zur Morphologie Synthetischer Calcium-Phosphat-Keramiken *In Vitro*," *A Zahnarztl. Implantol.* **III** (1987) 87.

84. G. Daculsi, R. Z. LeGeros, J. P. LeGeros and D. Mitre, "Lattice Defects in Calcium Phosphate Ceramics; High Resolution TEM Ultrastructural Study," *J. Appl. Biomat.* **2** (1991) 147-152.

85. S. Niwa, K. Sawai, S. Takahashi, H. Tagai, M. Ono and Y. Fukuda, "Experimental Studies on the Implantation of Hydroxylapatite in the Medullary Canal of Rabbits," *Biomat.* **1** (1980) 65-71.

86. J. Christofferssen and M. R. Christoffersen, "Kinetics of Dissolution of Calcium Hydroxyapatite. V. The Acidity Constant for the Hydrogen Phosphate Surface Complex," *J. Cryst. Growth* **57** (1982) 21-26.

87. L. L. Hench, R. J. Splinter, W. C. Allen and T. K. Greenlee, "Bonding Mechanisms at the Interface of Ceramic Prosthetic Materials," *J. Biomed. Mater. Res.* **2** (1971) 117-141.

88. R. Z. LeGeros, "Properties of Commercial Bone Grafts Compared to Human Bone and New Synthetic Bone Biomaterials," Ninth Annual Meeting of the Society for Biomaterials, Birmingham, Alabama (1983) Abstr. No. 86.

89. H. C. Margolis and E. C. Moreno, "Kinetics of Hydroxyapatite Dissolution in Acetic, Lactic and Phosphoric Acid Solutions," *Calcif. Tissue. Int.* **50** (1992) 137-143.

90. J. N. Kent, J. H. Quinn, M. F. Zide, L. R. Guerra and P. J. Boyne, "Augmentation of Deficient Alveolar Ridge with Non-Resorbable Hydroxylapatite Alone or With Autogenous Cancellous Bone," *J. Oral. Maxillofac. Surg.* **41** (1983) 429-435.

91. M. Ogiso, H. Kaneda, J. Arasaki and T. Tabata, "Epithelial Attachment and Bone Tissue Formation on the Surface of Hydroxyapatite Ceramic Dental Implants," *Biomaterials* **1** (1980) 59-66.

92. S. K. Doss, "Surface Property of Hydroxyapatite: The Effect on Various Inorganic Ions on the Electrophoretic Behavior," *J. Dent. Res.* **55** (1976) 1067-1075.

93.   P. Ducheyne, C. S. Kim and S. R. Pollack, "The Effect of Phase Differences on the Time-Dependent Variation of the Zeta Potential of Hydroxyapatite," *J. Biomed. Mater. Res.* **26** (1992) 147-168.

94.   S. R. Pollack, "Bioelectrical Properties of Bone: Endogenous Electrical Signals," *Orthop. Clin. N Amer.* **15** (1984) 3-13.

95.   D. F. Williams, "Review: Tissue-Biomaterial Interactions," *J. Mat. Sci.* **22** (1987) 3421-3445.

96.   M. Gorbunoff, "The Interaction of Proteins with Hydroxylapatite," *J. Anal. Biochem.* 136 (1984) 425-432.

97.   M. Spencer and M. Grynpas, "Hydroxyapatite for Chromatography. 1. Physical and Chemical Properties of Different Preparations," *J. Chromat.* 166 (1978) 423-434.

98.   A. L. Boskey, "Overview of Cellular Elements and Macromolecules Implicated in the Initiation of Mineralization," in *The Chemistry and Biology of Mineralized Tissues,* ed. W. Butler (Birmingham, Alabama, 1984) pp. 335-343.

99.   A. H. Reddi, "Implant-stimulated Interface Reactions During Collagenous Bone Matrix-Induced Bone Formation," *J. Biomed. Mater. Res.* **19** (1985) 233-239.

100. A. Beckham, T. K. Greenlee and A. R. Crebo, Jr. "Bone Formation at a Ceramic Implant Interface," *Calcif. Tiss. Res.* **8** (1971) 165-171.

101. J. Ganeles, M. . A. Listgarten and C. I. Evian, "Ultrastructure of Durapatite-Periodontal Tissue Interface in Human Intrabony Defects," *J. Periodntol.* **57** (1986) 133-140.

102. M. Jarcho, "Retrospective Analysis of Hydroxyapatite Development for Oral Implant Applications," *Dent. Clin. North. Amer.* **36** (1992) 19-26.

103. B. M. Tracy and R. H. Doremus, "Direct Electron Microscopy Studies of the Bone-Hydroxylapatite Interface," *J. Biomed. Mater. Res.* **18** (1984) 719-726.

104. M. Gregoire, I. Orly and Menanteau, "The Influence of Calcium Phosphate Biomaterials on Human Bone Cell Activities: An *In Vitro* Approach," *J. Biomed. Mater. Res.* **24** (1990) 163-177.

105. R. Z. LeGeros, I. Orly, M. Gregoire and J. Kazimiroff, "Comparative Properties and *In Vitro* Transformation of HA Ceramics in Serum," *J. Dent. Res.* **67** (1988) 177 (Abstr. # 512).

106. I. Orly, M. Gregoire, J. Menanteau and M. Dard, "Effects of Synthetic Calcium Phosphates on the H-Thymidine Incorporation and Alkaline Phosphatase Activity of Human Fibroblasts in Culture," *J. Biomed. Mat. Res.* **23** (1989) 1433-1440.

107. M. Heughebaert, R. Z. LeGeros, M. Gineste, A. Guilhem and G. Bone, "Physico-Chemical Characterization of Deposits Associated with HA Ceramics Implanted in Non-Osseous Sites," *J. Biomed. Mater. Res.* **22** (1988) 254-268.

108. G. Daculsi, R. Z. LeGeros and C. Deudon, "Formation of Carbonate-Apatite Crystals After Implantation of Calcium Phosphate Ceramics," *Calcif. Tissue Int.* **24** (1990) 471-488.

109. R. Baron, L. Neff, D. Louvard and P. J. Courtoy, "Cell-Mediated Extracellular Acidification and Bone Resorption: Evidence for a Low pH in Resorbing Lacunae

and Localization of 100-kD Lysosomal Membrane Protein at the Osteoclast Ruffled Border," *J. Cell. Biol.* **101** (1985) 2210-2222.

110. R. Z. LeGeros, G. Daculsi, I. Orly and M. Gregoire, "Substrate Surface Dissolution and Interfacial Biological Mineralization," in *The Bone-Biomaterial Interface*, ed. J. E. Davies (University of Toronto Press, 1991) pp. 76-88.

111. G. Daculsi, R. Z. LeGeros and D. Mitre, "Crystal Dissolution of Biological and Ceramic Apatites," *Calcif. Tiss. Int.* **45** (1989) 95-103.

112. U. Gross, V. Strunz, J. Bab and J. Sela, "The Ultrastructure of the Interface Between a Glass Ceramic and Bone," *J. Biomed. Mater. Res.* **15** (1981) 291-305.

113. C. M. Muller-Mai, C. Voigt and U. Gross, "Incorporation and Degradation of Hydroxyapatite Implants of Different Surface Roughness and Surface Structure in Bone," *Scan. Micros.* **4** (1990) 613-624.

114. M. Ogino, F. Ohuchi and L. L. Hench, "Compositional Dependence of the Formation of Calcium Phosphate Films on Bioglass," *J. Biomed. Mater. Res.* **14** (1980) 55-64.

115. F. B. Bagambisa, U. Joos and W. Schilli, "The Interaction of Osteogenic Cells with Hydroxylapatite Implant Materials *In Vitro* and *In Vivo*," *Int. Oral. Maxillofac. Implants* **5** (1990) 217-226.

116. H. S. Cheung and M. H. Haak, "Growth of Osteoblasts on Porous Calcium Phosphate Ceramic: An *In Vitro* Model for Biocompatibility Study," *Biomat.* **10** (1988) 63-67.

117. J. E. Davies, "The Use of Cell and Tissue Culture to Investigate Bone Cell Reactions to Bioactive Materials," in *Handbook of Bioactive Ceramics Vol. I.* eds. T. Yamamuro, L. L. Hench, and J. Wilson (CRC Press, Boca Raton, Florida, 1990):

118. P. Frayssinet, I. Primout, N. Rouquet, A. Autefate, A. Guilhem and P. Bonnevaille, "Bone Cell Grafts in Bioreactor: A Study of Feasibility of Bone Cell Autograft in Large Defects," *J. Mater. Sci. Mater. Med.* **2** (1991) 217-221.

119. M. H. Amler, "Osteogenic Potential of Non-Vital Tissues and Synthetic Implant Materials," *J. Periodnt.* **58** (1988) 758-761.

120. D. W. Holcomb and R. A. Young, "Thermal Decomposition of Human Tooth Enamel," *Calcif. Tiss. Int.* **31** (1980) 189-201.

121. J. F. Osborn and H. Neweseley, "The Material Science of Calcium Phosphate Ceramic," *Biomaterials* **1** (1980) 108-111.

122. M. Niki, G. Ito, T. Matsuda and M. Ogino, "Comparative Push-Out Data of Bioactive and Non-Bioactive Materials of Similar Rugosity," in *Bone-Material Interface*, ed. J. E. Davies (University of Toronto Press, 1991) pp. 350-356.

123. R. F. Ellinger, E. B. Nery and K. L. Lynch, "Histological Assessment of Periodontal Osseous Defects Following Implantation of Hydroxyapatite and Biphasic Calcium Phosphate Ceramics. A Case Report," *Int. J. Perio. Restor. Dent.* **3** (1986) 223-233.

124.   G. L. de Lange, C. de Putter and K. de Groot, "Histology of the Attachment of Gingival Fibres to Dental Root Implants of Ca-Hydroxyapatite," *Biomater. Biochem.* **5** (1983) 452-462.

125.   H. W. Denissen, W. Kalk, A.A.H. Veldhuis, A. van den Hooff, "Eleven-Year Study of Hydroxyapatite Implants," *J. Prosthet. Dent.* **61** (1989) 706-712.

126.   C. de Putter, J. de Groot and P.A.E. Sillevis-Smitt, "Transmucosal Implants of Dense Hydroxylapatite," *J. Prosthet. Dent.* **49** (1983) 87-95.

127.   K. Kuroshina, G. I. de Lange, L. de Putter and K. de Groot, "Reaction of Surrounding Gingiva to Permucosa Implants of Dense HA in Dogs," *Biomaterials* **5** (1984) 215-219.

128.   R. M. Frank, P. Widermann, J. Hemmerle and Freymann, "Pulp Capping with Synthetic Hydroxyapatite in Human Premolars," *J. Appl. Biomat.* **2** (1991) 243-250.

129.   J. Seibert and S. Nyman, "Localized Ridge Augmentation in Dogs: A Pilot Study Using Membranes and Hydroxyapatite," *J. Periodontol.* **61** (1990) 157-165.

130.   J. P. LeGeros and R. Z. LeGeros, "Characterization of Calcium Phosphate Coatings on Implants," The 17th Annual Meeting of the Society for Biomaterials, Scottsdale, Arizona. Abstr. #192 (1991).

Chapter 10

POROUS HYDROXYAPATITE

Edwin C. Shors and Ralph E. Holmes
Department of Plastic and Reconstructive Surgery
The University of California at San Diego
San Diego, CA 92110

INTRODUCTION

The rationale for using hydroxyapatite as a bone substitute material should be self evident: natural bone is approximately 70% hydroxyapatite by weight and 50% hydroxyapatite by volume. The rationale for making the material macroporous is not so obvious, and necessitates an appreciation for the architecture of tissues and its effect on regeneration and repair. All of our organs, such as liver, kidney and bone, have a parenchymal and a stromal component. The parenchyma is the physiologically active part of the organ; the stroma is the framework that supports the organization of the parenchyma. In soft tissues loss of parenchyma with maintenance of stroma allows a remarkable degree of regeneration and repair. A bone defect might regenerate more predictably if a stromal substitute is implanted to provide a framework for organization of the osteons. By providing the bone defect with a stromal substitute, containing spaces morphologically compatible with the osteons and their vascular interconnections, a partnership between biomaterials and biologic regenerative and repair responses can be encouraged.[1]

When examined closely, cortical bone consists of osteons or Haversian systems which are held together by a hard tissue stroma or interstitium (Fig. 1). The interstium does not have the Haversian canals and blood circulation of the osteons. Interosteonic communications, known as Volksmann canals, traverse the interstitial bone and permit blood to access the deepest osteons and maintain their osteocytes.

Interstitial bone accounts for one-third of the volume of long bones, like the femur and tibia, with the remaining two-thirds consisting of the parenchymal osteons.[2,3] To design an implant for osteoconduction it would seem logical to mimic the architecture of this interstitial or stromal bone. Since osteons average 190-230 μm in diameter, and intercommunicate through Volksmann canals, an idealized bone graft substitute would mimic osteon-evacuated cortical bone and have an interconnected porous system of channels of similar dimensions (Fig. 2). These pore dimension are in accordance with the classical studies of Klawitter and Hulbert,[4] which established a minimum pore size of 100 μm for bone ingrowth into ceramic structures.

Fig. 1. Microstructure of human cortical bone. The cylindrical osteons or Haversian systems represent the parenchymal component of bone. Blood flowing through the Haversian canals supply osteocytes contained between the osteonal lamellae. Interstitial or stromal bone occupies the space between osteons. Note fenestrations or Volksmann canals in the interstitial bone which permit interosteonic passage of blood supply.

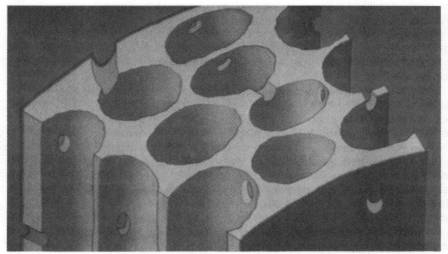

Fig. 2. Idealized microstructure for cortical bone regeneration. After evacuation of the osteons, there remains a structure with an interconnected porosity of 65% into which regenerated osteons can naturally fit.

Cancellous bone differs from cortical bone in being open-spaced and trabecular (Fig. 3). The trabeculae represent "unrolled" osteons on both surfaces which are in apposition to a central framework of interstitial bone. An ideal cancellous bone graft substitute would mimic osteon-evacuated cancellous bone and have a thin lattice interconnected by pores of 500-600 μm (Fig. 4). In this chapter, on porous hydroxyapatite, considerable emphasis will be devoted to the effects of porosity, since that is where bone regeneration takes place.

PROCESSING

Sintered HA

The most widely used process to fabricate porous hydroxyapatite implants utilizes isostatic compaction and sintering of calcium phosphate powders that contain naphthalene particles.[5] Volatilization of the naphthalene particles leaves a porosity which consists of spherical voids communicating by a narrow-mecked aperture wherever the particles were in contact. To permit bone ingrowth of any depth, these apertures must exceed 100 μm or they will represent blind ends and discontinuities in bone. Another sintering process for creating a macroporous structure utilizes pretreatment with hydrogen peroxide.[6]

HA Cement

More recently water-setting HA cements have been employed to create hydroxyapatite materials with various porosities.[7,8,9] The most completely characterized HA cement in this group[10] is made by reacting tetracalcium phosphate and calcium hydrogen phosphate in an aqueous environment.

$$Ca_4(PO_4)_2 + CaHPO_4 ---> Ca_5(PO_4)_3OH \qquad (1)$$

Under *in vitro* conditions at 37°C the HA cement sets in approximately 15 minutes and the isothermal chemical reaction is completed in 4 hours. Porosity is obtained by mixing the cement prior to set with sucrose granules and then dissolving the granules in water.

HA Conversion by Hydrothermal Exchange

During the early 1970's in the Material Research Laboratory at Pennsylvania State University, a process was developed that utilized the skeletal structure of marine invertebrates, especially reef building corals, as a template to make porous structures of other materials.[11] The calcium carbonate skeleton is reacted with dominium hydrogen phosphate and, by means of a hydrothermal exchange of carbonate and phosphate, is converted to hydroxyapatite:

$$10CaCO_3 + 6(NH_4)_2HPO_4 + 2H_2O ---> Ca_{10}(PO_4)_6(OH)_2 + 6(NH_2)CO_3 + 4H_2CO_3 \qquad (2)$$

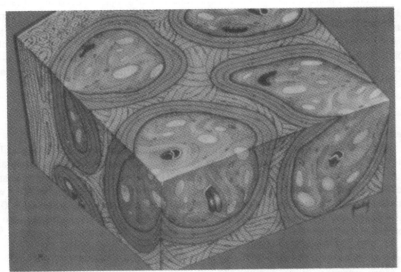

Fig. 3. Microstructure of human cancellous bone. The osteons appear as planar lamellae with interstitial bone filling the spaces in between. The thickness of these planar lamellae, similar to the radius of osteons, permits nutrition of its osteocytes from blood vessels in the large trabecular spaces.

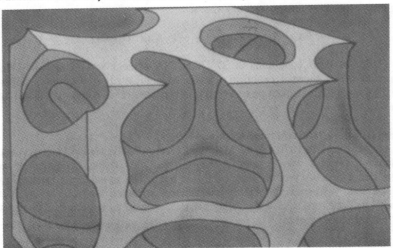

Fig. 4. Idealized microstructure for cancellous bone regeneration. The large interconnected pores can permit ingrowth of fibrovascular tissue, differentiation of osteoblasts and apposition of new bone against this framework.

Under suitable temperature and pressure conditions the exchange results in a nearly pure hydroxyapatite. The hydroxyapatite structure is an exact replicate of the porous marine skeleton, including its interconnected porosity. Prior to the exchange reaction, the organic component of these corals is removed with sodium hypochlorite. The extreme chemical and thermal conditions of the exchange reaction destroy any residual organic material. Of interest to present day environmental concerns is the lack of any impact due to harvesting these corals. Known as "nuisance coral" in the South Pacific, they are over-abundant and must be routinely removed from harbors and shipping lanes.

COMPOSITION

The chemical compositions of the different hydroxyapatite preparations are traditionally evaluated with x-ray diffraction. Preparations typically demonstrate a crystalline structure which is essentially pure hydroxyapatite with trace levels of beta-tricalcium phosphate, also known as beta-whitlockite (Fig. 5). This beta form of tricalcium phosphate is also present within human bone at similar low concentrations.

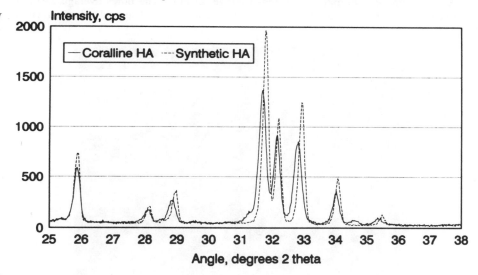

Fig. 5. X-ray diffraction scan of synthetic HA standard compared with HA from hydrothermal exchange of coral skeleton carbonate with phosphate.

PROPERTIES

The porosity of sintered and cement forms of hydroxyapatite is dependent on the numbers and dimensions of the volatilized or dissolved particles. The degree of particle compaction and contact helps to determine the interconnectivity of the porosity. For example, the sintered HA material studied by Klein[6] contained pores of 150-250 μm. The volume fraction porosity and the pore interconnectivity were not reported. The HA cement material studied by Costantino[10] had a volume fraction porosity of 10% and 20%. The pore dimensions and connectivity were not reported.

Early research identified two species of coral having suitable pore sizes for conversion into hydroxyapatite and subsequent use as a bone substitute.[12] The characteristics of the porosity in marine invertebrates were found to be genus and species dependent. Within a species and between members of that species, the average pore diameters were quite uniform, with a low range of variation throughout the structure. Accurate taxonomic identification thus assured a consistency of pore size and for the first time offered a legitimate educational reason for biomaterials students to visit and study in the West Indian and South Pacific littorals.

To mimic the osteon-evacuated stroma of cortical bone, the coral skeleton from the genus *Porites* was selected (Fig. 6). The solid framework and pore network are continuous and interconnected domains. The solid components of the implant framework average 75 μm and the interconnections average 95 μm. The pores average 230 μm diameter and their interconnections average 190 μm diameter. The void volume fraction is 65%.[13]

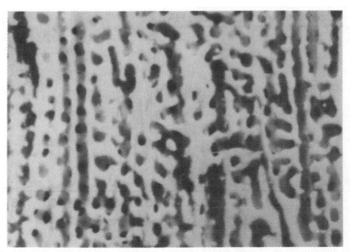

Fig. 6. Microstructure of *Porites* coral skeleton after conversion to hydroxyapatite (HA200). Fenestrations of pore channel walls resulting in high degree of interconnectivity are easily seen.

To mimic the osteon evacuated stroma of cancellous bone, the genus *Goniopora* was selected (Fig. 7). The solid components of its framework average 130 μm with interconnections that average 220 μm diameter. The pores average 600 μm diameter and their interconnections average 260 μm diameters. The void volume fraction is 63%.[13]

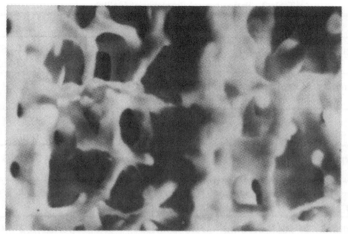

Fig. 7. Microstructure of *Goinopora* coral skeleton after conversion to hydroxyapatite (HA500). Trabecular framework with large interconnected pores is apparent.

For brevity and ease of association with pore size, the hydroxyapatite form of *Porites* is called HA200 while that for *Goniopora* is called HA500. Both structures are comparable with human bone, Table 1.

Table 1. Comparison of the Microstructure of HA500 and Human Cancellous Bone ($\pm$SE).

	HA500	**Iliac Bone**[14]	**Iliac Bone**[15]
Volume fraction (%)	35.1$\pm$1.5	20.5$\pm$0.4	20.3$\pm$0.4
Surface area (mm^2/mm^3	5.3$\pm$0.2	3.0$\pm$0.1	3.4$\pm$0.1
Ratio of surface area to volume fraction	15.3$\pm$0.6	14.6$\pm$0.6	17.3$\pm$0.2
Mean trabecular width (μm)	131.9$\pm$4.4	136.6$\pm$4.5	120.3$\pm$1.6
Mean pore width (μm)	245.0$\pm$9.0	529.6$\pm$22.9	468.2$\pm$27.2

From reference 17

Biomechanical Properties

The biomechanical properties of sintered and cement forms of porous HA depend on the degree of porosity and the paucity of published data suggests this needs further study. The biomechanical properties of HA 200 and HA 500 are presented in Tables 2 and 3.

Table 2. Biomechanical Properties of HA200.

Property	Test	Orientation	Mean	Range
Crush Strength (psi)	Compression	Parallel Perpendicular	1343 626	997-1675 257-963
Ultimate strength $(N\text{-}cm^{-1})$	Compression	Perpendicular	373	251-544
Stiffness $(N\text{-}cm^{-1})$	Compression	Perpendicular	8300	3310-11470
Energy absorption (N-cm)	Compression	Perpendicular	9.9	4.5-13
Tensile strength $(gm\text{-}cm^{-2}x10^4)$	4-point bending	Not reported	Not reported	2.4-3.3
Young's modulus $(gm\text{-}cm^{-2}x10^4)$	4-point bending	Not reported	Not reported	5.2-6.0
Elastic modulus $(dynes\text{-}cm^{-2}x19^{10})$	Resonance frequency	Parallel Perpendicular	4.8 2.6	3.6-5.8 1.9-3.2

From reference 17

Table 3A. Failure Stress Values for HA500 ($\pm$SD).

Direction	Tensile	Compressive	Transverse	Torsional	Bending
Parallel	188.7$\pm$---	851.7$\pm$319.7	143.0$\pm$41.3	41.3$\pm$11.2	403.3$\pm$97.5
Perpendicular	95.4$\pm$49.4	489.0$\pm$160.3	268.0$\pm$98.5	40.2$\pm$15.1	529.5$\pm$224.7
Diagonal	122.0$\pm$67.2	661.3$\pm$267.0	495.7$\pm$51.9	46.3$\pm$18.0	332.8$\pm$137.2

Table 3B. Moduli of HA500 ($\pm$SD).

Direction	Tensile	Compressive	Flexure	Rigidity
Parallel	56.0$\pm$---	80.2$\pm$54.4	110.0$\pm$29.3	0.9$\pm$0.4
Perpendicular	32.7$\pm$11.7	36.5$\pm$14.5	159.8$\pm$69.4	0.9$\pm$0.4
Diagonal	23.6$\pm$14.1	56.2$\pm$31.1	111.7$\pm$35.2	1.1$\pm$0.8

From reference 36

The biomechanical properties of HA200 and HA500 are distinctly anisotropic, i.e., different in one direction than in the others, as a result of the growth characteristics of most scleractinian corals. The relationship between compressive strength and channel axis of the pores in HA200 has been well defined.[16] In the absence of any bone ingrowth the ultimate compressive strength of HA200 is significantly less than cortical bone while that of HA500 is similar to cancellous bone. The stiffness of HA200 as defined by conventional destructive testing is comparable with bone graft, but the implant material still fails like a ceramic with a brittle fracture. Like its dense form, porous HA materials have a low fatigue strength relative to most metals and polymers. Although not well studied, crack propagation has been considered to be partially ameliorated by the presence of pores.[17]

Good
Ref:
later

Surface Chemistry

The surface chemistry of the porous sintered and cemented HA are presumable not different from their dense forms. In the converted HA it has been noted that the crystallite size of the hydroxyapatite is significantly smaller than the crystallite size within the original coral. Although this reaction has been studied in detail, the explanation for this polycrystalline morphology is not well defined.[18,19] A high power scanning electron photomicrograph (Fig. 8) demonstrates this high surface area which may be a factor in the osteoconductive behavior of this porous HA material.

Fig. 8. High power scanning electron micrograph of HA200 showing polycrystalline surface. Reference line indicates 15 μm.

TISSUE RESPONSE

Ingrowth

The tissue response to porous HA implants is inherently different from dense HA because of the opportunity for ingrowth. Porosity and interconnectivity are key determinants of amount and type of ingrowth. In highly porous and interconnected implants like HA200 and HA500, fibrovascular tissue ingrowth starts by day 3 or 4. By 28 days this ingrowth is completed throughout the implant and apposition of bone against the pore walls has begun.[20-22] The spatial apposition of bone then progresses temporally from the implant surface towards the center.[23] A transient appearance of multinucleated giant cells has been reported.[24]   The eventual bone-HA bonding within pores is considered to be like that documented for dense HA. Studies in several dog models have found bone ingrowth to be nearly complete by 3 months (equivalent to 6 months in man). Data from HA200 in Table 4 and HA500 in Table 5 summarize these studies.

Table 4. Tissue Volume Fractions (%) of HA200 Implants Retrieved
From the Dog Radius After 3, 6, 12, 24 and 48 Months ($\pm$SE).

Months	n	Soft Tissue	Bone	HA Matrix
3	3	12.1$\pm$2.4	49.4$\pm$2.4	38.6 $\pm$1.5
6	3	13.7$\pm$1.8	46.1$\pm$2.1	40.2$\pm$3.6
12	3	11.6$\pm$2.3	52.7$\pm$3.1	35.7$\pm$1.2
24	3	7.7$\pm$0.5	54.8$\pm$1.0	37.6$\pm$1.5
48	2	6.2$\pm$1.1	54.3$\pm$4.2	39.6$\pm$3.2

From reference 26.

Table 5. Tissue Volume Fractions (%) of HA500 Implants Retrieved
From the Dog Proximal Tibial Metaphysis After 2, 4, 6 and 12 Months ($\pm$SE).

Months	n	Soft Tissue	Bone	HA Matrix
2	2	57.3$\pm$1.4	10.3$\pm$0.6	32.6$\pm$1.3
4	2	50.1$\pm$1.6	11.6$\pm$0.7	38.3$\pm$1.5
6	2	51.2$\pm$1.6	13.0$\pm$0.7	35.9$\pm$1.4
12	2	49.0$\pm$1.7	17.3$\pm$0.9	33.8$\pm$1.6

From reference 27

The histologic appearance of the initial bone ingrowth demonstrates an irregular and unorganized orientation of the collagen fibers and distribution of osteocytes that is characteristic of immature woven bone.[23] This initial bone is subsequently replaced with mature parallel fiber lamellar bone. This sequence is similar to normal bone formation. The woven bone of the embryo and newborn skeleton is gradually replaced with a parallel fiber lamellar bone which will remodel throughout life.[25] In dog studies normal physiologic remodelling of the mature bone, reflecting the influence of Wolffs law, can be observed by 12 months. In diaphyseal radius implants of HA200 the intramedullary portion showed a substantial remodelling with removal of the bone ingrowth (and preservation of bone ingrowth within the intracortical portion).[26] In metaphyseal tibial implants of HA500 the HA pores within the cancellous region contained trabecular bone ingrowth while the pores within the cortical shell regions contained osteonal bone ingrowth.[27] The findings of these normal physiologic behaviors of bone (maturation sequence and response to Wolffs law) within the pores of HA characterizes the implant composition and microstructure (pore shape, size and interconnectivity) as highly biocompatible.

Union and Incorporation

In orthopedic terminology, the tissue response of bone ingrowth into graft or implant surface pores is called union. An implant or graft is said to be united when sufficient bone ingrowth (approx. 200-500 μm depth) has occurred to unite host bed and implant as measured by histometry and shear testing. When bone ingrowth throughout the pores is present, it is conceptually conceivable that the bone could occupy the pore center without contacting the pore walls, or it could contact the pore walls and thicken into the pore center. Apposition of bone ingrowth against the walls of the implant or graft matrix is called incorporation. An implant or graft is said to be incorporated when the matrix is substantially coated with bone ingrowth. In addition to traditional measures of osteocompatibility,[28] studies of animal specimens and human biopsies should ideally utilize an easy, accurate and unbiased system of image acquisition and analysis which includes the measurement of specific implant matrix surface area and fraction of this area covered by bone ingrowth.[29] As exemplified by the data in Table 6, the tissue response to porous HA implants can result in a high degree of incorporation.

Table 6. Porous HA Matrix Surface Areas (mm^2/mm^3) and Surface Fractions (%) Covered by Bone Ingrowth in Implant Specimens Retrieved After 11 to 17 Months ($\pm$SD).

Months	Surface Area	Surface Fraction
11	9.6$\pm$0.8	88.8$\pm$8.8
12	9.3$\pm$0.9	90.2$\pm$6.9
14	9.2$\pm$0.4	89.3$\pm$8.4
15	9.3$\pm$0.4	96.6$\pm$1.8
16	9.0$\pm$0.6	92.2$\pm$5.6
17	9.2$\pm$0.7	90$\pm$5.6

From reference 37

Effect on Strength

In porous HA the tissue responses result in an implant-bone composite that significantly changes the original biomechanical properties. A high correlation ($r=0.92$) was reported between the bending strength of porous HA and the amount of pore space occupied by bone ingrowth.[30] In HA200 compressive strength was found to increase 3.5-7.2 fold and anisotropy of the original matrix was neutralized after bone ingrowth (Table 7). In HA 500 compressive strength was found to increase 2.7-6.8 fold after bone ingrowth.[12]

Table 7. Crush Strength (psi) in Compression of HA200 Before and After Bone Ingrowth.

Orientation	Before		After	
	Mean	Range	Mean	Range
Parallel	1343	997-1675	4776	2750-8479
Perpendicular	626	257-963	4529	2475-7562

From reference 38

Biodegradation

Two mechanisms - cell mediated and dissolution - participate in the biodegradation and resorption of HA in the body. The activity of both of these mechanisms is directly related to implant surface area. Because of its lower surface area, dense HA has demonstrated very low rates of biodegradation. Porous HA, on the other hand, can undergo a significant degree of resorption. Scanning electron microscopy of HA cement after setting revealed that it is composed of small petal-like crystals with an interconnected microporosity averaging 2-5 nm in diameter. Experimental study of a macroporous form of this microporous HA cement has not been reported. However when a microporous-only form of HA cement was used to obliterate the frontal sinus of cats, only 27% of the HA cement remained after 18 months.[31] HA converted by hydrothermal exchange has much larger micropores 1-5 μm in diameter, resulting in a much smaller total surface area. Perhaps not surprisingly, converted HA only resorbs 1-2% per year.

The different sequences of biodegradation are probably important from a functional point of view. When placed in the cat sinus, a retreating front of HA cement resorption was replaced by an advancing front of new bone repair. The lack of mechanical strength at this front, of no consequence in the frontal sinus, might have serious consequences in a weight bearing long bone. In contrast, HA200 and HA500 demonstrate no bulk front of resorption and replacement. The individual HA matrix members of converted HA become thinner by 1-2% per year, apparently replaced by bone, resulting in maintenance of the bulk dimensions of the implant and its mechanical strength, until it is finally "thinned" out of existence. The mechanisms of biodegradation of these different forms of HA, along with the biomechanical consequences, need further study.

CLINICAL APPLICATIONS

Of the different forms of porous HA materials, only the converted HA forms, HA200 and HA500, have undergone major clinical trials. Clinical applications in maxillofacial and orthopedic surgery continue to be evaluated. A representative study will be reviewed from each field of application. These studies are small and a larger data base is required before all indications and contraindications can be determined.

Maxillofacial Surgery

In 92 consecutive patients undergoing orthognathic surgery a total of 355 HA200 implants were placed in the maxilla (294), mandible (4) and midface (20).[32] In the 47 patients who had maxillary surgery, 202 implants were positioned directly adjacent to the maxillary sinus. Of these 202, 58 were placed in a maxillary step osteotomy, 99 in the lateral maxillary wall osteotomy, and 45 were placed between the pterygoid plate and maxillary tuberosity. The remaining maxillary implants were placed interdentally (53) and midpalatally (36) after segmentalization and expansion. Of the 41 mandible implants, 28 were positioned in the buccal cortical defect left by a sagittal split osteotomy, 10 were used for chin onlay, and 3 were used for chin interpositional inlay. Of the 20 midface implants, 12 were used for the lateral orbital rim, 7 for infraorbital only and one for nasofrontal interposition.

These procedures, representing a broad spectrum of maxillofacial surgery, were associated with long term complications in 4 of the 92 patients (4.3%). One patient developed bilateral maxillary sinusitis two months post-surgery, responded well to antibiotics and had no further problems. Two patients developed intranasal exposure of midpalatal implants which were removed at 6 and 14 months with no further problems. The fourth patient had persistent drainage from an interdental implant which was removed 21 months after surgery with no further problems. Cephalometric measurement and analysis of post-surgery facial bone position revealed stability equivalent to that seen after the use of autogenous bone grafts. Biopsies taken from 9 patients at 4 to 16 months post-surgery revealed structurally intact implants that were both united and incorporated.[33] The biopsy specimens were composed of 48.5% HA200 matrix (range: 36.5-56.7%), 18% bone (range: 6.7-31%), and the remainder was soft tissue and vascular space. Up to 9 months woven bone was still apparent, with longer term biopsies showing only parallel-fiber lamellar bone. Clinical studies of these applications continue.

Another maxillofacial application of interest is the use of HA200 in the form of 2-3 mm diameter granules placed in the floor of the maxillary sinus. The ingrowth of bone into these granules provides more bone stock into which titanium cylinders and screws may be placed for use in dental restoration. In a series of 4 patients receiving implants in 5 maxillary sinuses, 5 biopsies were retrieved.[34] Histometric analysis demonstrated a mean bone ingrowth of 23.1%, soft tissue ingrowth of 44.9% and HA matrix of 31.9%. After biopsy confirmation of bone ingrowth, a total of 12 dental implants were placed and subsequently used for dental restoration. Clinical studies of

HA200 alone, mixed with cancellous autografts, and infiltrated with bone marrow aspirate continue.

Orthopedic Surgery

A series of 46 patients with traumatic defects of their long bones underwent reconstruction with HA500 and stabilization with plate and screw fixation.[35]   The mechanism of injury was motor vehicle accident in 32 cases, falls in 10, and gunshot injuries in 4.   There were 25 men and 212 women, with an average age of 34.5 years (range: 17-67 years). All operations were performed between 6 hours and 5 days from the time of injury.   No supplemental autogenous bone graft was used in any of these 46 cases.

Metaphyseal (end-shaft) defects arising from axial compression injuries to adjacent joint surfaces constituted 34 of the cases.   The location of the defect was tibial plateau in 23, distal tibia in 4, distal radius in 3, and distal femur in 4.   All but three cases had displaced osteochondral fragments impacted into the crushed cancellous bone of the metaphysis. Following reduction and rigid internal fixation of all major components of the fracture, the resultant metaphyseal defects varied in volume from 1 to 120 cm^3 (mean: 9.5 cm^3; median: 2.5 cm^3).   A satisfactory press fit of the HA 500 block was achieved in all cases.   An illustrative case is drawn in Fig. 9.

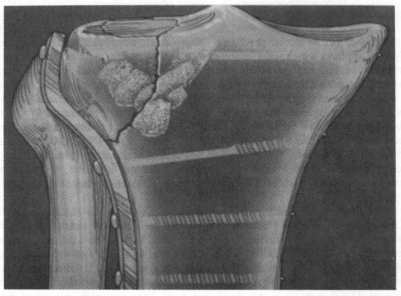

Fig. 9. Illustration of tibial plateau fracture treated with porous HA implant. After returning the fragment to its anatomic position, a defect remained in the cancellous interior of the metaphysis which was grafted with blocks of porous hydroxyapatite.

Diaphyseal (mid-shaft) defects, secondary to high energy bending forces, constituted the remaining 12 cases. The location of the defect was tibia in 3, femur in 2, humerus in 3, radius in 3 and ulna in 2. The volume of cortical defects implanted with HA500 ranged from 1 to 4 cm^3. An illustrative case is drawn in Fig. 10.

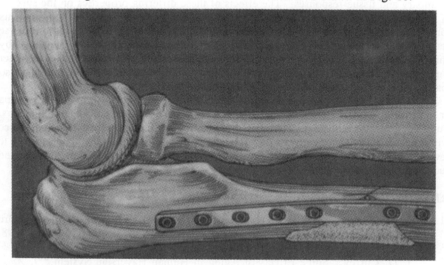

Fig. 10. Illustration of radius fracture in forearm treated with porous HA implant. After stabilization with plate and screws, a segment of cortical bone was missing which was grafted with a block of porous hydroxyapatite.

Fracture union was achieved in all cases. As judged by disappearance of all cortical and cancellous fracture lines, fracture union occurred at an average of 28 weeks, identical with that noted in a comparable group of historical autograft controls. Four complications occurred in these 46 patients. These consisted of one early loosening of a humeral plate, one soft tissue slough, one contiguous septic arthritis, and one loss of reduction followed by a late onset infection. None of these complications were attributable to the porous hydroxyapatite implant.

Biopsy of the HA500 implant was performed in 10 patients at the time of elective hardware removal. Fluoroscopic imaging was used to ensure that the biopsy was taken from the center of the implant. The time from surgery to biopsy averaged 11 months (range: 7-18 months). The biopsies cases included 7 proximal tibial metaphyseal fractures, 2 distal tibial metaphyseal fracture, and 1 humeral diaphyseal fracture. Histologic evaluation showed compact bone in the superficial (cortical) portions of the biopsy specimens with regenerated osteons filling the HA pores. The deeper (cancellous) sections demonstrated normal appearing trabeculae in apposition to the HA matrix. Histometric analysis was performed on all biopsies and revealed an average bone volume fraction of 40.7%, HA matrix of 31.8% and soft tissue and vascular space of 27.5%.

SUMMARY

A variety of possibilities exits for the fabrication of porous hydroxyapatite materials with differing porosities, interconnectivities, mechanical properties, surface chemistry and tissue responses. The effects of these differing properties on the success of clinical applications needs more study. The diversity of reconstructive requirements for clinical defects of the skeleton is great and there will be a need for equally diverse ceramic implant materials. Because HA materials are so brittle their acceptance by clinicians remains low. New forms of HA materials that incorporate collagen or other similar polymers must be developed so clinician can easily shape and fix the implants so that patients can walk without fear of failure while bone repair is taking place.

The cellular partnership between host repair and an implant providing a stroma or interstitium represents an exciting challenge to biomaterials scientists and clinicians. Future optimizations of these material and microstructural components will be combined with impregnation of bone growth factors and population by cultured osteoblast cells to reward the patients, clinician and biomaterials scientist with even more predictable and successful surgical outcomes.

REFERENCES

1.   R. E. Holmes, "Alloplastic Implants," in *Reconstructive Plastic Surgery*, 3rd edition, ed. J. G. McCarthy (W. B. Saunders Co., Philadelphia, 1990) pp. 698-731.

2.   J. D. Currey, "Some Effects of Aging in Human Haversian System," *J. Anat. [Lond]* **98** (1964) 69-75.

3.   F. G. Evans and S. Bang, "Physical and Histological Differences Between Human Fibular and Femoral Compact Bone," in *Studies on the Anatomy and Function of Bone and Joints,* ed. F. G. Evans (Springer-Verlag, New York, 1966) pp. 142-150.

4.   J. J. Klawitter and S. F. Hulbert, "Application of Porous Ceramics For The Attachment of Load Bearing Orthopedic Applications," *J. Biomed. Mater. Res. Symp.* **2** (1971) 161.

5.   W. Hubbard, *Physiological Calcium Phosphates As Orthopedic Biomaterials*, (PhD. thesis, Marquette University, 1974).

6.   C. Klein, P. Patka and W. den Hollander,"Macroporous Calcium Phosphate Bioceramics in Dog Femora: A Histological Study of Interface and Biodegradation," *Biomaterials* **10** (1989) 59-62.

7   W. E. Brown and L. C. Chow, "A New Calcium Phosphate, Water-Setting Cement," in *Cements Research Progress* (American Ceramic Society, Westerville, Ohio, 1986) pp. 352-379.

8.   P. J. Capano *The Chemical Synthesis, and Biomedical and Dental Applications of The First Truly Successful, in vivo Replacements for Bones, Teeth, and Similar Materials* (PhD Thesis, University of Texas at Austin, 1987).

9.   B. Constantz, S. W. Young, H. Kienapfel, B. L. Dajhlen, D. R. Sumner, T. M. Turner, R. M. Urban, J. O. Galante, S. B. Goodman and S. Gunasekaran, "Pilot Investigations of a Calcium Phosphate Cement in A Rabbit Femoral Canal Model and A Canine Humeral Plug Model," *Transactions of the 17th Annual Meeting, Society for Biomaterials,* Scottsdale, Arizona, May 1-5, (1991) 92.

10. P. D. Costantino, C. D. Friedman, K. Jones, L. C. Chow, H. J. Pelzer and G. A. Sisson, "Hydroxyapatite Cement: I. Basic Chemistry and Histologic Properties," *Arch Otolaryngol Head Neck Surg* **117** (1991) 379-384.

11. D. M. Roy and S. K. Linnehan, "Hydroxyapatite Formed from Coral Skeletal Carbonate by Hydrothermal Exchange," *Nature* **247** (1974) 220-222.

12. R. Holmes, V. Mooney, R. Bucholz and A. Tencer, "A Coralline Hydroxyapatite Bone Graft Substitute," *Clin. Orthop.* **188** (1984) 252-262.

13. W. M. Hanusiac *Polymeric Replamineform Biomaterials and A New Membrane Structure* (PhD Thesis, Pennsylvania State University, 1977).

14. H. H. Malluche, W. Meyer, D. Sherman and S. G. Massry, "Quantitative Bone Histology In 84 Normal American Subjects: Morphometric Analysis and Evaluation of Variance in Iliac Bone," *Calcif Tissue Internat.* **34** (1982) 449-455.

15. W. A. Merz and R. K. Schenk, "Quantitative Structural Analysis of Human Cancellous Bone," *Acta. Anat.* **75** (1970) 54-66.

16. A. F. Tencer, V. Mooney, K. L. Brown and P. A. Silva, "Compressive Properties of Porous Coated Hydroxyapatite for Bone Grafting," *J. Biomed. Mater. Res.* **19** (1985) 957.

17. E. White and E. C. Shors, "Biomaterials Aspects of Interpore-200 Porous Hydroxyapatite," *Dent. Clin. North Am.* **30** (1986) 49-67.

18. W. Eysel and D. M. Roy, "Hydrothermal Flux Growth of Hydroxyapatite By Temperature Oscillation," *J. Crystal Growth* **20** (1973) 245-250.

19. W. Eysel and D. M. Roy, "Topotactic Reaction of Aragonite to Hydroxyapatite." *Z. Kristallogr* **141** (1975) 11-24.

20. R. A. Finn, W. H. Bell and J. A. Brammer, "Interpositional Grafting with Autogenous Bone and Coralline hydroxyapatite," *J. Maxillofac. Surg.* **8** (1980) 217-227.

21. J. F. Piecuch, R. G. Topazian, S. Skoly, et al., "Experimental Ridge Augmentation with Porous Hydroxyapatite Implants," *J. Dent. Res.* **62** (1983) 148-151.

22. T. E. Butts, L. J. Peterson and C. M. Allen, "Early Soft Tissue Ingrowth Into Porous Block Hydroxyapatite," *J. Oral Maxillofac. Surg.* **47** (1989) 475-479.

23. R. E. Holmes, "Bone Regeneration Within A Coralline Hydroxyapatite Implant," *Plast. Reconstr. Surg.* **63** (1979) 626-633.

24. C. Itatani and G. J. Marshall, "Cellular Responses to Implanted Replamineform Hydroxyapatite," *Trans. Orthop. Res. Soc.* **9** (1984) 123.

25. J. A. Ogdon, "Chondro-osseous Development and Growth," in *Fundamental and Clinical Bone Physiology*, ed. M. R. Urist (J. B. Lippincott, Philadelphia, 1980) pp. 108-171.

26. R. E. Holmes, R. W. Bucholz and V. Mooney, "Porous Hydroxyapatite as a Bone Graft Substitute in Diaphyseal Defects: A Histometric Study," *J. Orthop. Res.* **5** (1987) 114-121.

27. R. E. Holmes, R. W. Bucholz and V. Mooney, "Porous Hydroxyapatite as a Bone Graft Substitute in Metaphyseal Defects," *J. Bone Joint Surg.* **68A** (1986) 904-911.

28. A. F. Tencer, R. E. Holmes and K. D. Johnson, "Osteocompatibility Assessment," in *Handbook of Biomaterials Evaluation: Scientific, Technical, and Clinical Testing of Implant Materials,* ed. A.F. von Recum (Macmillan Publishing Company, New York, 1986) 321-338.

29. R. E. Holmes, H. K. Hagler and C. A. Coletta, "Thick-section Histometry of Porous Hydroxyapatite Implants Using Backscattered Electron Imaging," *J. Biomed. Mater. Res.* **21** (1987) 731-739.

30. R. B. Martin, M. W. Chapman, R. E. Holmes, D. J. Sartoris, E. C. Shors, J. E. Gordon, D. O. Heitter, N. A. Sharkey and A. G. Zissimos, "Effects of Bone Ingrowth on the Strength and Non-Invasive Assessment of a Coralline Hydroxyapatite Material," *Biomaterials* **10** (1989) 481-488.

31. C. D. Friedman, P. D. Costantino, K. Jones, L. C. Chow, H. J. Pelzer and G. A. Sisson, "Hydroxyapatite Cement:II. Obliteration and Reconstruction of the Cat Frontal Sinus," *Arch. Otolaryngol Head Neck Surg.* **117** (1991) 385-389.

32. L. M. Wolford, R. W. Wardrop and J. M. Hartog, "Coralline Porous Hydroxylapatite as a Bone Graft Substitute in Orthognathic Surgery," *J. Oral Maxillofac. Surg.* **45** (1987) 1034-1042.

33. R. E. Holmes, R. W. Wardrop and L. W. Wolford, "Hydroxylapatite as a Bone Graft Substitute in Orthognathic Surgery: Histologic and Histometric Findings," *J. Oral Maxillofac. Surg.* **46** (1988) 661-671.

34. D. G. Smiler and R. E. Holmes, "Sinus Life Procedure Using Porous Hydroxyapatite: A Preliminary Clinical Report," *J. Oral Implantology* **13** (1987) 17-32.

35. R. W. Bucholz, A. Carlton and R. E. Holmes, "Hydroxyapatite and Tricalcium Phosphate Bone Graft Substitute," *Orthop. Clin. North Am.* **18** (1987) 323-334.

36. A. F. Tencer, E. C. Shors, P. L. Woodard and R. E. Holmes, "Mechanical and Biological Properties of a Porous Polymer-Coated Coralline Ceramic." In *Handbook of Bioactive Ceramics, Vol. II, Calcium Phosphate and Hydroxylapatite Ceramics* (CRC Press, Boca Raton, Florida 1991) 209-221.

37. R. E. Holmes and S. M. Roser, "Porous Hydroxyapatite as a Bone Graft Substitute in Alveolar Ridge Augmentation: A Histometric Study," *Int. J. Oral Maxillofac. Surg.* **16** (1987) 718-728.

38. J. F. Piecuch, A. J. Goldberg, C. V. Shastry, et al., "Compressive Strength of Implanted Porous Replamineform Hydroxyapatite," *J. Biomed. Mater. Res.* **18** (1984) 39-45.

Chapter 11

STABILITY OF CALCIUM PHOSPHATE CERAMICS AND PLASMA SPRAYED COATING

C.P.A.T. Klein, J.G.C. Wolke and K. de Groot
Department of Biomaterials, University of Leiden
The Netherlands

INTRODUCTION

Insertion of a biomaterial in a living tissue creates an artificial interface between living tissue and biomaterial. An ideal interface should behave in the same way as a theoretical plane present at the same place in the healthy tissue. When an implant and tissue are brought into contact, a mutual interaction is initiated, the implant may induce changes in the biological system, which may in turn, via surface reactions, induce changes in the implant material. At the interface, primary reactions will take place on molecular scale, such as dissolution of ions from the material, protein adsorption and desorption, denaturing of proteins, etc. The processes may induce secondary reactions "far" away from the interface. Uncertainties still exist on the initial events occurring around an implant.[1-3]

In hard tissue replacement, permanent attachment of limb prostheses, or total tooth implants, a biomaterial must interface with bone. The biocompatibility of implant materials is optimal when the material elicits the formation of normal tissues at its surface and, establishes a contiguous interface capable of transferring the loads which normally occur at the implantation site. To what extent bone-plus-implant will be able to function as an integrated mechanical unit depends on: the mechanical and physiological characteristics of the living bone; the chemical, mechanical and physical properties of the implant, and the interaction between bone and implant.

The main inorganic phases of bones and teeth of vertebrates as well as hard tissues of humans, although chemically quite complex, appear to be predominantly in a single structural state closely resembling that of hydroxylapatite, $Ca_{10}(PO_4)_6(OH)_2$, [HA]. Biological apatites are known for their occurrence in non-stoichiometric form, usually with low Ca/P ratios, and contain, besides structural imperfections and defects, substantial amounts of foreign ions such as CO_3^{2-}, citrate, Mg^{2+}, and Na^+, and trace amounts of Cl^-, F^-, K^+, Sr^{2+}, and other metal ions[4-6] (see also Chapter 9).

The synthetic form has been shown to be chemically and crystallographically similar, although not identical to, naturally occurring hydroxylapatite and has thus received a great deal of attention for use as bone graft substitute.[7,8]

Calcium phosphate implant materials are composed of the same ions which make up the bulk of the natural bone mineral. Because of this, these materials when implanted in bone are capable of participating in calcium phosphate solid-solution equilibria at their surfaces. The required calcium and phosphate ions needed to establish their equilibria may be derived from the implant, the surrounding bone, or both.[9-11]

The biocompatibility of synthetic HA is not only suggested by its composition but also by results of *in vivo* implantation, which has produced no local and systemic toxicity, no inflammation, and no foreign body response. Many investigators have demonstrated direct bone apposition of new bone to hydroxylapatite. This contact appears to be direct, without an intervening fibrous layer. During mechanical testing, fracture often occurs through the bone and/or HA, rather than at the bone-HA interface. The intimate bonding of new bone to HA is the main advantage of using HA as a bone graft substitute.[12-16]

Calcium phosphate ceramics can be made with properties resembling those of hard tissues. Dense ceramics with a compressive strength >500 MPa, as well as porous ceramics allowing bony ingrowth, can be prepared. Another approach to influence the processes at the biomaterial-bone interface is the use of biomaterials with controlled chemical breakdown characteristics, i.e., bioactive/biodegradable materials. With time, such biomaterials should totally be resorbed by the body and replaced by tissues. Consequently, the function of totally bio-degradable biomaterials is merely to serve as a scaffolding or filler of space, thereby permitting tissue infiltration and replacement. Essentially, this is the same function as that of bone grafts. A major advantage of the use of resorbable bioceramics over autologous bone grafts is a ready supply, controlled variations in size, and elimination of a second surgical procedure. However, a disadvantage of this type of bioceramic is the serious strength reduction that occurs during the resorption process. Consequently, mechanical design factors must be seriously taken into account to prevent fracturing of the resorbable ceramic during the intermediate stages of healing. Biodegradable implants are preferable when they are used ultimately to restore the normal function of bone. These implants should encourage bone growth and facilitate integration of the implant with bone, the rate of its resorption should match the rate of the bone formation, and the reduction in implant strength should closely match the increase in strength of the healing tissues.

So-called tricalcium phosphates [TCP] appear to be the most suitable bioceramics of this type.[17] With a nominal composition of $Ca_3(PO_4)_2$, this material has a Ca/P ratio of 1.50. TCP is found in two different whitlockite crystallo-graphic configurations, α-TCP, and the more stable β-TCP.[8,13] Due in part to its crystalline structure the biodegradation rate of TCP has been shown to be much greater than that of HA.[8,18] While the exact mechanism of biodegradation remains unclear, some researchers suggest that when placed in an acidic environment, TCP dissolves *in situ*.[13] Other investigators noted osteoclast-like cells attached to the surface of the implanted TCP and suggest a cellular breakdown by macrophages.[19,20]

As the material resorbs, new bone fills the area once occupied by the TCP implant. While much of the TCP implant is resorbed within the first months, some material remains in the defect site for extended periods, perhaps years in humans.[20,21,22] The material that is not resorbed appears to be incorporated within the new bone structure. Resorbable TCP or ß-whitlockite ceramics give rise to more bone remodelling activity than hydroxylapatite ceramics. Depending on the Ca/P ratio, an irregular or more planar bone formation is present at the ceramic surface. [13,24-26]

Data on biodegradability of the different calcium phosphate ceramics (HA or TCP, dense or porous) are conflicting. The reported experimental conditions show a variety of surgical procedures and animal models. Also manufacturing conditions, crystal structure, Ca/P ratio, impurities, degree and type of porosity were not clearly defined.[27]

Because the mechanical properties of calcium phosphate bioceramics are limited they should either be unloaded or loaded only in compression. To achieve the high strength necessary for implants, metal alloys can be coated with calcium phosphate particles. The bone bonding capacity of these coatings may help cementless fixation of orthopaedic prostheses. It has been shown that skeletal bonding is enhanced immediately after implantation.[13,28,29] Coatings on metals and other substrates (ceramics, polymers, composites) have been applied by a variety of methods including, dip, plasma spraying, electrophoretic deposition, sputter coating, hot isostatic pressing and ion assisted sputtering[30,31] (see also Chapter 12).

Variations of the material properties of calcium-phosphate coatings affect on the bone bonding mechanism and the rate of bone formation. Local supersaturation in the constituent ions of the bone mineral phase, arising from enhanced solid-solution exchange at the coating surface, could be a cause of bone tissue growth enhancement.[32] Variation in Ca/P ratios (2.0-1.5), i.e., tetracalcium phosphate, hydroxylapatite and α/ß-TCP lead to differences in degradability and to differences in bone contact.[20,33-35] Composition changes by addition of fluor to apatite or magnesium to ß-whitlockite will also influence the stability of the coating.[36,39] Understanding the cause of biodegradation of calcium phosphates and enhancement of bone tissue ingrowth or bonding requires knowledge of the characteristics of the calcium-phosphate coating itself and the coating process.

CALCIUM PHOSPHATE BIOCERAMICS

Density

Dense ceramics are usually made by compressing a powder into a pellet, which is then subjected to a heat treatment that causes the powder particles to fuse by means of solid-state diffusion. Such process is called sintering and is schematically depicted in Fig. 1. Depending on variables such as sintering temperature, time, and particle size distribution, a dense shape can be produced. 'Dense' is defined as less than 5% (in volume) porous.

Several methods are used to introduce macropores into a bioceramic. Holes can be drilled into the fired body, but a more appropriate technique is based on mixing the starting powder with appropriately sized organic powders, threads, or sponges. These organic additives burn out and leave behind their replica as voids when the green body is heated up to sintering temperatures. Alternatively, a powder may turned into a slurry with hydrogen peroxide, dissociation of which at higher temperatures also leaves pores in the green body.[41,42]

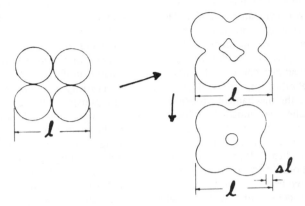

Fig. 1. Schematic of the sintering process.

Both the material tensile and compressive strength depend on the material volume portion occupied by interstices. These interstices or porosities are usually classified as comprising either micropores (having a diameter of several microns due to the incomplete sintering of the particles) or macropores (having a diameter of several hundred microns allowing bone ingrowth).

Stability

Although calcium phosphate bioceramics are usually obtained by sintering at high temperatures, sometimes with the exclusion of water vapor, it is the stability at ambient and body temperatures that determines their fate after implantation. Since solid-state reactions hardly ever occur at room temperatures, solid, unstable phases will only react at their surfaces. If the surface continually dissolves, the whole implant may dissolve. If surface reactions lead to the formation of a thin layer of a second stable phase, the virtual absence of solid-state reactions causes the unstable solid to be stabilized. As Driessens[43] showed, there are only two calcium phosphate materials that are stable at room temperature when in contact with aqueous solutions, and it is the pH of the solution that determines which one is stable. At a pH lower than 4.2, the component $CaHPO_4 \cdot 2H_2O$ (dicalcium phosphate) is the most stable, while at higher pH (> 4.2), hydroxylapatite $[Ca_{10}(PO_4)_6(OH)_2]$ is the stable phase (Fig. 2). High-temperature stability of calcium phosphates is best illustrated by the phase diagrams shown in Figs. 2-4. These focus on temperatures at which sintering processes usually take place, 1000 to 1500°C.

Figure 3 shows that when the ambient atmosphere contains no water, various calcium phosphates can be found at high temperatures, such as tetracalcium phosphate ($= C_4P$), α-tricalcium-phosphate (α-C_3P), monetite ($= C_2P$), and mixtures of calcium oxide (CaO) and C_4P. Hydroxylapatite is not stable under these conditions. If the

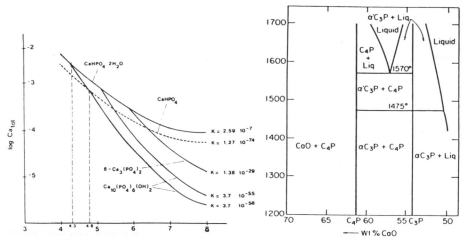

Fig. 2. Solubility of various phases in the system CaO P_2O_5 H_2O as function of pH (horizontal axis).

Fig. 3. Phase diagram of the system CaO P_2O_5 at high temperature (vertical axis °C). No water present.

partial water pressure is increased from 0 to 50 mmHg, Fig. 4, then the situation is quite different: hydroxylapatite can be found (=Ap). If the ratio Ca/P is not exactly equal to 10/6, a wide range of apatite containing mixtures is thermo-dynamically stable, e.g., tetracalcium phosphate, triphosphate and calcium oxide (CaO).

The two phase diagrams, Figs. 4 and 5 stress the importance of temperature, exact Ca/P ratio, and partial pressure of water vapor in the ambient atmosphere in the determination of stable phases (ß-triphosphate turns into α-tri phosphate around 1200°C; the latter phase is considered to be stable in the range 700 to 1200°C).

The importance of partial water pressure is shown more clearly in Fig. 6 an enlarged part of Fig. 4. This diagram shows, that for a Ca/P ratio higher than 10/6, at temperature of 1300°C (10^4/T= 6.4, if T is expressed in °K), the stable phase is C_3P + C_4P if the vapor pressure is 1 mmHg (log P_{H2O} = 0). The stable phases are Ap + C_4P at 10 mmHg. Mixtures of Ap + CaO are stable at pressures of around 100 mmHg. Thus, with a Ca/P ratio exceeding that of apatite by only a few percent, stable phases can vary from C_3P + C_4P(log P_{H2O} = 0), Ap + C_4(log P_{H2O} = 1), and Ap + CaO (log P_{H2O} = 2). Control over temperature, Ca/P ratio and vapor pressure during sintering provides the stability to produce a wide range of well defined calcium phosphate products.

If these conditions are not controlled a less well defined end product may result.[44] Both Ca^{2+} and PO_4^{3-} ions, as well as the OH⁻ group in hydroxylapatite, can be replaced by other ions, several of them present in physiological surroundings, for example fluor, magnesium or carbonate. In synthetic apatite, carbonate can partially substitute for OH⁻,

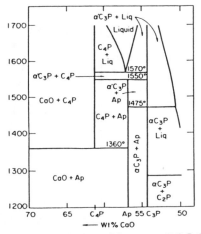

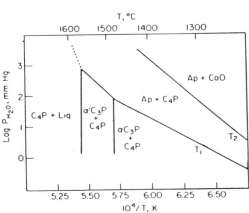

Fig. 4. Phase diagram of the system CaO P₂O₅ at high temperature (vertical axis °C). Water vapor P H₂O=500 mmHg.

Fig 5. Influence of ambient water vapor pressure (vertical axis P H₂O in mmHG) on phase composition.

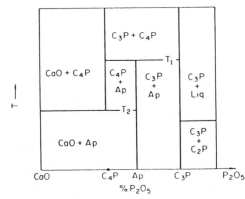

Fig. 6. Enlarged part of Fig. 4. Instead of 1360°C, we use T_2, and instead of 1475°C we use T_1, to indicate that these values hold only for P H₂O = 500 mmHG.

and PO_4^{3-} ions in the crystal lattice, while magnesium, is incorporated only to a very limited extent. Further, these ions decrease the crystallinity of synthetic apatite and promote the formation of amorphous calcium phosphate. Magnesium ions stabilize ß-triphosphate (see Chapter 9 for details).

Biodegradation of Calcium Phosphate Bioceramics

Biodegradation data on different calcium phosphate ceramics are contradicting. Many authors [18,28,45-50] report rather fast degradation of ß-whitlockite, while others [51,53] report a minimal or very slow resorption. Many investigators[18,54-56] found no degradation of HA, where as another described resorption.[57] It is thought that the chemical composition of the ceramic determines whether or not calcium phosphate materials with $1 <$ Ca/P ratio > 2 are degradable[17,50]. This question has been investigated by Klein et al.[35,43,58,59] A well-defined series of hydroxylapatite and ß-whitlockite crystal structures were investigated, with microporosity ranging from 2 to 55% and with macroporosity of either 0 or 30%. These experiments show that the composition, Ca/P ratio, impurities like F⁻ or Mg^{++} and structural relationships (micro-/macrostructure) are important factors associated with biodegradation. Ceramics with a Ca/P ratio 1.67 are more stable. Microporosity plays a more dominant role than the macroporosity. Microporosity determines the geometry of "necks" between sintered particles, while macroporosity determines the amount of necks in contact with the environment. The "neck" formation depends on preparation technique, sintering temperature, pressure applied to compress the powder into a tablet before sintering. Varying the surface chemistry by addition of F⁻ or Mg^{++} ions decreased the biodegradation rate.

LeGeros et al.[8] studied the influence of the ß-TCP/HA ratios on biodegradation and *in vivo* transformation. The higher the ratio, the greater the extent of dissolution. ß-TCP will dissolve earlier than HA. The processes of dissolution and precipitation appeared to occur simultaneously and correlated with the ratio. The resorption of ß-TCP/HA appears to occur simultaneously with bone formation. For the same material (HA) and the same pH, dissolution rates vary considerably in different buffers.

Table 1 shows that the dissolution rate of HA at pH 7.2 varies from 97.4 when buffered in citrate, to 44.3 in Gomori's buffer. Without buffers, dissolution studies yield quite different results. In addition to Klein et al.,[35] Bauer et al.,[60] showed that, in deionized water, the pH may change from 8.6, when tri-phosphate is incubated, to 12.3 for tetracalcium phosphate. Solubilities decrease rapidly with increasing pH; one expects to find a very low apparent dissolution rate for tetracalcium phosphate under such conditions. This explains the finding by Adam et al.[61] that dissolution of tetracalcium phosphate is lower than that of hydroxylapatite. When the same buffer is used, hydroxylapatite has a lower solubility rate than both tri- phosphate and tetracalcium phosphate; however, it is uncertain that this is relevant *in vivo* where the fluids are probably saturated with respect to calcium and phosphate.

PLASMA SPRAYED COATINGS OF CALCIUM PHOSPHATE

Plasma Spray-Technique

The plasma spray technique is described in Chapter 12.

The very high temperature of plasma spraying can lead to dehydroxylation or phase transformations in the coating. The transition temperature of tricalcium phosphate is around 1200°C, hydroxylapatite decomposes at 1300°C, tetracalcium phosphate and fluorapatite are stable above 1300°C. Hence the capacity of the plasma flame to induce phase transitions decreases in this order. However hydroxylapatite show dehydroxylation, ß-whitlockite changes to α-whitlockite and tetracalcium phosphate will turn into a hydroxylapatite crystalline structure. Plasma spraying decreases crystallinity.

Table 1. Concentration of Ca and P (ppm) After 1 Week Incubation in Various Buffers.

Material	Citrate		Gomori's		Deionized H_2O	
	Ca	P	Ca	P	Ca	P
TCP[a]	85.0	45.9	48.9	19.6	4.6	1.7
HA[a]	97.4	43.8	44.3	17.6	5.1	2.2
Tetra[a]	70.3	47.9	77.6	18.6	9.7	2.2
TCP[b]	153	82.5	17.1	8.0	3.3	1.5
HA[b]	44.0	19.8	10.8	4.3	44.4	2.3
Tetra[b]	351	127	94.4	9.0	8.8	0.2

[a]30 mg sintered powder particles/30 ml buffer
[b]coated plug(=15 mg coating)/30 ml buffer

Ideally, only a thin outer layer of each powder particle should become molten and in the plastic state in which phase transition is unavoidable. This plastic state is necessary to ensure dense and adhesive coatings, but should comprise a negligible volume fraction of the calcium phosphate particles. By choosing an optimum relation between particle size type of gas (heat content of a plasma, and thus ability to increase the temperature of a particle, depends strongly on the gas used), speed of the plasma (the longer a particle resides in a plasma, the higher its temperature), and cooling process of the coated surface, one obtains coatings with the desired calcium phosphate(s) and crystallinity.

Figure 7 shows the influence of different plasma gases on the coating crystallinity. Plasma gas argon mixed with hydrogen gives a higher degree of crystallinity but without hydrogen the powder particles cannot enter the gas. This is because the high velocity and viscosity of the argon gas cause the particles to bounce back from the flame, instead of entering.

Figure 8 shows that the plasma gas nitrogen gives a thicker coating layer compared to argon. An important criterion is the calcium oxide formation which can react with water and destroy the coating layer.

Figure 9 shows that using a higher hydrogen content results in more calcium oxide than use of only nitrogen. Hydrogen leads to more enthalphy in the flame and a lower flame velocity, thus the particles undergo more melting and decomposition.

Physico-Chemical Properties

An important aspect of plasma sprayed ceramic coatings is their thickness: mismatch in thermal properties coupled with the fast coating rate of the coating material during the plasma spray process gives rise to stresses in the coating and substrates; these stresses increase with the thickness of the coating. The compressive stress at the coating-

Plasmaspraying of HA
difference in plasmagas

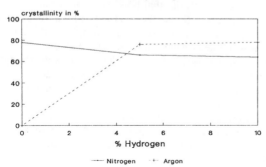

Fig. 7. The influence of different plasma gases on the crystallinity of the coating.

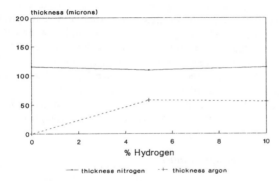

Fig. 8. The influence of different plasma gases on the thickness of the coating.

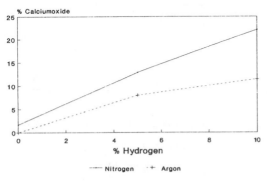

Fig. 9. The influence of different plasma gases on the % calcium oxide in the coating.

pp2/ 10gr/min

substrate interface weakens the bond strength. Therefore, the thinner the coating, the higher its bond strength.

It has been shown, by finite element analysis, that bond strength usually suffers from significant stress concentration effects. Since these effects are a function of variation in elastic modulus, the testing method, shape of testing sample, and thickness of the brittle layer, bond strength measurements have only a relative value.

For plasma-sprayed coatings, two methods are used to determine bond strength:

1. Scratch testing, in which a sharp needle under a given weight necessary to reach the underlying metallic surface is an indication of the comparative "bond strength".

2. Tensile and shear testing, in which, by means of a glue, loads normal and parallel, respectively, are applied to the ceramic coating, the loads expressed in force per area being called the tensile and shear strength, respectively.

HA coatings have only been subjected to the second test: tensile and shear strength determination. We have coated rods with a layer of 50 μm. For tensile strength measurement we coated the cross section (having an area of 0.95 cm^2), and for shear strength, the cylindrical surface. Araldite® AV 118 glue was used to apply a force. For tensile strength the glue was put on the coated cross section and then subjected to a normal force (in tension). The values were in the order of 70 MPa, and the failure mode occurred at the glue-HA interface. The measured values therefore represent more the adhesive strength in shear, the coated rod was plotted into the glue and then pulled out of it. We found values of around 30 MPa and, due to the fact that shearing results in wear attrition, it was not possible to determine the mode of failure, although usually more than half of the coating still seemed to be present.

Table 2. Tensile Strength of Different Glues.

	Strength	Failure mode	n
Araldite AV 118	60.8 MPa	100% Glue/HA	5
3M	38.3 MPa	100% Glue/HA	5
Lee Insta-Bond	5.4 MPa	100% HA/Ti	3
Concise	11.7 MPa	100% HA/Ti	3

Table 3. Tensile Strength of Different Coating Thickness Glued with Araldite AV 118.

	Strength	Failure mode	n
40 μm	66.8 MPa	100% Glue/HA	5
80 μm	60.7 MPa	100% Glue/HA	5
120 μm	45.3 MPa	100% HA/Ti	2

Our values are about twice as high as those reported by Kay et al.[50] Since we found that our strength actually represented the adhesive strength of the glue, it might well be that Kay et al. used another type of glue, and hence also measured an "apparent" bond strength instead of the true one, where the failure mode should have been at the right interface. The roughness of the coating is important if the main failure mode is at the glue-ceramic interface: increasing the roughness increases the contact area, and hence, the "apparent" bond strength.

Bone Attachment of Plasma Sprayed Calcium Phosphate Coatings (Push-Out Test)

The thickness of the coating is a primary influence on the strength of the coated device. A thick coating will have mechanical properties that are somewhat similar to that of the bulk material. If the coating is too thick, it spalls from the metal.

Mechanical and histological evaluations of uncoated and plasma-sprayed hydroxylapatite were performed. The attachment characteristics of interface shear strength were determined by mechanical push-out testing. The results showed variability, probably due to different implantation procedures (cortical or cancellous bone, loaded or unloaded), different coating techniques (grit-blasted titanium surface, particle size), and/or treatment of the tissue-implant sample before the push-out test (formaldehyde fixation or fresh wet bone). However, with identical specimen preparation, there still exists a large scatter in data (Table 4).

Dhert et al.[36] has studied, with finite element analysis, the push out test and the effect of varying conditions of the push-out model on the interfacial shear strength. Distance between implant and support jig is very critical for the occurrence of peak stresses in the interface. The clearance of the clearance of the hole in the support jig should be at least 0.7 mm, to give the most uniform distribution of stresses along the interface. Many researchers have reported that clearance should be minimal however this will lead to unrealistically low apparent push-out strength values. Variations of the

Table 4. Mean Push-Out Strength (MPa) + SD of Various Studies on HA Coatings.

Week	Boone	Cook	Dhert	Geesink	Geesink	Klein	Poser	Verheyen
3								$2.9 \div 0.3$
4	$5.9 \div 0.8$							
5		$7.0 \div 3.2$						
6				$49.1 \div 2.3$				$5.8 \div 3.3$
10		$7.3 \div 2.2$						
12	$8.2 \div 2.8$		$13.3 \div 2.1$	$54.8 \div 2.6$	$34.5 \div 6.5$	$8.2 \div 1.1$	$9.8 \div 3.8$	
25			$17.3 \div 6.1$					$3.3 \div 0.4$
104					29.7			
108						44.0		

Young's modulus of the implants resulted in a wide range of interface shear stresses. A low Young's modulus results in considerably higher stresses at the medial site of the cortex, from which the load is applied on the implant. A high modulus results in slightly higher stresses at the lateral site of the cortex where the jig edge supports the bone. In this situation the interface stress distribution is much more uniform compared to the situation with a low Young's modulus implant. Only materials with similar Young's modulus can be compared. Variation of the cortical thickness showed a reciprocal relationship between cortical thickness and interface shear strength. The diameter of the implant hardly affected the interface stress distribution.

Biodegradation of Calcium Phosphate Coatings

Variations of the material properties of calcium phosphate coatings affect the bone bonding mechanism and the rate of bone formation. Local supersaturation in the constituent ions of the bone mineral phase, arising from enhanced solid-solution exchange at the coating surface, may be a cause for the bone growth enhancement.

Variation in Ca/P ratios (2.0 - 1.5), i.e., tetracalcium phosphate, hydroxylapatite and α/β-TCP lead to differences in degradability and differences in bone bonding. Hydroxylapatite (Ca/P ratio 1.67) and tetracalcium phosphate (Ca/P ratio 2.0) give a strong, intimate bone contact. α-TCP and uncoated titanium evoked remodeling of bone and less bone contact.

Composition changes by addition of fluor to apatite or magnesium to ß-whitlockite also influence the stability of the coating. Hydroxylapatite showed lower push out data (not significant) and higher degradation of the coating than fluorapatite. Magnesium whitlockite showed significant lower bone contact and push-out data and higher degradation. Parallel *in vitro* solubility studies showed that the solubility of tetracalcium phosphate and α-TCP is much higher than that of hydroxylapatite. Fluorapatite showed lower solubility than hydroxylapatite while magnesium whitlockite shows a very high solubility. Understanding the causes of enhancement of bone tissue ingrowth bonding requires knowledge of the characteristics of the calcium phosphate coating and the coating process.[20,36,37]

The particle size distribution which should be not too broad otherwise the particles do not melt uniformly, some will be overheated or vaporized. Differences in particle size of the sintered powder used for plasma spraying gives differences in crystalline/amorphous structure of the coating. A particle size distribution of 1-45 μm gives a coating with an almost totally amorphous structure due to the complete melting of the very small particles. However, fluorapatite coating with this particle size distribution (1-45 μm) gives a crystalline coating comparable with a distribution of 1-125 μm. The powder port used may influence the crystallinity and stability of the coating. A heat treatment after the plasma spraying can have an effect on the coating properties, because recrystallization occurs.

In vitro and *in vivo* experiments were done to study effect of factors which may influence coating stability, solubility, crystallinity and bone tissue response. Factors involved with the plasma spray coating procedure, such as starting powder compound

(FA, HA, Mg-TCP or Tetracalcium phosphate), powder particle distribution (1-45 μm, 1-125 μm) of hydroxylapatite versus fluorapatite coatings [HA_{45}, HA_{125}, FA_{45} and FA_{125}], the plasma spray powder port factor (2 or 6) of hydroxylapatite coatings [$HA_{45/2}$, $HA_{45/6}$, $HA_{125/2}$ and $HA_{125/6}$] and the effect of post heat treatment of 1 hour at 600°C were examined. In the *in vitro* study the materials were compared using solubility tests, X-ray diffractometry and scanning microscopy. The different coatings were incubated in buffer solutions for 3 months and at different time intervals the Ca and P concentrations were measured. Before and after incubation the non heat-treated and heat treated coatings were examined by X-ray and SEM. Solubility (Table 5) and crystallinity (Fig. 10) depended on Ca/P ratio, particle distribution and post heat treatment.

XRD patterns of FA-45 and FA-125 plasma sprayed coatings showed a well crystallized material. XRD patterns of HA-45 coatings compared with HA-125 coatings showed almost entirely amorphous phases of HA-45 and a more crystalline HA-125 coating.

After heat treatment of the different plasma sprayed coatings, FA coatings showed no significant change in crystallinity, HA-45 and HA-125 showed a higher crystallinity. Difference in powder port factor did not effect the degree of crystallinity. XRD patterns of fluorapatite coatings [FA-45, FA-125] demonstrated both in non-heat treated and heat treated coatings a crystalline structure. The crystallinity of fluorapatite was hardly altered by factors as temperature (1hr, 600°C) or particle size distribution (1-45 μm, 1-125 μm). However, during plasma spraying HA decomposes. XRD patterns

Table 5. Solubility of Different Calcium Phosphate Coatings.

	Ca^{++} citrate	Ca^{++} gomori	P^- citrate	P^- gomori	
FA45	325	150	153	56	No
	150	80	83	35	H.T.
FA125	325	120	150	47	No
	170	70	82	31	H.T
HA45/2	400	330	184	127	No
	360	200	184	80	H.T
HA45/6	400	240	175	99	No
	400	225	195	93	H.T
Mg-TCP	500	800	152	275	No
	500	800	210	300	H.T
TETRA	400	450	118	75	No
	220	150	76	27	H.T

No=no heat treatment, H.T. =heat treatment

CRYSTALLINITY
Peak Height

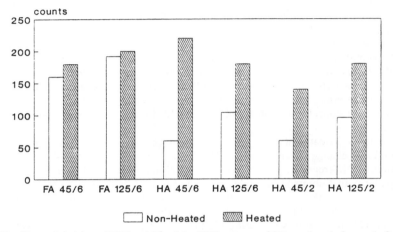

Fig. 10. The peak height at 25.7° and 2Θ of the XRD patterns of HA coatings before and after a heat treatment are shown. HA$_{125}$ coatings are more crystalline than HA$_{45}$. Heat treatment increased the crystallinity for all HA coatings.

of HA coatings showed less crystallinity, especially HA-45 showed an amorphous structure probably because of the complete melting of the very small particles. HA-coatings became more crystalline after a heat treatment, probably because of recrystallization (Fig. 11). The post heat treatment influenced both crystallinity and degree of solubility. The plasma spray powder port factor examined for HA coatings was not very significant. Incubation of different calcium phosphate coatings showed precipitation of Ca^{++} and P$^-$ions at the surfaces of nearly all non-heat treated coatings (HA, Mg-TCP,Tetra) except FA coatings. In all heat treated coatings no precipitation was observed (Figs. 12 A,B).

In *in vivo* experiments the same factors were studied in relation to coating stability and bone tissue response. Different calcium phosphate plasma sprayed coatings were implanted in goat femora for 3 months. After the implantation period the samples were prepared for histology and histometry.

An enhancement of the coating stability of HA-45 and HA-125 because of a heat treatment after the plasma spray procedure was demonstrated. Probably the increased crystallinity, caused by the recrystallization during the heat treatment, promoted the stability. The enhanced stability of HA coatings leads to increased direct bone contact. It was found that α-TCP coatings evoked more remodelling activity than the more stable HA coatings during the first weeks of implantation, after a longer period the difference

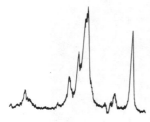

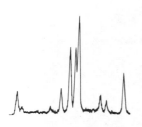

Fig. 11. XRD patterns of; (above) non-heat treated HA_{45} coating after incubation in Gomori's buffer, (middle) heat treated HA_{45} coating, (below) non-heat treated HA_{45}.

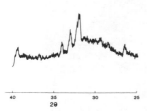

in bone response is less. Also, this study demonstrated that the stable FA and heat treated HA coatings were incorporated into the bone with the formation of lamellar bone next to the implants and with little remodelling, while the more unstable non-heat treated HA coatings showed remarkable remodelling and cellular response. In the first months of implantation less stable coatings can induce high bone remodelling activity with fewer bone contacts (Figs. 13 and 14).

There were few effects of powder port (2 or 6) on HA coatings *in vivo* However, a heat treatment enhanced the coating stability of both HA coatings prepared with powder port 2 and 6. Comparing the XRD pattern data of HA coatings with coating stability

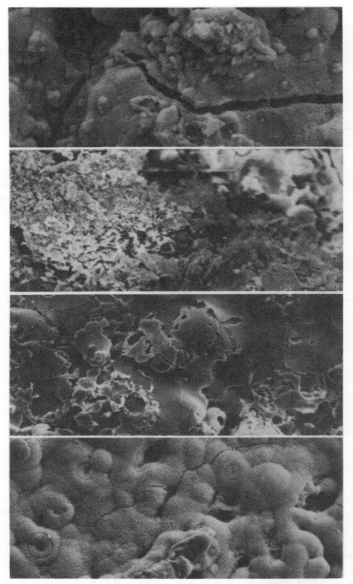

Fig. 12A. SEM photographs of different calcium phosphate coatings without a heat treatment and after 3 months incubation in Gomori's buffer. From top to bottom; TETRA, Mg-TCP, FA_{45} and HA_{45}. All coatings except FA showed precipitation of calcium and phosphate.

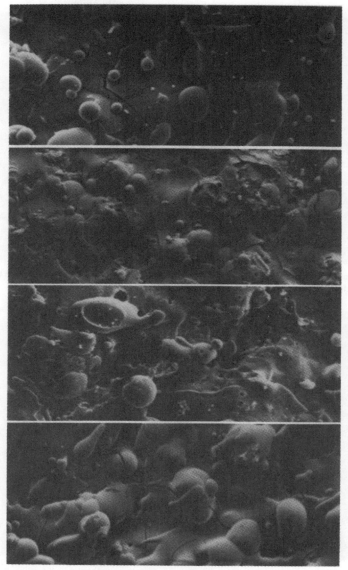

Fig. 12B. SEM photographs of different calcium phosphate coatings with a heat treatment and after 3 months incubation in Gomori's buffer. From top to bottom; TETRA, Mg-TCP, FA_{45} and HA_{45}. No coating showed precipitation of calcium and phosphate.

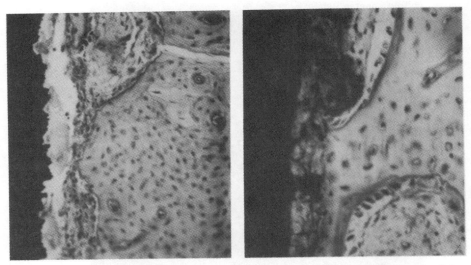

Fig. 13. HA$_{45}$ coating implanted in goat femur for 3 months, signs of degradation give rise to high cellular response (left). Heat treated HA$_{45}$ coatings were more stable (right).

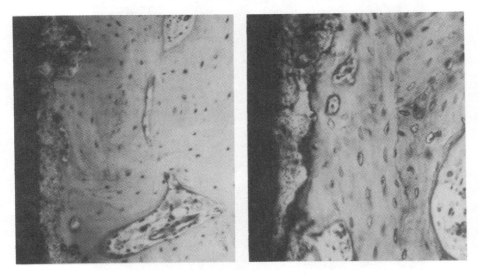

Fig. 14. Both FA$_{45}$ coating (left) and FA$_{45}$ heat treated (right) showed a highly stable, less cellular response and more direct bone contact.

data it seemed that *in vitro* heat treatment and particle size were factors that influenced the crystallinity degree, while *in vivo* only the heat treatment factor is clearly detectable. Comparing data of an *in vitro* solubility study of HA coatings it seemed that also the heat treatment is the most significant factor influencing the stability/solubility of the coating. FA-45 and FA-125 non-heat-treated coatings showed a significantly higher stability than the HA coatings, which were not further improved by a heat treatment. The degree of crystallinity of FA coatings was also not altered by a heat treatment. FA coatings are very stable and not subjected to factors such as particle size distribution or temperature. Comparing the FA data of *in vitro* study with the *in vivo* study it appeared that after a heat treatment the solubility was decreased significantly, while crystallinity, coating stability and bone tissue response for non heat treated and heat treated FA coatings remained similar.

REFERENCES

1. J. D. Andrade, "Interfacial phenomena and biomaterials," *Med. Instrum.* **7** (1973) 110.
2. A. E. Clarke, L. L. Hench, and H. A. Paschall "The Influence of Surface Chemistry on Implant Interface Histology. A Theoretical Basis for Implant Materials Selection," *J. Biomed. Mater. Res.* **10** (1976) 161.
3. P. Ducheyne and J. Lemons in *Bioceramics, Materials Characterization Versus In Vivo Behavior* (NY. Acad. Sci., 1987).
4. A. Engstrom, in *The Biochemistry and Physiology of Bone*, ed. G. H. Bourne (Academic Press, New York, 1972) pp. 237.
5. L. L. Meyer and B. O. Fowler, *Inorg. Chem.* **21** (1982) 3029.
6. D. McConnel in *Applied Mineralogy*, Vol 5 (Springer-Verlag, Wien, 1973).
7. H. Alexander, J. R. Parsons, J. L. Ricci, P. K. Bajpai, A. B. Weiss, "Calcium-Based Ceramics and Composites in Bone Reconstruction" in *CRC Critical Reviews in Biocompatibility*, ed. D. F. Williams (CRC Press, Boca Raton, Florida, 1987) pp. 43-77.
8. R. Z. LeGeros, J. R. Parsons, G. Daculsi, F. Driessens, D. Lee, S. T. Liu, S. Metsger, D. Peterson and R. M. Walke "Significance of the Porosity and Physical Chemistry of Calcium Phosphate Ceramics: Biodegradation-Bioresorption," *Ann. N.Y. Acad. Sci.* **523** (1988) 286-271.
9. J. F. Osborn and H. Newesely "Dynamic Aspects of the Implant Bone-Interface" in *Dental Implant*, ed. G. Heimke and Carl Hanser (Verlag, Munchen, 1980) p. 111.
10. H. U. Cameron, I. MacNab and R. M. Pilliar "Evaluation of Biodegradable Ceramic" *J. Biomed. Mater. Res.* **11** (1977) 179.
11. K. Kato, H. Aoki, I. Tabota and M. Ogiso "Biocompatibility of Apatite Ceramics in Mandibles" *Biomater. Med. Devices Artif. Organs.* **7(2)** (1979) 291.
12. M. Jarcho, J. F. Kay, K. I. Gumaer, R. H. Doremus and H. P. Drobeck "Tissue Cellular and Subcellular Events at a Bone Ceramic Hydroxylapatite Interface" *J. Bioeng.* **1** (1979) 79.

13. M. Jarcho "Calcium Phosphate Ceramics as Hard Tissue Prosthetics," *Clin. Orthop.* **157** (1981) 259-278.

14. B. M. Tracy, R. H. Doremus, "Direct Electron Microscopy Studies of the Bone-Hydroxylapatite Interface," *J. Biomed. Mater. Res.* **18** (1984) 719-726.

15. J. Ricci, H. Alexander, R. H. Parsons, R. Salbury, P. K. Bajpai "Partially Resorbable Hydroxylapatite-Based Cement for Repair of Bone Defects," in *Biomedical Engineering: Recent Developments,* ed. S. Saha (Pergamon Press, New York, 1986) pp. 469-474.

16. J. R. Parsons, J. L. Ricci, H. Alexander, P. K. Bajpai, "Osteoconductive Composite Grouts for Orthopaedic Use," *N. Y. Acad. Sci.* **523** (1988) 190-207.

17. K. Koster, H. Heide and R. Konig, "Resorbable Calcium Phosphate Ceramics Under Load," *Langenbecks Arch. Chir.* **343** (1977) 174.

18. C.P.A.T. Klein, A. A. Driessen, K. de Groot and A. van den Hooff "Biodegradation Behavior of Various Calcium Phosphate Materials in Bone Tissue," *J. Biomed. Mater. Res.* **17** (1983) 769-784.

19. P. S. Eggli, V. Muller and R. K. Schenk, "Porous Hydroxyapatite and Knead Phosphate Cylinders with Two Different Pore Size Ranges Implanted in the Cancellous Bone in Rabbits," *Clin. Orthop.* **232** (1988) 127-137.

20. C.P.A.T. Klein, P. Patka, H.B.M. van der Lubbe, J.G.C. Wolke, and K. de Groot, "Plasma-Sprayed Coatings of Tetracalcium Phosphate, Hydroxylapatite, and α-KNIGHT on Titanium Alloy: An Interface Study," *J. Biomed. Mater. Res.* **25** (1991) 53-65.

21. S. F. Hulbert, L. L. Hench, D. Forbers and L. S. Bowman, "History of Bioceramics," *Ceramics Internat.* **8** (1982) 131-140.

22. J. W. Boretos, "Advances in Bioceramics," *Advanced Ceramic Materials* **2** (1987) 15-30.

23. G. M. Bowers, J. W. Vargo, B. Levy, J. R. Emerson and J. J. Berquist, "Histologic Observations Following the Placement of Knead Phosphate Implants in Human Intrabony Defects." *J. Periodontol* **57** (1986) 286-287.

24. C. A. Van Blitterswijk, J. J. Grote, W. Kuypers, W.T h. Daems and K. de Groot, "Macropore Tissue Ingrowth: A Quantitative Study on Hydroxylapatite Ceramic,: *Biomaterials* **7** (1986) 137.

25. C.P.A.T. Klein, P. Patka and W. den Hollander, "Macroporous Calcium Phosphate Bioceramics in Dogs Femora; a Histological Study of Interface and Biodegradation," *Biomaterials* (1989).

26. B. J. Fellows, G. Bauer, G. Gottschalk, H. J. Oel and K. Donath, "Investigation into the Nature of Bioactivity in the Calcium Phosphate System," in *Ceramics in Clinical Applications* (Materials Science Monograph No.39) ed. P. Vincenzini (Elsevier, New York, 1987) p. 283.

27. C.P.A.T. Klein and K. de Groot, "Implant Systems Based on Bioactive Ceramics," in *Osseo-Integrated Implants Vol II, Implants in Oral and ENT Surgery*, ed. G. Heimke (CRC Press, Boca Raton, Florida, 1990) pp. 193-208.

28.   R.G.T. Geesink, K. de Groot and C.P.A.T. Klein, "Chemical Implant Fixation Using Hydroxyapatite Coatings," *Clin. Orthop.* **225** (1987) 147-170.

29.   S. D. Cook, K. Thomas, J. F. Kay and M. Jarcho, "Hydroxylapatite Coated Titanium for Orthopaedic Implant Applications," *Clin. Orthop. Rel. Res.* **232** (1988) 225-243.

30.   W. R. Lacefield, "Hydroxyapatite Coatings," in *Bioceramics; Material Characteristics Versus In Vivo Behavior.* eds. P. Ducheyne and J. E. Lemons. (Ann. NY. Acad. Sci., 1988), Vol. 523, pp. 72-80.

31.   P. Ducheyne, J. F. McGuckin, "Composite Bioactive Ceramic-Metal Materials," in *Handbook of Bioactive Ceramics,* eds. T. Yamamuro, L. L. Hench and J. Wilson (CRC Press, Boca Raton, Florida, 1990) pp. 175-186.

32   K. de Groot, C.P.A.T. Klein, J.G.C. Wolke and J.M.A. de Blieck-Hogervorst, "Plasma-Sprayed Coatings of Calcium Phosphate," in *Handbook of Bioactive Ceramics. Vol II.* eds. T. Yamamuro, L. L. Hench and J. Wilson (CRC Press Boca Raton, Florida, 1990) 133-142.

33.   C.P.A.T. Klein, J.M.A. de Blieck-Hogervorst, J.G.C. Wolke and K. de Groot, "A Study of Solubility and Surface Features of Different Calcium Phosphate Coatings *In Vitro* and *In Vivo*: A Pilot Study," in *Ceramics in Substitutive and Reconstructive Surgery,* ed. P. Vincenzini (1991) pp. 363-374.

34   C.P.A.T. Klein, J.M.A. de Blieck-Hogervorst, J.G.C. Wolke and K. de Groot, "Solubility of Hydroxyapatite, Tricalcium Phosphate and Tetracalcium Phosphate Coatings on Titanium *In Vitro*," in *Advances in Biomaterials, Clinical Implant Materials,* ed. G. Heimke **9** (1990) 277-282.

35.   C.P.A.T. Klein, J.M.A. de Blieck-Hogervorst, J.G.C. Wolke and K. de Groot, "Studies of the Solubility of Different Calcium Phosphate Ceramic Particles *In Vitro*," *Biomaterials* (1990).

36.   W.J.A. Dhert, C.P.A.T. Klein, J.G.C. Wolke, E.A. van der Velde, K. de Groot. and P. M. Rozing, "A Mechanical Investigation of Fluorapatite, Magnesium-Whitlockite and Hydroxylapatite Plasma Sprayed Coatings in Goats," *J. Biomed. Mater. Res.* in press 1991.

37.   W.J.A. Dhert, C.P.A.T. Klein, J.G.C. Wolke, E.A. van der Velde, K. de Groot, and P. M. Rozing, "Fluorapatite, Magnesium-Whitlockite, and Hydroxyapatite Coated Titanium Plugs: Mechanical Bonding and the Effect of Different Implantation Sites," in *Ceramics in Substitutive and Reconstructive Surgery,* ed. P. Vincenzini (1991) pp. 385-394.

38.   S. D. Davis, D. F. Gibbons, R. L. Martin, S. R. Levitt, I. Smith and R. V. Harrington, "Biocompatibility of Ceramic Implants in Soft Tissue," *J. Biomed. Mater. Res* **6** (1972) 425-449.

39.   L. Heling, R. Heindel and B. Merin, "Calcium Fluorapatite; A New Material for Bone Implants," *J. Oral Implantol.* **9** (1981) 548-555.

40.   C.P.A.T. Klein, J.G.C. Wolke, J.M.A. de Blieck Hogervorst and K. de Groot, "Heat-Treatment Influences Plasma Sprayed Calcium Phosphate Coatings," (Proc. Fourth World Congress 1992 Berlin).

41. A. A. Driessen, C.P.A.T. Klein and K. de Groot, "Preparation and Some Properties of Sintered Kneed-Whitlockite," *Biomaterials* **3** (1982) 113-116.

42. W. van Raemsdonk, P. Ducheyne and P. de Meester, "Calcium Phosphate Ceramics," in *Metal and Ceramic Biomaterials, Vol.2*, eds. P. Ducheyne and G. W. Hastings (CRC Press, Boca Raton, Florida, 1984) p. 143.

43. F.C.M. Driessens, "Formation and Stability of Calcium Phosphates in Relation to the Phase Composition of the Mineral in Calcified Tissues," in *Bioceramics of Calcium Phosphate*, ed. K. de Groot (CRC Press, Boca Raton, Florida, 1983) p. 1.

44. K. de Groot, C.P.A.T. Klein, J.G.C. Wolke and J.M.A. de Blieck-Hogervorst, "Chemistry of Calcium Phosphate Bioceramics," in *Handbook of Bioactive Ceramics, Vol II*, eds. T. Yamamuro, L. L. Hench and J. Wilson (CRC Press Boca Raton, Florida, 1990) pp. 3-16.

45. S. N. Bhaskar, J. M. Brady, L. Getter, M.P. Growen and T. Driskell, "Biodegradable Ceramic Implants in Bone," *Oral Surg.* **32** (1971) 336.

46. M. P. Levin, L. Getter and D. E. Cutright, "A Comparison of Iliac Marrow and Biodegradable Ceramic in Periodontal Defects," *J. Biomed. Mater. Res.* **9** (1975) 183.

47. P. J. Boyne, T. J. Strunz and J. P. Shaghat, "Three Year Long Term Study of Implants of Hydroxylapatite in Dog Alveolar Bone," *8th ANN. Meet. Soc. Biomaterials* (1982).

48. K. de Groot, "Dental and Other Head and Neck Uses of Calcium Phosphate Bioceramics," in *Bioceramics, Materials Characterization Versus In Vivo Behavior*, eds. P. Ducheyne and J. Lemons (NY. Acad. Sci., 1987) p.272.

49. K. Janke, "Ceramic Implants in Ear, Nose and Throat Surgery," in *Ceramics in Clinical Applications (Materials Science Monogr. 39)* ed. P. Vincenzini (Elsevier, N.Y., 1987) p. 381.

50. K. A. Thomas, J. F. Kay, C. D. Cook and M. Jarcho, "The Effect of Surface Macrostructure and Hydroxylapatite Coating on the Mechanical Strength and Histologic Profiles of Titanium Implant Materials," *J. Biomed. Mater. Res.* **21** (1987) 1395.

51. E. B. Nery, L.K.L. Pflughoeft and G. E. Rooney, "Functional Loading of Bioceramic Augmented Alveolar Ridge: A Pilot Study," *J. Prosthet. Dent.* **43**(3), (1989) 338.

52. P. Duquette, *"An Evaluation of the Osteogenic Potential of a Porous, Single Phase, Knead Phosphate Ceramic,"* (PhD Thesis, Indiana, 1973).

53. Wagner, "Tierexperimentelle Untersuchungen zur Knochen Regeneration Genormter Defekte nach der Implantation einer Trikalziumphosphat Keramik," *Dtsch. Zahnarztl. Z.*, **36** (1981) 82.

54. P. Patka, G. den Otter, K. de Groot, A. A. Driessen, "Reconstruction of Large Bone Defects with Calcium Phosphate Ceramics. An Experimental Study," *Neth. J. Surg.* **37** (1985) 38.

55. C.A. van Blitterswijk, H. K. Koerten and J. J. Grote, "Biological Performance of Whitlockite," in *Advances in Biomaterials, Vol.6,* eds. P. Christel, A. Meunier and A.J.C. Lee (1986), p. 27.

56. H. Aoki, K. Kato and T. Tabata, "Osteocompatibility of Apatite Ceramics in Mandibles," *Rep. Inst. Med. Det. Eng.* **11** (1979) 33.

57. J. R. Strub and T. W. Gaberthuel, "Trikalziumphosphat und desen Biologisch Abbaubare Keramik in der Parodontale Knochen Chirurgie, Eine Literaturubersicht," *Schweiz. Monatsschr. Zahnheilkd* **88** (1978) 798.

58. C.P.A.T. Klein, K. de Groot, A. A. Driessen and H.B.M. van der Lubbe, "Interaction of Biodegradable Kneed-Whitlockite Ceramics with Bone Tissue: *In Vivo* Study," *Biomaterials* **6** (1985) 189.

59. C.P.A.T. Klein, K. de Groot, A. A. Driessen and H.B.M. van der Lubbe, "A Comparative Study of Different Kneed-Whitlockite Ceramics in Rabbit Cortical Bone with Regard to Their Biodegradation Behavior," *Biomaterials* **7** (1986) 144.

60. G. Bauer, B. J. Fellows, H. Gottschalk, J. Dumbach, W. Spitzer and K. Donath "Auswirkung des pH-werts Verschiedener Calciumphosphat Keramiken auf die Biologische Umgebung DVM-AK 'Implantate'," *Bioactive Werkstoffe* **6** (1985) 103.

61. P. Adam, A. O. Nebelung and M. Vogt, "Loslichkeit und Umwandlungen Biokeramischer Schichten," *Z. Zahnarztl. Implantol.* **IV**, 15, 19.

Chapter 12

HYDROXYLAPATITE COATINGS

William R. Lacefield
Department of Biomaterials, University of Alabama at Birmingham
School of Dentistry, Birmingham, Alabama

INTRODUCTION

Ceramic coatings are used on metallic substrates in a variety of applications, including enhancement of corrosion resistance of a metal or the creation of a more refractory surface for high temperature service. In the biomedical field, coatings have been used to modify the surface of implants, and in some cases to create an entirely new surface which gives the implant properties which are quite different from the uncoated device. Because of its similarity to the inorganic component of bone and tooth structure, synthetic hydroxylapatite $[Ca_{10}(PO_4)_6(OH)_2]$ was one of the first materials considered for coating metallic implants. As bulk hydroxylapatite (HA) is brittle and relatively weak when compared with common implant metals and alloys and high strength ceramics like aluminum and zirconium oxides (Chapter 1), the best use of HA in load bearing implant applications is as a coating on one of these stronger implant materials. In spite of the relatively good tissue response to metallic implant surfaces, such as the passive titanium oxide layer present on titanium, with the use of calcium phosphate materials as coatings it is possible to present a surface which is conducive to bone formation.

One reason for the use of HA or a similar calcium phosphate surface is to cause earlier stabilization of the implant in surrounding bone. This is the case for example in a dental implant where the healing time is reduced and the prosthetic attachment can be placed earlier. Another reason to use an HA coating is to extend the functional life of the prosthesis, as in the case of a cementless hip prosthesis stabilized by the HA coating in the surrounding femur without the use of polymethylmethacrylate bone cement. Under the proper conditions a cementless prosthesis should remain functional longer than a cemented device in which stability is threatened by fracture of the bone cement after a limited number of years in service.

The specific function of an implant should be considered when determining what properties are desired in a coating. For example, a carbon coating on a heart valve prosthesis increases wear resistance and provides a non-thrombogenic surface to the device (Chapter 14). Where HA coatings are used currently, these two factors are not of primary importance. Of major importance is that the HA coating enables the implant to present a surface to the surrounding bone or soft tissue which will elicit the optimum tissue response. Not only does the surface need to be of a composition that is conducive to the proper tissue response, but the breakdown of the coating or the release of ions from the coating should not cause an adverse reaction. Dense HA has a very low dissolution rate in neutral and alkaline aqueous solutions. If there is a decrease in density or crystallinity or if non HA material is created during the coating process, the dissolution rate can be considerable greater. This does not rule out the possibility that some coating dissolution can take place without compromising biocompatibility; in fact, the presence of Ca and PO_4 ions in the area around the implant may be more conducive to bone formation than a surface which does not release ions, although this is speculation at present.

Another important factor when considering the desirable properties of an implant coating is that the coating should be strongly bonded to the metal substrate to maintain implant integrity as well to facilitate proper transmission of load from the implant to the surrounding bone. An HA coating which separates from the implant *in vivo* would provide no advantage over an uncoated implant and may be less desirable than no coating at all. In a worse case condition, a weakly bonded HA coating may separate from the implant and fragments of the coating would be in close proximity to the bare metal surface. Any movement of the loose ceramic coating fragments on the substrate surface could result in disruption of the passive oxide layer on metal surfaces, such as TiO_2 on titanium, and Cr_2O_3 on cobalt-chromium alloys. Although reformation of the oxide layer occurs quite rapidly, a momentary increase in metallic ion release will take place. Also, the oxide which forms under *in vivo* conditions existing at the time of reformation may not be as passive or protective as was the previous oxide layer.

PROCESSING

Industrial and laboratory techniques used for coating HA and other ceramic materials onto metallic substrates include plasma spraying, electrophoretic deposition, sputtering and hot isostatic pressing. The plasma spray method will be discussed in detail, as it is currently the only widely used method for coating commercially available implant devices with HA. A number of alternative ways of coating implants with HA are briefly reviewed, as the use of some of these provide coatings which are essentially 100% crystalline HA or are potentially advantageous in particular applications such as for coating porous surfaced implants.

The first consideration in evaluating methods for producing HA coatings is whether the composition and properties of the starting material are altered so that the *in vivo* performance of the coating is compromised. This alteration can be either contamination of the coating by foreign materials, or changes in the basic structure and chemistry of the starting material due to exposure to high temperatures during processing. One source of contamination of HA coatings can occur due to the breakdown of the equipment used, such as the copper anode nozzle in the plasma spraying process. Other deposition techniques have different sources of contamination, such as the electrolyte solution used in the electrophoretic deposition method.

Another potential problem area, in addition to the alteration of the HA or other coating material during deposition, is that changes in the substrate itself may occur if it is subjected to high temperatures for an extended period. This may occur, either during deposition of the coating, or during a subsequent heat treatment which may be used to increase density or crystallinity of the coating. It is possible that during high temperature heat treatment or sintering any of the available metallic substrate materials can be adversely affected by excessive time and temperature combinations. A decrease in mechanical properties of metallic substrates may result from a number of microstructural changes, including grain growth in wrought alloys of titanium and stainless steel, changes in the α - ß structure of titanium alloy, and embrittlement of cast cobalt-chromium alloys by carbide precipitation at the grain boundaries. These changes are not likely to occur with ceramic substrate materials, such as aluminum oxide.

Plasma Spraying

Plasma spraying, the most common means of applying HA coatings to implant devices, employs a plasma, or ionized gas, partially to melt and carry the ceramic particulate onto the surface of the substrate. In flame spraying, another thermal spraying technique, the carrier gasses are not ionized and the temperatures generated are considerably lower than in plasma spraying.

A schematic of a typical plasma spraying process is shown in Fig. 1. The carrier gas is usually argon, which is ionized as it passes within the high temperature discharge zone as the current arcs across the gap between the anode and the cathode. The nozzle of the plasma gun is kept from melting by water cooling, as temperatures developed in the plasma may exceed 10,000°C. As the ceramic particulate remains in the heated plasma zone for only a fraction of a second, usually only partial melting of the powder takes place. Distance of the substrate from the plasma is one of the critical factors controlling the degree of melting of the particle and the ability of the particulate to flow into a dense coating. One advantage of the plasma spraying process is that during the coating process the substrate remains at a relatively low temperature (generally less than

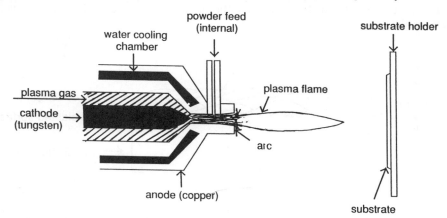

Fig. 1. Typical plasma spraying operation with powder fed into the plasma stream internally.

300°C) so the mechanical properties of the metallic implant materials are not compromised.

Other factors which influence the degree of melting of the particulate during plasma spraying include variables which control the temperature of the plasma, such as current, anode-cathode gap distance, and gas mixture. The carrier gas may be pure argon, or a hotter plasma is produced by small additions of hydrogen or other gases. A gas composition of 90% Ar, 10% H gives a significantly hotter plasma than 100% Ar, other conditions remaining constant.

Another factor influencing the degree of melting of the particulate is the position at which the material enters the plasma stream. If the material enters within the nozzle near the start of the plasma zone, it is known as an internal feed system. If the powder is fed into the plasma outside the nozzle, it is an external feed system. As the material stays longer in the plasma for the internal feed system, lower plasma temperatures can usually be to achieve the same degree of melting as the external feed system.

Plasma spraying of HA usually takes place under normal atmospheric conditions, as opposed to the plasma spraying of some metallic powders during which a vacuum or an inert atmosphere is used to minimize oxidation. The higher heat transfer conditions present during atmospheric plasma spraying results in greater deposition efficiency for HA and other ceramic materials when compared with plasma spraying in a vacuum.

Pure, 100% crystalline HA particles in the 20-40 μm range are typically used as the starting material for plasma spraying. When the softened particulate impinges on the substrate surface, individual particles deform into characteristic shapes called "splats". Figure 2 shows a schematic of the formation of a plasma sprayed HA coating on a metal substrate surface. Three passes of the HA spray are typically made for any given area of the implant surface. Deformation and spreading of individual particles takes place on impact with the substrate, and the final coating thickness typically averages 40 -60 μm. The flow of the softened HA is usually sufficient to form a dense coating with less than 2% residual porosity.

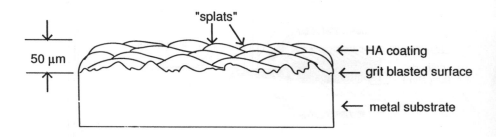

Fig. 2. Formation of plasma sprayed HA coating on surface of grit blasted metal substrate.

Hydroxylapatite coatings produced by the plasma spraying process typically contain considerable amorphous calcium phosphate material and small amounts of crystalline phases other than HA (see Chapter 9). It is possible to increase the crystallinity and in some cases the bond strength of plasma sprayed HA coatings by a post deposition heat treatment. However, this extra step is usually not feasible commercially because of factors, which include the adverse effects of the annealing temperature on the mechanical properties of the substrate metal or alloy, the time and expense of the additional operation and the possibility of contamination of the HA surface.

Other Coating Techniques

Because of the difficulty in producing highly crystalline coatings and reproducibly high bond strength using the plasma spraying technique, other coating methods have been investigated for commercial applications:

1. An electrophoretic deposition process in which HA particulate is suspended in an alcohol or other suitable solution and then subjected to an electric field is a method which deposits the HA on an implant surface with minimal alteration of the starting material. This is a useful technique for placing HA on porous surfaces which can not be completely coated with line-of-sight techniques such as plasma spraying. However, as the HA is only weakly deposited and the individual particles are not bonded together, high temperature sintering is necessary after deposition. Because of low bond strength, electrophoretically deposited coatings are perhaps best used for porous implant designs where the presence of an HA coating is necessary for only a limited time period.

2. Hot isostatic pressing (HIP'ing) can be used to densify HA powder placed on the surface of a metallic implant. In order to achieve a uniform application of pressure on the particulate mass, an encapsulation material (e.g., a noble metal foil) is necessary. The advantage of the method is that lower sintering temperatures (less than 900°C) are required to attain densification and bonding of the HA coating, thus the chances of altering the microstructure or mechanical properties of the metal substrate are reduced.

3. Ion beam sputtering and radio frequency (RF) sputtering are thin film deposition techniques in which a target material is bombarded with an ion beam in a vacuum chamber, and atomic sized fragments of sputtered material form coatings on suitably placed substrates. The typical coatings sputtered from an HA target are amorphous on deposition, as the sputtered components from the HA target (Ca, P, O and H) do not posses enough energy to recombine into HA. A heat treatment on the order of 500°C is usually sufficient to provide enough thermal energy to form a crystalline coating which is predominantly HA. Although sputter deposited coatings generally have better bond strength and mechanical

properties than thick coatings, the durability of thin 1 μm coatings in the body has not yet been demonstrated.

4.   Thermal spray techniques other than plasma spraying are also potential candidates for the production of commercial HA coatings on implants. One approach involves the use of the high velocity oxy-fuel (HVOF) technique. In this method the much higher velocity of the particles causes them more easily to fuse and flow into irregularities on the metal surface in spite of being subjected to much lower temperatures than plasma spraying, thus the initially high crystallinity of the HA can be maintained.

COMPOSITION

The composition of plasma sprayed HA coatings is somewhat different from that of the starting material, as might be expected because of the high degree of melting experienced by the ceramic particulate. There are several analytical techniques which are useful for evaluation and analysis of HA coatings. Observation of the coating surface by scanning electron microscopy (SEM) or light microscopy (LM) can be useful prior to the use of other techniques which determine the composition and physical properties.

Figure 3 is a SEM showing a typical, plasma sprayed surface of HA on a Ti substrate. The structure appears to be highly melted, with little evidence of the crystalline nature of the coating at this magnification, although comparison of x-ray diffraction patterns showed the crystallinity on this particular specimen to be on the order of 50%. Other features of the surface include very low porosity levels, another indication of the high degree of melting experienced by the HA particulate. Also, microcracks can be observed scattered throughout the structure of plasma sprayed HA coatings, perhaps as a result of a very rapid cooling rate.

Fig. 3. SEM micrograph of plasma sprayed HA coating surface showing partially melted structure with some porosity and microcracks (500X).

X-ray diffraction (XRD) has been widely used as a means of determining the composition and structure of plasma sprayed, HA coatings, as well as for estimating the percentage of crystallinity and identifying secondary crystalline phases generated as a result of the high temperature spraying process. Figure 4 shows a diffraction pattern of HA powder prior to plasma spraying. In the plot of the intensity vs 2θ, there are numerous sharp peaks and a low background indicative of highly crystalline HA material. Figure 5 is an XRD pattern of the resulting plasma sprayed coating made using the powder with the diffraction pattern given in Fig. 4. Several changes in the diffraction pattern can be noted, including an increase in the background area under the peaks. Truly amorphous materials exhibit only a "glass bulge" with no sharp peaks discernable. A slight bulge in combination with peaks indicates a material which is partly crystalline and partly amorphous. A second change in the pattern would be the appearance of new peaks as indicated by arrows to the left. These new peaks are indicative of one or more crystalline phases which have been generated by the plasma spraying process, such as α and ß-tricalcium phosphate, calcium oxyphosphate, calcium pyrophosphate, or calcium oxide. Matching of diffraction patterns to determine other crystalline phases has to be made with care because of possible overlapping of some of the major peaks.

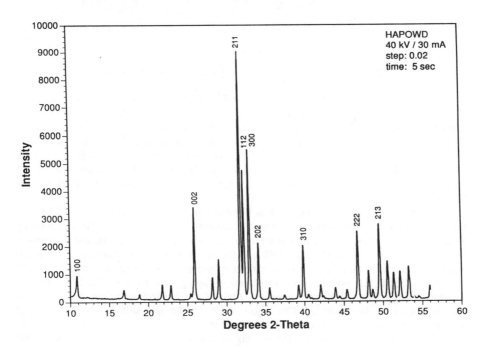

Fig. 4.   X-ray diffraction pattern of HA powder.

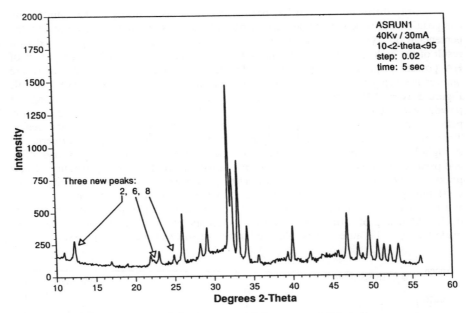

Fig. 5. X-ray diffraction pattern of plasma sprayed HA coating.

Lattice parameter determination by XRD is another way to find out whether changes have taken place in the hydroxylapatite as a result of the coating process or subsequent exposure to a physiologic solution. Hydroxylapatite normally possesses a hexagonal structure with a $P6_3/m$ space group with lattice parameters of a = b = 9.42 A and C = 6.88 A. If the HA is pure and free of vacancies a monoclinic form with the space group $P2_1/b$ can exist. Any changes in the lattice composition due to the substitution of fluorine, carbonates, etc., alters the a and c lattice parameters (see Chapter 9 for details).

Other analytical techniques such as Fourier transform infrared analysis (FTIR) and Raman spectroscopy can provide structural information in addition to that obtained by XRD analysis. FTIR is a technique which is sensitive to the asymmetrical vibrational modes of groups such as PO_4 and OH and therefore is useful in determining changes which take place in the relative concentrations of those groups. Figure 6(A) is the FTIR spectrum of pure HA powder prior to plasma spraying, and Figure 6(B) is the plasma sprayed coating using the same powder. One of the absorption bands for phosphate groups occurs at 1090 cm^{-1}. The larger bulge in the left side of the spectrum is indicative of amorphous material or the presence of moisture. The decrease in the absorption bands at 3572 cm^{-1} for the coating indicates that some of the OH groups were driven off during the high temperature process. The FTIR spectrum can be also be used

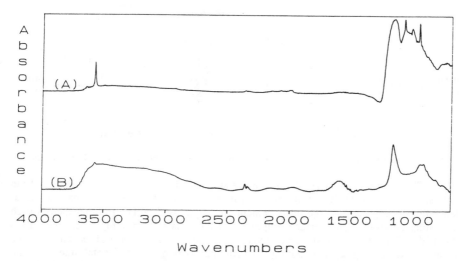

Fig. 6. (A) FTIR spectrum of HA powder. (B) FTIR spectrum of plasma sprayed HA coating.

to determine the presence of carbonate in the coatings and other crystalline phases such as calcium pyrophosphate ($Ca_2P_2O_7$). Raman spectroscopy, a technique which is sensitive to symmetrical vibrational modes, has also been valuable for evaluating certain changes which occur in plasma sprayed HA coatings, such as the loss of OH, which are not readily determined by XRD. Additional information on the characterization of calcium phosphate phases is given in Chapters 8 and 9.

PROPERTIES

Two important properties of plasma sprayed HA coatings are bond strength to the implant and dissolution rate in solution. Bond strength measurements are often used by manufacturers for purposes of quality control, as small changes in the conditions of the plasma spraying process can cause major alteration of the adherence of coating to substrate.

The most commonly used method of determining tensile bond strength involves the coating of a metallic disc followed by testing utilizing a variation of ASTM C633. To test the adherence of HA coatings to an implant material such as titanium, the surface of the substrate disc is first prepared in the same way as the actual surface on an implant, using all cleaning and grit blasting steps. The coated disc is then fixed to a stainless steel rod and bonded to another rod using a heat cured epoxy, as shown schematically in Fig. 7. The two rods are pulled in tension until failure occurs, with exact alignment retained during the test. The tensile strength is then calculated by dividing the load at failure by the cross sectional area of the disc.

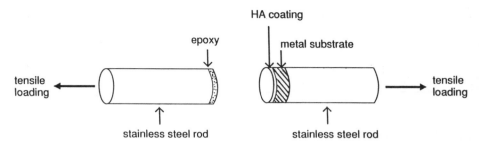

Fig. 7.   Tensile bond testing of HA coated specimens using a modified ASTM C633 procedure.

The tensile bond strengths of plasma sprayed HA coatings on various substrate metals, reported in the literature, range from less than 7 MPa to more than 80 MPa, depending in large part on the measurement technique used, although other factors such as coating conditions and surface roughness of the substrate certainly affect the values obtained.   The use of heat cured epoxy is somewhat controversial because of the possibility of the epoxy penetrating through the coating to the metal interface and artificially strengthening the bond. Plasma sprayed HA is more completely melted, and therefore more dense, than the typical thermal sprayed ceramics such as Al_2O_3 and BeO, so penetration of epoxy is not as much of a problem.   Factors such as compressive stresses generated during cooling of the heat cured epoxy are additional sources of error in bond test measurements of plasma sprayed HA coatings.

The roughness of the substrate is of primary importance in achieving high bond strength of a plasma sprayed HA coating, and coarse aluminum oxide grit is generally used to roughen the implant surface prior to deposition.   The bonding of the plasma sprayed HA coatings to metal appears to be entirely mechanical in nature, as there is no evidence of any degree of chemical bonding in as-deposited coatings.

The dissolution rate of HA coatings is of interest because a rapidly dissolving coating may not remain on the implant for a sufficient time to allow full stabilization in bone or the desired tissue response *in vivo*. A coating which is breaking down quickly *in vivo* will release a higher quantity of Ca and PO_4 ions to the surrounding tissues than will a more stable coating with a lower dissolution rate. The quantity of ions released (typically in the ppm range) can be determined by placing the coated specimen in a simulated physiologic solution and monitoring the Ca ion content for given time periods. One means of measuring ion release is by atomic absorption, in which the quantity of ions in solution is determined by lamps which measure the absorbance of a particular species such as Ca.

Plasma spraying of HA powder produces calcium phosphate coatings with a crystalline structure which is primarily HA usually with small amounts of other crystalline phases. However, there is also a considerable amount of amorphous material generated during the plasma spraying operation, as the crystallinity typically ranges between 40 and 80%. The dissolution rate of plasma sprayed HA coatings is quite variable and typically is considerably higher than fully dense, 100% crystalline HA, either in the human body or in simulated physiologic solutions. This is mainly due to the presence of amorphous material and other more soluble crystalline calcium phosphate and calcium oxide phases present in plasma sprayed HA coatings. High dissolution rates are generally seen as undesirable for HA coatings, although another calcium phosphate material - tricalcium phosphate - is used with the objective of complete dissolution within a predictable time period. Although ongoing development and optimization of the plasma spraying of HA is aimed at higher crystallinity coatings without sacrificing bond strength, there is no evidence that very high crystallinity (approaching 100%) is better *in vivo*. Further *in vitro* and *in vivo* research which correlates the crystallinity, secondary phase composition and ion release rate with tissue response to an HA coated implant is necessary before optimum coating conditions can be established.

SURFACE CHEMISTRY

The surface chemistry of HA coatings is mainly dependent on the coating technique used as well as the composition of the starting material. For plasma sprayed HA coatings, the amorphous content of the coating as well as the presence of more soluble crystalline phases, such as tricalcium phosphate, result in a surface which is actively releasing Ca and PO_4 ions. Techniques such as electrophoretic deposition produce HA coatings which are essentially 100% crystalline and therefore may be less active as far as ion release, depending in part on the final density of the material after sintering. Sputter deposited coatings have different surface chemistries and ion release rates depending on the degree of crystallization and the presence of other elements such as fluorine. More radical surface modifications under investigation include the incorporation of agents which induce the formation of bone, such as bone morphogenic proteins, into HA coatings.

One concern with any coating method is whether the starting material is contaminated during the coating process. Surface analytical techniques such as Auger electron spectroscopy (AES) and x-ray photoelectron spectroscopy (XPS) are useful for determining trace amounts of contaminants introduced during the coating, handling or sterilization operations (see Chapter 18). For plasma sprayed HA, the main contaminant introduced during the process is copper from the anode nozzle or tungsten from the internal cathode. Higher plasma temperatures, used in conjunction with an external feed system, cause more rapid deterioration of the plasma spray gun nozzle, and may result in more Cu contamination of the coating. However, with the internal feed system there is the increased risk of abrasion of the inside of the nozzle by the ceramic particulate carried by the plasma stream. In actual practice, minute quantities of Cu as well as Cu particulate have been identified on the surface of plasma sprayed HA coatings by AES and energy dispersive spectroscopy (EDS) from both internal and external feed systems.

The surface chemistry of the portion of the metal implant covered by HA should not be a factor except in cases where the coating dissolves or separates from the substrate. In cases of delamination or dissolution of the coating, the substrate must have an acceptable surface which was not contaminated or adversely altered during the coating operation. Surface chemistry of the metal substrate may also be altered by the operations used to texture the surface. In the grit blasting of metallic substrates for plasma sprayed HA coatings, the embedding of aluminum oxide particles is unavoidable. These particles are difficult to remove, but may pose no problem because of the relative inertness of aluminum oxide.

Alteration of the implant surface may occur during sterilization with gamma irradiation indicated by a light tan color observed in some HA coatings after treatment. This color change may be a result of displacement of electrons from their proper locations in the electron shells, and the white color characteristic of the coating prior to sterilization can be restored by annealing if esthetics are a concern. Other surface changes which may occur as a result of sterilization procedures include contamination of the coating in a steam autoclave and formation of cytotoxic products on rough HA surfaces sterilized by ethylene oxide. Potential problems of surface contamination with wet heat (autoclave) and chemical sterilization (ethylene oxide) is the reason that HA coated implants are now sterilized by gamma irradiation or dry heat.

TISSUE RESPONSE

The area of primary concern with the use of any implant system is the body's response to the device. Animal studies have been done on HA coated implants to predict the response in humans. Most investigations matched HA coated implants against identical uncoated implants to determine if the coating enhances stabilization of the implant in the newly developing bone. Some studies found no significant difference between HA coated and uncoated implants; in a majority it appears that bone forms sooner around the implant when an HA coating is present. Some long term animal studies found that there was little difference between HA-coated and uncoated implants after 6 months or more *in vivo*. Thus, the main advantage of HA coatings may be in short term stabilization of the implant.

Animal studies often involve histological examination of the tissues surrounding an implant after killing the animals at preset time periods. The presence of bone directly against the implant surface is seen as evidence of proper tissue response as opposed to the presence of fibrous tissue. Several techniques including high voltage SEM have been used to determine if cell attachment is present or if a very thin layer of fibrous tissue exists between the bone and implant surface. An example of bone formed against an HA coated screw implant in a canine study is given in Fig. 8 (implant removed). Whether a chemical bond, between an HA surface (or any other implant surface such as TiO_2) and bone, is formed is still controversial which will require high resolution TEM to resolve.

Push-out tests have been used to measure the ability of the implant to resist forces tending to shear the bond between the implant surface and bone. Some tests have indicated that HA coatings enhance fixation of the implant in the surrounding bone. However, these tests have been performed on smooth sided implants without grooves or other features designed for retention, and the bond strength of the implant to bone is typically quite low (e.g., 7 MPa). It is not practical to rely on the bond of an HA

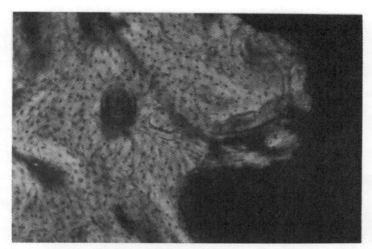

Fig. 8. Bone formed around an HA-coated screw type implant on a dog (original magnification X190).

coating to bone to provide long term stability of an implant in the absence of any other retentive features.

There are reports, from practitioners, of HA coated implants being removed where the coating has fractured and fragments remain bonded to the surrounding bone. This indicates that the bond strength of the HA coating may be stronger to bone, in some cases, than to the metal substrate after a time *in vivo*. Figure 9 shows an HA coated cobalt-chromium subperiosteal dental implant in which part of the coating was stripped off when the device was removed from the surrounding bone. Most of the HA coated implants removed clinically were plasma spray coated >5 years ago, and progress has been made in the quality and bond strength of the coatings since then.

CLINICAL APPLICATIONS

The use of HA coatings is an attempt to enhance the response of the surrounding bone or soft tissue to a metallic or ceramic implant. The widest use of HA coatings is on metallic dental endosseous and subperiosteal implants, and to a lesser extent on metallic orthopaedic devices such as total hips and knees. There are currently very few commercially available HA coated ceramic or polymeric implant devices.

On dental and orthopaedic implants which have plasma sprayed HA coatings, the coating itself covers only a portion of the device. In a one or two stage root form dental implant, only the portion to be placed within the bone is HA coated, with the exception of some devices in which the coating extends onto the neck, i.e., transgingival area of the implant. Subperiosteal dental implants are usually entirely coated except for the posts to which the superstructure (e.g., full denture) will be attached. For orthopaedic

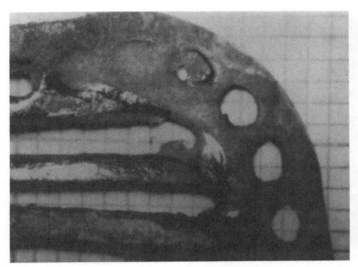

Fig. 9. Removed HA-coated subperiosteal implant showing remaining bone and sections of bare metal where coating has separated.

implants, generally only a portion of the contact surface between the implant and bone is HA coated. In some total hip designs only the proximal end of the shaft portion is HA coated, based on the premise that some mobility of the distal end is desirable.

There are few, if any, long term, controlled clinical studies which use plasma sprayed HA coated implants, which is one reason why some feel that the widespread use of these devices is not yet warranted. There are, however, a number of observations made by clinicians who have observed first hand the results of HA coated implants and feel they are superior to uncoated devices. In the case of dental implants, it is commonly felt that the use of HA coatings is more advantageous in the maxilla than in the mandible, so some practitioners will use HA coated Ti root form implants in the maxilla and uncoated Ti in the mandible. Similarly, for cast cobalt-chromium subperiosteal implants, plasma sprayed HA coated devices are more likely to be used in the maxilla than the mandible.

Some of the coating techniques, other than plasma spraying, under consideration for the deposition of HA on medical and dental implants produce coatings which are 100% crystalline HA. Other techniques may lend themselves more easily to coating porous surfaced implants than plasma spraying. Although these alternative means of producing HA coated implants may appear to be satisfactory for producing commercial products, years of animal and clinical testing will be needed before general use.

Reports from the early use of HA dental implants by practitioners and the results of some animal studies indicate that plasma sprayed HA coatings cause faster adaptation of bone to an implant device than do uncoated titanium or other implant alloys, although the long term advantages of the use of these coatings in either dental or orthopaedic devices has not been established. The rapidly increasing popularity of HA coatings, especially for endosseous dental implants, is perhaps due to the perception that an implant coated with a substance which is similar to bone should result in a more optimum response from either the surrounding bone or soft tissue. In that respect, HA coatings continue to interest both the clinician and the researcher as work continues to improve both the design of implants which utilize HA coatings and to optimize such factors as crystallinity and ion release rate to attain the proper tissue response in each particular orthopaedic and dental application.

READING LIST

K. de Groot, R. G. T. Geesink, C. P. A. T. Klein, and P. Serekian, "Plasma Sprayed Coatings of Hydroxyapatite," *J. Biomed. Mater. Res.* **21** (1987) 1375-1387

K. A. Thomas, J. F. Kay, S. D. Cook, and M. Jarcho, "The Effect of Surface Macrotexture and Hydroxyapatite Coating on the Mechanical Strengths and Histological Profiles of Titanium Implant Materials," *J. Biomed. Mater. Res.* **21** (1987) 1395-1406.

W. R. Lacefield, "Hydroxylapatite Coatings," in *Bioceramics: Materials Characteristics Versus In Vivo Behavior*, eds. P. Ducheyene and J. Lemons (NY Acad. Sci., 1988) pp. 72-80.

J. Koeneman, J. Lemons, P. Ducheyne, W. Lacefield, F. Magee, T. Calahan, and J. Kay, "Workshop on Characterization of Calcium Phosphate Materials," *J. Applied Biomaterials* **1(1)** (1990) 79-90.

M. I. Kay, R. A. Young and A. S. Posner, "Crystal Structure of Hydroxyapatite," *Nature* **204** (1964) 1050-1052.

S. Radin and P. Ducheyne, "The Effect of Plasma Sprayed Induced Changes in the Characteristics on the *In Vitro* Stability of Calcium Phosphate Ceramics," *Transactions 16th Annual Meeting of the Society for Biomaterials* **XIII** (1990) 128.

B. Koch, J. G. C. Wolke, and K. de Groot, "X-ray Diffraction Studies on Plasma Sprayed Calcium Phosphate-Coated Implants," *J. Biomed. Mater. Res.* (1990) 655-667.

P. Ducheyne, S. Radin, M. Heughbaert, and J. C. Heughebaert, "Calcium Phosphate Ceramic Coatings on Porous Titanium: Effect of Structure and Composition on Electrophoretic Deposition, Vacuum Sintering and *In Vitro* Dissolution," *Biomaterials* **11** (1990) 244-254.

P. Ducheyne, W. van Raemdonck, J. C. Heughebaert, and M. Heughebaert, "Structural Analysis of Hydroxyapatite Coatings on Titanium," *Biomaterials* **7** (1986) 97-103.

R. G. T. Geesink, K. deGroot, and C. P. A. T. Klein, "Chemical Implant Fixation Using Hydroxyl-Apatite Coatings," *Clin. Orthop.* **2** (1987) 147-170.

M. J. Filiaggi and R. M. Pilliar, "Interfacial Characterization of a Plasma-Sprayed Hydroxylapatite/Ti-6A1-4V Implant System," *10th Annual Conference Canadian Biomaterials Society* (1989) pp. 23-25.

J. F. Kay, M. Jarcho, G. Logan, and S. T. Liu, "The Structure and Properties of Hydroxylapatite Coatings on Metal," *Transactions 12th Annual Meeting of the Society for Biomaterials* **IX** (1986) 13.

S. D. Cook, K. A. Thomas, J. F. Kay, and M. Jarcho, "Hydroxyapatite-Coated Porous Titanium for Use as an Orthopedic Biologic Attachment System," *Clin Orthop.* **230** (1988) 303-312.

R. Z. Legeros, J. P. Legeros, O. R. Trantz, and W. P. Shirra, "Conversion of Monetite, $CaHPO_4$, to Apatites: Effect of Carbonate on the Crystallinity and the Morphology of the Apatite Crystallites", *Adv. X-ray* **14** (1979) 57-66.

Chapter 13

BIOACTIVE GLASS COATINGS

Larry L. Hench and *Örjan Andersson
University of Florida, Gainesville, Florida
*Abo Akademi, Turku, Finland

INTRODUCTION

Bioactive glasses are not strong enough to be used for load-bearing applications. One approach to solving this problem is to combine the glass with a fracture tough phase, such as a metal or a polymer, to produce a composite (Chapters 15 and 16). The other alternative is to apply the glass as a coating on a mechanically tough substrate, the subject of this chapter. Coating a substrate with a bioactive glass serves three purposes:

1) The substrate is protected from corrosion and degradation of properties.

2) Tissues are protected from corrosion products which may induce systemic effects.[1] Figure 1a shows tissues adjacent to a bone plate and metallic screw after only eighteen months following implantation in a 58 year old woman. An extensive concentration of corrosion products is present in the tissue. Figure 1b shows tissues adjacent to a bioactive glass coated stainless steel implant, after 18 months in a monkey. No corrosion products are present.

3) The bioactive glass coating provides interfacial attachment to bone, e.g.,bioactive fixation, thereby eliminating the need for polymeric bone cements.

The metals commonly used as load-bearing prostheses in orthopedics and dentistry have been used as substrates for bioactive glasses including: 316L stainless steel, Ti-6-4 alloy, and Co-Cr-Mo alloys. Table 1 summarizes the composition and properties of these metals. Medical grade alumina has also been used as a substrate. Chapter 2 summarizes the properties of alumina.

Three methods are used to apply bioactive glasses as coatings.

1) **Enamelling or Glazing**, using glass frits to provide a protective first layer or ground coat applied between the substrate and the bioactive glass.

2) **Flame-spray coating**, where the glass is applied to the substrate as a stream of molten particles. The particles fuse to the substrate upon impact.

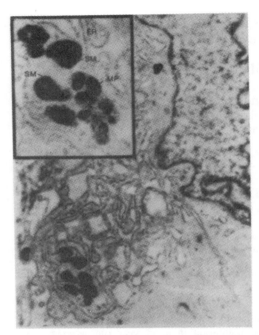

Fig. 1a. Fibroblast-like cell containing aggregates of metallic particles in cytoplasmic vacuoles (23,000X). Inset: High magnification electron micrograph to demonstrate metallic particles within membrane-bound vacuoles (55,000X).

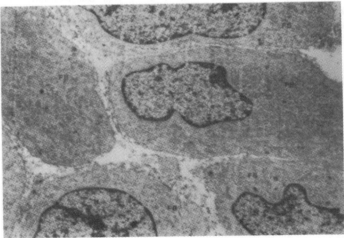

Fig. 1b. Representive field from tissue surrounding Bioglass®-coated implants.

Table 1. Orthopedic and Dental Implant Materials.

Material	Composition (wt%)																		Young's modulus (GPa)	Ultimate tensile strength (MPa)	Fracture Toughness K_{IC} ($MNm^{-3/2}$)	Coefficient of Thermal Expansion (K^{-1})
	Al	C	Co	Cr	Fe	H_2	Mg	Mn	Mo	Ni	N_2	O_2	P	S	Si	Ti	V	W				
Stainless steel		0.3 max		17.00-20.00	Bal*			2.00 max	2.00-4.00	10.00-14.00			0.03 max	0.03 max	0.75 max				200	207-1,160	~100	20×10^{-6}
Co-Cr alloy		0.05-0.15	Bal	19.00-21.00	3.00 max			2.0 max		9.00-11.00					1.0 max			14.0-16.0	230	430-1,028	~100	$13\text{-}16 \times 10^{6}$
Ti-6Al-4V alloy	5.5-6.5	0.08 max			0.25 max	0.013 max					0.05 max	0.13 max				Bal	3.5-4.5		105	700-1,050	~80	9×10^{-6}

*Bal, balance of element in the material.

3)   **Rapid-immersion coating**, where the metallic substrate is heated to form an
       oxide layer of critical thickness and is then rapidly inserted into a container of
       molten glass. The oxide layer dissolves in the glass forming an adherent bond
       between the substrate and the glass coating.

       The following sections review the principles, results and problems of these coating
methods.

THE PROBLEM

       The problem of obtaining a bioactive glass coating with high mechanical integrity
is the chemical reactivity of this type of glass. Compositions of silicate-based glasses
that form a bond with tissues have < 60 mole % SiO_2 (Chapter 3). Such glasses have
a random, two-dimensional, sheet-like, network structure with many open pathways for
ion transport. It is this structure that results in the rapid formation of a calcium phosphate
(CaP) and hydroxy-carbonate apatite (HCA) layer on the glass which provides binding
sites to collagen (Chapter 3). The open network also makes it easy for other cations,
such as Fe, Cr, Ni, Co, Mo, Ti, or Ta, to pass through the glass or react with the
surface. If these cations are present they rapidly react with the surface and prevent
formation of the CaP layer and its crystallization to HCA and thereby inhibit or eliminate
the bioactivity of the glass. Only a few percent of multi-valent cations are needed to
make a glass non-bioactive. See Hench[2] and Gross et al.[3] for a review of these
compositional factors.

ENAMELLING

       Glass coatings have been applied to metallic substrates for millennia in a process
called enamelling. The glass is melted and homogenized and quenched into either water
or air, forming a frit, and then ground into a powder (Chapter 18). The metal is coated
with the powder by painting, spraying, or dipping. An aqueous slurry of the glass
powder is often used. After drying, the coated part is then heated to above the softening
temperature of the glass (400-600°C) where the glass fuses to an oxide layer on the metal
forming a mechanical-chemical bond at the interface. The outer layer of the powder
sinters together (Chapter 1) and forms a coherent layer of glass fused to the metal.
       In order to facilitate adherence of the glass to the metal the traditional enamelling
method usually uses an intermediate layer, called a ground coat, containing oxides of Co
or Ni which reacts chemically with the metal, usually steel. Efforts by Hench and
colleagues to use commercial enamel ground coats followed by a top layer of bioactive
glass, such as 45S5 in Table 1 in Chapter 3, failed. Failure was due to the rapid
diffusion of metal ions from the ground coat through the bioactive glass layer
contaminating it and destroying bioactivity.
       A solution is to use two layers of glass on the substrate. The first layer of glass
provides interfacial bonding of glass to the metal, achieves an appropriate match of
coefficients of thermal expansion (CTE) of the metal and the glass, has a high durability

and a suitably low glass transition temperature, T_g, and low viscosity and still retains bioactivity of a top layer of glass fused to it. The T_g of the inner layer of glass must not be lower than that of the outer layer since it would soften too much during firing of the top layer.

Compositional Optimization

Optimization of glass compositions to produce and control the range of thermal and chemical properties listed above has been achieved by Karlsson[4,5] and Andersson[6,7] and co-workers at Abo Akademie University, Turku, Finland. A phenomenological model for glass optimization was developed using the following steps:

1) Define the compositional ranges of interest.

2) Design an experimental plan (select compositions) that allows regression analysis of the results and avoids cross-correlation.

3) Measure the properties of interest for the selected compositions.

4) Conduct a computer-based regression analysis to describe the relationship between composition and property.[6,7]

Example: T_g dependence on glass soda content.

As an example, the glass transition temperature (T_g) is considered. The glass transition temperature is traditionally determined by dilatometry, but also differential thermal analysis (DTA) can be used. T_g is seen as a downward shift in the baseline of a DTA curve. In this example 16 compositions were chosen within the following ranges: SiO_2, 38.9-65.5; Na_2O, 15.0-30.0; CaO, 10.0-25.0; P_2O_5, 0.0-8.0; B_2O_3, 0.0-3.0; Al_2O_3, 0.0-3.0 weight-%. T_g was then determined by DTA. By regression analysis a model with a regression coefficient, R^2, of 96.5%, and a residual standard deviation, ω, of 6.7°C was obtained. Thus,

$$T_g \ (^\circ C) = 44.3136 + 6.2716 \ (SiO_2) + 6.9072 \ (CaO) + 6.3210 \ (P_2O_5) + 6.7516 \ (Al_2O_3) \tag{1}$$

See Andersson, 1992, for details, ref. 8.

With slightly lower regression coefficient and higher residual standard deviation ($R^2 = 93.4\%$, $\omega = 8.2°C$) a simple linear dependence of the Na_2O content is obtained.

$$T_g \ (^\circ C) = 674.483 + 6.26843 \ (Na_2O) \tag{2}$$

Since this equation only contains one component it can be visualized as in Fig. 2. The figure shows T_g versus soda content for compositions within or close to the bioactive

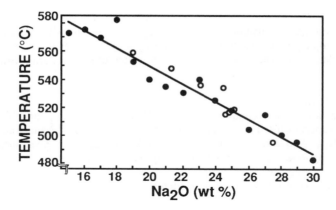

Fig. 2. T_g as a function of Na_2O-content for the A-series (●) and the B-series (o). The line corresponds to Equation 2 which was developed on basis of the A-series. The T_g for glasses in the B-series are only indicated for comparison.

range in the SiO_2-Na_2O-CaO-P_2O_5-B_2O_3-Al_2O_3-system. The 16 compositions used in developing the model are indicated as dots, and eight other compositions as circles. Due to the number of components in model (1), which gives a better fit to the experimental data, the relationship is not easily visualized. From a practical point of view this is also not necessary.

Factors in Developing a Model

A major problem in developing models is to design experiments that accurately measure or describe the property of interest. A good example is solubility of a glass. Up to a certain limit, a higher solubility results in a higher level of bioactivity. The high solubility (low leaching resistance), however, reduces the long-term reliability (Chapter 3). Therefore, it is desirable to design a bioactive glass coating with sufficient bioactivity and as low solubility as possible. It is difficult to develop models for the solubility since the complex reactions at the surface are strongly time dependent. Some reactions remove material from the glass whereas others add components (Chapter 3). Therefore, depending on the testing time, different results are obtained. Models that give at least a rough estimate of the solubility can, however, be developed.

When attempting to develop a model that quantitatively describes the host response to an implanted material other problems are encountered. If the percentage of bone at the interface is taken as a measure of the performance of the glass, the Index of bioactivity, see Chapter 3, the evaluation becomes difficult due to large scatter in the results. The scatter is caused by variables such as implant fit, immobilization, and individual variations in the animal. Andersson et al. developed a phenomenological description that is based on the *in vivo* behavior of glasses in the SiO_2-Na_2O-CaO-P_2O_5-B_2O_3-Al_2O_3-system[9]. Both the tissue response and the reactions within the glass were considered. A classification of the glasses was made based on a SEM/EDXA study. The glasses were qualitatively classified in order of increasing reactivity (nearly unchanged surface, Si-rich layer, HA-layer) and bone response (no contact, contact, bonding). By this classification it was possible to develop a model that qualitatively predicts the reactions within the glass as well as the type of interface developed in bone.

If reliable models for the properties of interest are available these may be used for simple prediction of the properties of given compositions. As discussed above, however, when developing coatings the opposite problem is more interesting, i.e. to specify the properties and find the composition with the closest fit to the properties. Therefore, this can be done by computer supported optimization. Usually some deviation in the properties can be tolerated. All properties are also not of equal importance. Therefore, the optimization program should preferably allow specifying the upper and lower limits as well as giving the different properties different weights according to their importance. It should be remembered that the min and max limits for the compositions are set by the experimentally investigated range.

The Alumina Effect

It is desirable to produce bioactive glasses of reduced solubility. The traditional way to control the solubility of glass is by Al_2O_3-additions. Gross and Strunz showed[3] that bone mineralization was disturbed in the presence of glasses containing a combination of Al_2O_3 and Ta_2O_5. The Al_2O_3-content was 7.5-15 wt-%. Osteoid formed but the mineralization of the osteoid was disturbed. Andersson et al. found that about 1.5 wt-% Al_2O_3 could be used without interfering with the mineralization of osteoid.[6] At higher Al_2O_3-content, a great amount of osteoid was found at the implant surface. This suggests that the alumina retards or inhibits the transformation of osteoid into bone. This is essentially the same result as reported by Gross and Strunz for glasses of high alumina-contents.[3]

Hench attributed the destructive effect of ions such as Al^{3+} on bone bonding to an increased corrosion resistance and to the precipitation of the multivalent ions as oxides, hydroxides or carbonates.[2] Andersson et al. suggested the following ways of loss of bioactivity due to alumina additions:[9]

(a)   Considerable decrease of the solubility of the glass, so that no silica-rich layer forms.

(b)   Stabilization of the silica gel in glasses of higher solubility. Thus, the silica gel forms but calcium phosphate formation is limited or inhibited. This may be attributed to a decreased flexibility of the gel (cross-linking) or to Al occupying calcium phosphate bonding sites in the gel. In order for aluminum ions to bond to the silica structure, the structure does not need to be as flexible and hydrated as for calcium phosphate formation.

(c)   Adsorption of Al^{3+} to the calcium phosphate rich surface layer of formation of Al-containing compounds, the crystals of which are not compatible with bone. Also release of aluminum into the tissue might contribute to the lack of bonding.[3]

Thus, in alumina-containing glasses the surface compositional changes depend on the composition of the glass. For glasses of relatively low solubility, the alumina may stabilize the structure enough to completely inhibit Ca,P-accumulation (case a). In glasses of higher solubility, a partial stabilization may take place. A certain flexibility of the silica-structure and a sufficient number of non-bridging oxygens still exist. The alumina ties up the hydrated silica gel which results in a limited or inhibited Ca, P-formation (case b). Figure 3a shows a cross-section of a fairly soluble glass (52.0 st-% SiO_2) after 8 weeks in rabbit tibia.[9,10] The glass contains 3.0 wt-% Al_2O_3, and it is seen that bone contact, but no bonding, has resulted. Figure 3b shows normalized compositional profiles, obtained by EDXA, over the interface. A thick silica-rich layer but no calcium phosphate has formed. It is clearly seen that alumina is accumulating in the hydrated SiO_2 gel. Naturally the amount of added alumina also affects the stabilization. If small amounts of alumina (approx. up to 1.5 wt-%) are added to a fairly soluble bioactive glass, a tendency towards alumina-enrichment can be seen in the silica-rich layer. However, the alumina content might not be enough to stabilize the structure, and as a result a calcium phosphate-rich surface forms. In case (c) the glass is characterized by formation of a non-bonding calcium phosphate surface. It is important to notice that the formation of the calcium phosphate surface layer *in vivo* or *in vitro* is not a sufficient criterion for bioactivity if the glass (or the substrate) contains alumina or other multivalent cations.

Use of the phenomological model outlined above makes it possible to add a small amount of component such as Al_2O_3 to the glass to improve durability without losing bioactivity. Likewise, the optimization formula makes it possible to match thermal expansion coefficients and substrate.

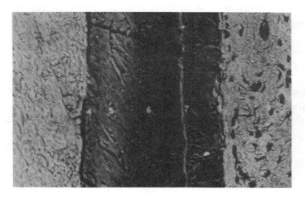

Fig. 3a. SEM micrograph of glass S52P3 showing contact, but no bonding, after 8 weeks in rabbit tibia.

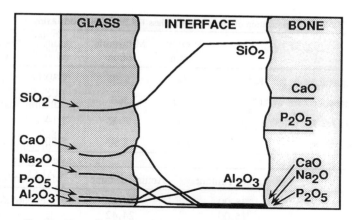

Fig. 3b. Normalized compositional profiles over the interface in Fig. 3a.

DEVELOPMENT OF COATING FOR Co-Cr-Mo ALLOY

A procedure for developing a bioactive dual glass coating by composition optimization follows:

1. Choice of alloy and specification of required glass properties.
2. Computer-based optimization
3. *In vitro* bioactivity test
4. Development of optimal firing scheme and testing for porosity, diffusion, metal-glass adhesion, etc.
5. *In vitro* bioactivity test
6. *In vivo* test.

As an example consider the development[11] of a dual coating for a Co-Cr-Mo alloy (similar to that in Table 1). The CTE of the alloy used is 1.47×10^{-5} K^{-1}. The compression strength of glass is good, whereas the tensile strength is poor. In order to avoid tension in the coating the CTE of the glass must not be higher than that of the metal substrate. The CTE of the coatings were chosen to be somewhat below that of the substrate in order to induce compression in the coating. This is advantageous since a higher load can be applied before tensile stresses occur. The glass optimization data for the bioactive outer coating is summarized in Table 2.

The optimized compositions for the inner and outer layers are:

	SiO_2	Na_2O	CaO	P_2O_5	B_2O_3	Al_2O_3	
HI-4	59.62	25.84	10.24	1.30	0.00	3.00	(inner layer)
HI-2	55.86	24.42	12.99	4.90	1.83	0.00	(outer layer)

Table 2. Optimization of Outer Layer, Glass HI-2 for Co-Cr-Mo Alloy.

Table 2. Optimization of Outer Layer, Glass HI-2 for Co-Cr-Mo Alloy.

Properties	Weight factor	Minimum value	Actual value	Maximum value
Thermal expansion in 10^{-5} K^{-1}	100	1.30	1.39	1.39
Transition temperature in °C	100	500	520	520
Solubility as ml HCl/200 mg glass	10	0.00	2.14	3.00
Reaction Number	100	5.80	5.80	8.00

Oxides	Minimum content	Actual content	Maximum content
SiO_2	38.00	55.86	65.50
Na_2O	15.00	24.42	30.00
CaO	10.00	12.99	25.00
P_2O_5	0.00	4.90	8.00
B_2O_3	0.00	1.83	3.00
Al_2O_3	0.00	0.00	3.00
Total		100.00	

Note: Example of optimization of bioactive outer layer for a Co-Cr-Mo alloy. The minimum and maximum limits for the compositions are set by the ranges used in developing the models, the limits for the properties by the requirements on the enamel. For the thermal expansion coefficient Appen's method is used, for the transition temperature Equation 1 in the text, and for the biological behavior the reaction number model discussed in the text. The solubility is expressed as the amount (ml) of 0.01 M HCl consumed to maintain the pH at 7.4 in 100 ml aqueous 15 mM Na-citrate solution, during 60 minute immersion of 200 mg glass grains (297-500 μm). In the present optimizations the viscosity was not considered. The content was not fixed for any component.

The glasses were tested *in vitro*. Both the inner layer (HI-4) and the outer one (HI-2) has slightly higher solubility than predicted by the optimization. A calcium phosphate surface layer developed at the surface of HI-2 within 24 hours in a simulated body fluid (SBF). (See Chapter 18 for details of SBF composition.) Thus, the glass was concluded to be bioactive. The glasses were pulverized and dispersed in ethanol. Co-Cr-Mo cylinders were then coated by dipping. A range of different firing time-temperature combinations were tested in order to avoid bubbles and achieve minimal diffusion of aluminum from the inner layer into the outer layer. The Co-Cr-Mo cylinders could be coated by firing the inner layer at 680°C for 20 minutes and subsequently the outer layer at 640°C for 30 minutes. A nearly bubble-free coating was obtained. The mixing of the two layers (aluminum diffusion) was limited to a 5 μm thick zone. No cracks appeared which demonstrates the close match in CTE between the layers. The coated cylinders were then immersed in SBF for 72 hours. No calcium phosphate formed. Following the same procedure, the same dual coating was applied on a Au-Pd alloy. The coating retained bioactivity. Thus, some component of the Co-Cr-Mo alloy disturbs the bioactivity even if no contamination of the coatings was observed by EDXA.[11]

A phosphate-free dual coating has also been developed.[11] Rabbit hip implants made of the Co-Cr-Mo alloy were coated. No cracks appeared. The outer coating remained somewhat porous after firing, but induced apatite formation *in vitro*. The prostheses were tested 2 months under load-bearing conditions in rabbit hip.[12] The prostheses were well fixed, whereas uncoated controls were loose. Thus, bone bonded to the coating. However, the coating was slowly resorbing, probably due to porosity and partial crystallinity.

It can be concluded that the composition optimization approach has been successful with respect to the properties considered (CTE, T_g, durability, bioactivity) but that phase separation in the glass, reactions between glass and substrate, elimination of porosity and ion release from the coated substrate require more work before the coatings are reliable in an *in-vivo* environment.

GLAZING

Glass coatings fused to polycrystalline ceramic substrates, called glazes, have been used for thousands of years to improve properties and appearance. Bioactive glasses have also been fused to medical grade alumina to achieve bioactive fixation.[13,14] The problem to be overcome in applying a bioactive glass to alumina is the very large difference in coefficients of thermal expansion (CTE); the glass has a very high CTE ($1.3-1.6 \times 10^{-5}/C$) and alumina has a much lower CTE ($8 \times 10^{-5}/C$). Traditional glazes have their composition adjusted to eliminate CTE mismatch. Unfortunately, changing glass composition to lower its CTE to match that of alumina will also eliminate its bioactivity. Greenspan showed that only 3% Al_2O_3 addition to a 45S5 bioactive glass will prevent hydroxyapatite film formation and bioactive bonding to bone.[13]

The solution is to use two coats of the bioactive glass. The first layer is applied as a fine powder dispersed in an organic binder and solvent. After binder burn out at 650°C the glass is fused to the alumina at 1350°C for 15 minutes. During cooling the large CTE mismatch results in grazing of the coating to produce "micro-islands" of glass fused to the alumina substrate with a chemical gradient of Al_2O_3 diffused into the glass.[13] A second coating is applied and fired at only 1150°C for 30 minutes followed by a slow furnace cool to anneal the stresses developed between the layers. The second coating has limited diffusion of Al_2O_3 because of the lower firing temperature and therefore the top layer retains its bioactivity. The difficulty with this process is achieving reliability of adherence of the layers. If cracks develop between the two layers the glass will react rapidly with body fluids in the cracks, leading to spalling of the coatings from the substrate.

FLAME SPRAY COATING

Flame spray coating has been used for nearly seventy-five years to apply metallic coatings and for forty years to apply ceramic and glass coatings to substrates. Bioactive glasses (45S5, 45B$_{15}$S5, and 45S5F) were applied as coatings to 316L surgical stainless steel by use of flame spraying as early as 1974.[15]   Figure 4 is a schematic of a flame spray apparatus which consists of an oxy-hydrogen torch in conjunction with a powder feed hopper. The glass powder is entrained in the oxygen stream and passes through the torch   where the oxygen and hydrogen react to form a flame with a theoretical temperature in the range of 3000°C. The powder is rapidly heated, becomes molten and impinges on the substrate where it quenches back into a glass.

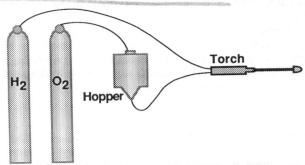

Fig. 4. Schematic representation of flame spray apparatus.

In order to achieve a smooth, bubble free coating, without loss of volatile components, such as Na$_2$O or P$_2$O$_5$, it is essential to optimize the softening range, quench rate, and viscosity of the glass. The relationship between these variables is shown in Fig. 5. The volume decrease associated with cooling of the molten droplets is shown as the dark curves in the figure. If the liquid does not crystallize upon cooling to T$_m$, it will retain its random, liquid-like structure through a super-cooled liquid regime, B, and solidify into a glassy material within the transformation range, T$_g$.   The transformation temperature of a flame sprayed glass depends upon its quench rate. A rapid quench produces a high T$_g$ and a less dense glass; i.e., larger volume for the same mass, line C in Fig. 5. A slower quench rate results in higher densities and lower transformation temperatures, curves D and E in Fig. 5.

When a glass has a high viscosity before the glass transformation occurs on quenching, such as the dashed curve (1), the flame sprayed particles cannot flow and create a uniform glassy coating on the surface.   When the viscosity-temperature dependence of the glass is low, curve (2), spreading and interparticle flow occurs before solidification of the glass.   This condition must be achieved to produce a smooth, bubble-free, adherent coating. The glass composition must be modified to achieve this condition.

The principles of glass composition optimization outlined above can be used here as well.   There is a limit on compositional adjustment for glasses to be used in flame spraying.   Crystallization at T$_m$ will occur if there are insufficient network formers (Chapter 1) in the molten glass particles. Crystallization must be avoided since the large

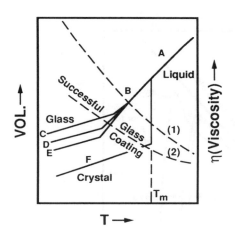

Fig. 5. Volume-temperature reactions for a liquid (A), supercooled liquid (B), glasses (C,D,E), and crystal (F), all shown as solid lines.

volume change that results at T_m (Fig. 5) prevents formation of a smooth coherent pore-free coating.

Compositional changes in the 45S5 bioactive glass formula shifted the viscosity-temperature behavior from curve (1) in Fig. 5 to curve (2). The results are illustrated in the dilatometer curves, thermal expansion vs temperature, shown in Fig. 6. The two compositions developed for flame spraying are shown in Table 3. Composition $45B_{15}S5$ replaced 15 weight % SiO_2 with 15 weight % B_2O_3 and composition 45S5F substituted CaF_2 for half of the CaO and 2 weight % of SiO_2. Both altered compositions decreased the viscosity, as shown by the lowering of the softening points in Fig. 6 from 550°C to 470°C. Both modified glass compositions had higher coefficients of thermal expansion which were much closer matches to that of 316L stainless steel. The fluorine containing glass had a better high temperature fluidity and produced a superior flame spray coating.

Studies of fluoride-containing bioactive glasses by the Florida group[16,20] show that they have an equivalent level of bioactivity but more durability than the 45S5 composition. Therefore, the fluoride-containing glasses are preferable for some clinical applications where small powders are required, in addition to being the preferred composition for flame spraying.

Pre clinical testing of flame spray coatings of bioactive glass (45S5F) was conducted using monkey femoral models. Load bearing femoral stem replacements of 316L stainless steel with flame spray coatings were designed by G. Piotrowski, implanted by W. C. Allen and evaluated at an EM and TEM level by H. A. Paschall, as described in Hench et al., 1975.[15] The results showed that the coatings protected the metallic devices from corrosion and no metallic corrosion products could be detected adjacent to the prostheses, in contrast to tissues in contact with uncoated stainless steel (Fig. 1).

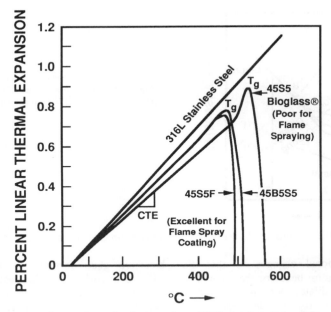

Fig. 6. Thermal expansion and softening points of 316L stainless steel and three bioactive glasses.

Table 3. Bioglass® Compositions Developed for Coating Orthopaedic Prostheses.

Glass	Use	Weight percentage					
		SiO_2	Na_2O	CaO	CaF_2	P_2O_5	B_2O_3
45S5	Bulk, powder	45.0	24.5	24.5		6.0	
45S5F	Flame Spray	43.0	23.0	12.0	16.0	6.0	
$45B_{15}S5$	Flame Spray	30.0	24.5	24.5		6.0	15.0
52S4.6	Immersion coating	52.0	21.0	21.0		6.0	

Mechanical testing of the interfaces between the monkey femur and the flame spray coated stem of the prostheses was conducted six months after implantation using an Instron testing machine in tensile mode. The femoral head replacements were almost totally surrounded by bone and tight within the femur, even after all the bone around the proximal end of the implants had been resected. At a force of 258 lb-f, the distal condyles were torn off the bone with no loosening of the implant. Three point bend tests causing fracture of the bone at 257 lb-f also did not loosen the implant. The 45S5F bioactive glass-coated femoral head of the monkey partial hip prosthesis still attached to bone (B) after mechanical testing is shown in Fig. 7. A portion of the bone has been removed by an osteotome to expose the implant surface (G). Fragments of bone are still attached to the glass surface.

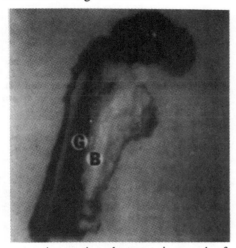

Fig. 7. Bioglass®-coated femoral head of monkey hip prosthesis showing (B) attached bone and (G) osteotome-exposed implant surface (2X).

in vivo Hench

see Ref ← *previous page* *Coating*

A scanning electron micrograph of a cross section of the implant and monkey femur (Fig. 8a) shows three morphological regions: metal (M), bioactive glass coating (BG), and bone (B).[16] The microporosity present at the BG-M interface is difficult to *D is.* eliminate in flame spraying. Such porosity can lead to low values of interfacial strength[15] because the bond between metal and coating will deteriorate with time. When there is little interfacial porosity, failure occurs within bone, as discussed above, rather than at the bone-glass interface or the glass-metal interface. Energy dispersive X-ray (EDX) analysis[21] of three regions that are 50 μm from each other across the bioactive glass-bone interface are shown in Fig. 8b. The EDX data show a decrease in Si concentration and a variation in the Ca/P ratio across the interface from glass to bone. The compositional gradients occur across a region that is contiguous and morphologically uniform.

These results show that flame spray coatings of bioactive glasses can protect metallic substrates from corrosion and also provide an interfacial bond between a functional load-bearing prosthesis that is as strong or stronger than bone. The studies also show that it is difficult to obtain high reliability in metal-glass coatings and if microporosity is present at the interface low failure strengths occur.

RAPID IMMERSION COATING

The rapid immersion process of coating a bioactive glass onto a metal eliminates many of the problems discussed above for enamelling or flame spray coating.[22] The process involves immersion of a pre-oxidized metal substrate into a molten batch of

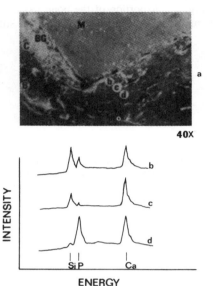

40X

Fig. 8. (a) Cross-sectional view of monkey hip implant and bone with the metal (M), Bioglass® coating (BG), and bone (B) indicated. Interfacial regions (b,c,d) are entified by corresponding EDX spectra.

glass. By controlling the temperature of the metal prior to immersion and the temperature of the melt it is possible to match the total thermal contraction of the substrate and the glass layer that fuses to it. The solubility of the oxide layer on the metal within the glass is controlled by the immersion time, which is only a few seconds. A compositional gradient of metal oxide is produced during the immersion. This results in chemical bonding at the metal-glass interface. The outer layer of the glass is not contaminated by the substrate metal oxides, since diffusion times are so short, and therefore has the same bioactivity as the bulk composition of the glass.

The rapid immersion method has been applied to 316L stainless steel[22] and recently optimized for clinical use.[23] It has been applied to a Co-Cr surgical alloy,[24] and a Ti alloy.[25] Results of the three systems are briefly summarized.

316L Stainless Steel

Thickness and composition of the oxide layer grown on the steel substrate are critical variables in the rapid immersion coating process. Temperature, time and oxygen partial pressure control both variables. Oxidation in air for 15-30 minutes at 900°C, Fig. 9, produces an intimate mixture of two metal oxides on the steel, magnetite (Fe_3O_4) and eskolite (Cr_2O_3), based upon X-ray diffraction, SEM-EDS, electron microprobe (EMP), and Auger electron spectroscopy (AES). The AES results show that the thickness of the oxide layer is in the range of 1 μm.

The pre-oxidized metal devices are immersed in a batch of molten 45S5 Bioglass® held at 1350°C. The glass temperature falls to the range of 1050°C as the device is rapidly inserted and removed (Fig. 9). A slow cool and annealing soak at 450°C is used to eliminate stresses. SEM results show that there is intimate contact between the metal substrate and the glass coating without any bubbles, pores or cracks. The EMP analysis shows a diffusional zone of approximately 4 μm across the metal-glass interface. There is counter diffusion across the interface indicating formation of a chemical bond between the glass and the oxidized metal surface. Fourier transform IR (FTIR) spectroscopy of the glass coatings indicates that no compositional change has occurred in the glass other than in the 4 μm diffusional zone. *In-vitro* bioactivity tests of the rate of hydroxy-carbonate apatite (HCA) formation shows that the coating behaves the same as the bulk 45S5 glass (Fig. 10).

Indentation fracture toughness analysis indicates that the coatings are equivalent to the bulk 45S5 glass and other soda-lime-silica glasses with values of 0.8 MPa·m$^{1/2}$. Interfacial failure strengths of the glass-metal bond were determined to be 126-130 Kg/cm^2. These values are higher than the 40-80 Kg/cm^2 values reported by Lacefield and Hench[24] for 45S5 Bioglass® coatings on a Co-Cr alloy using a similar test method. Push-off tests of the coatings, using a procedure similar to that used to measure the bone-bonding strength of bioactive glasses, produced values of 11-12 MPa. These values are similar to the post *in vivo* push-out values reported by West et al.[25] and Andersson and et al[10] using similar test methods.

Co-Cr Alloys

Lacefield and Hench[24] showed that the rapid immersion process could be used to produce mechanically strong coatings of bioactive glasses on a commercial cobalt-chromium alloy (Vitallium®). The 52S4.6 Bioglass® composition (Table 3) was used for the optimization study because of the close match of its CTE to that of the metal alloy ($13-16\times10^{-6}$/C). The thickness of oxide layer grown on the alloy is a critical variable in the adherence strength of the coating. Time, temperature, and oxygen partial pressure control the rate of oxide growth. An excessively thick oxide layer is weak and shear will occur within the oxide or between the oxide and the metal. An oxide layer that is too thin will be completely dissolved by the glass causing poor adhesion between the coating and the substrate. Figure 11 illustrates the importance of time and temperature, at a fixed partial pressure of oxygen, on interfacial shear strength of the glass coating on the Co-Cr alloy. The results for the 45S5 coating on 316L stainless steel are compared in Fig. 11. The steel requires a higher temperature oxidation to produce a high strength interface.

Adherent coatings of about 1 mm thickness can be applied to Co-Cr alloys and stainless steel by the rapid immersion process. An advantage of the process is that the metal is exposed to high temperature for only 3-15 seconds which does not affect the microstructure and strength of the alloys.

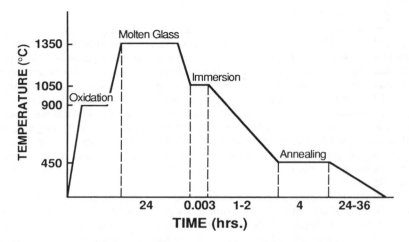

Fig. 9. Process schedule for rapid immersion coating of 45S5 Bioglass® on 316L stainless steel (based on ref. 18).

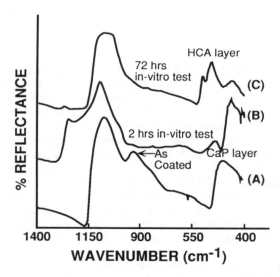

Fig. 10. FTIR diffuse reflection spectra of 45S5 Bioglass® coating on 316L stainless steel. (A) As coated (same as bulk glass spectrum). (B) Coating after 2 hrs *in vitro* test. Note formation of Ca,P layer. (C) Coating after 72 hrs.

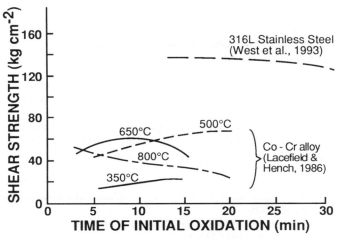

Fig. 11. Glass-metal bond strength as a function of time and temperature of oxidation.

Ti Alloys

West et al.[25] describe efforts to coat a Ti-6Al-4V alloy with bioactive glass. Earlier work had shown that deleterious chemical reactions occurred at the 45S5 glass interface with the Ti alloy. A precoat of boric acid in a reducing flame produced a protective Ti-B-O layer on the metal which made it possible to use the rapid immersion process to apply a top layer of bioactive glass. Implants with the bioactive glass coating were compared with HA-coated Ti alloy implants prepared commercially by plasma spraying and bulk 45S5 Bioglass® rods of the same dimensions. Coated cylinders were placed transcortically in the mid-diaphyseal region of the femurs of dogs. Push out strengths were determined using the method of Bobyn et al.[26] All implants bonded to bone with no statistically significant difference in push-out strengths of 15-17 MPa at four weeks. All implants increased in bond strength with time to 40-50 MPa after 24 weeks. The glass coated implants had greater variation in interfacial strength than HA coated samples due to bubbles at the glass-metal interface. The Ti-B-O precoat had too much variability to use clinically.

SUMMARY

Numerous systems for applying bioactive glass coatings on metals and ceramics have been investigated. All have deficiencies and problems with reliability of the glass-metal interface. The system that is closest to meeting a >95% reliability criterion is rapid immersion coating of bioactive glass on 316L stainless steel. Because of the previous history of unpredictability of *in vivo* performance of bioactive glass coatings, extensive *in vitro* testing is necessary before clinical trials can be justified for load bearing prostheses.

REFERENCES

1.   J. Black, "Systemic Effects of Biomaterials," *Biomaterials*, **5** (1984) 11.
2.   L. L. Hench, "Bioactive Glasses and Glass-Ceramics: A Perspective," in *Handbook of Bioactive Ceramics, Vol. I*, eds. T. Yamamuro, L. L. Hench, and J. Wilson (CRC Press, Boca Raton, FL, 1990) pp. 7-23.
3.   U. Gross, R. Kinne, H. J. Schmitz, V. Strunz, "The Response of Bone to Surface Active Glass/Glass-Ceramics," *CRC Critical Reviews in Biocompatibility*, **4** (1988) 2.
4.   T. Westerland, L. Hatakka, and K. H. Karlsson, "A Model for Optimizing Glass Batch Compositions," *J. Am. Ceram. Soc.* **66[8]** (1983) 574-579.
5.   L. Hatakka and K. H. Karlsson, "Optimization of Batch Costs for an Industrial Furnace," *Glass Technology* **27[1]** (1986) 17-20.
6.   Ö. H. Andersson, K. H. Karlsson, K. Kangasniemi, and A. Yli-Urpo, "Models for Physical Properties and Bioactivity of Phosphate Opal Glasses," *Glastech. Ber.* **61** (1988) 300-305.
7.   Ö. H. Andersson and K. H. Karlsson, "On the Bioactivity of Silicate Glass," *J. Non-Cryst. Solids* **129** (1991) 145-151.
8.   Ö. H. Andersson, "Glass Transition of Glasses in the SiO_2-Na_2O-CaO-P_2O_5-Al_2O_3-B_2O_3 System," *J. Mater. Sci.: Materials in Medicine* **3** (1992) 326-328.
9.   Ö. H. Andersson, G. Liu, K. H. Karlsson, L. Niemi, J. Miettinen and J. Juhanoja, "*In Vivo* Behaviour of Glasses in the SiO_2-Na_2O-CaO-P_2O_5-Al_2O_3-B_2O_3 System," *J. Mater. Sci.: Materials in Medicine* **1** (199) 219-227.
10.   Ö. H. Andersson, G. Liu, K. Kangasniemi, and J. Juhanoja, "Evaluation of the Acceptance of Glass in Bone," *J. Mater. Sci.: Materials in Medicine* **3** (1992) 145-150.
11.   Ö. H. Andersson, et al., to be published, 1993.
12.   K.J.J. Pajamäki, T. S. Lindholm, Ö. H. Andersson, K. H. Karlsson, E. Vedel, A. Yli-Urpo, and R-P. Happonen, "Bioactive Glass and Glass-Ceramic Coated Hip Endoprosthesis: Experimental Study in Rabbit," submitted 1993.
13.   D. C. Greenspan and L. L. Hench, "Chemical and Mechanical Behavior of Bioglass Coated Alumina," *J. Biomed. Maters. Res.* **10[4]** (1976) 503-509.
14.   J. E. Ritter, D. C. Greenspan, R. A. Palmer, and L. L. Hench, "Use of Fracture Mechanics Theory in Lifetime Predictions for Alumina and Bioglass-Coated Alumina," *J. Biomed Mater. Res.* **13** (1179) 251-263.
15.   L. L. Hench, H. A. Paschall, W. C. Allen, and G. Piotrowski, "Interfacial Behavior of Ceramic Implants," *National Bureau of Standards Special Publication*, **415** (1975) 19-35.
16.   L. L. Hench and H. A. Paschall, "Direct Chemical Bonding Between Bio-Active Glass-Ceramic Materials and Bone." *J Biomed. Maters. Res. Symp.* **4** (1973) 25-42.
17.   L. L. Hench, D. B. Spilman and D. R. Nolletti, "Fluoride Bioglasses," in *Biological and Biomechanical Performance of Biomaterials*, eds. P. Christel, A.

Meunier and A.J.C. Lee (Elsevier Science Publishers B.V., Amsterdam, 1986) pp. 99-104.

18.   J. Wilson, S. Low, A. Fetner and L. L. Hench, "Bioactive Materials for Periodontal Treatment: A Comparative Study," in *Biomaterials and Clinical Applications*, eds. A. Pizzoferrato, P. G. Marchetti, A. Ravaglioli and A.J.C. Lee, (Elsevier Science Publishers, Amsterdam, 1987) pp. 223-228.

19.   June Wilson and D. Nolletti, "Bonding of Soft Tissues to Bioglass®," in *Handbook of Bioactive Ceramics*, eds. T. Yamamuro, L. L. Hench, and J. Wilson (CRC Press, Boca Raton, FL, 1990) pp. 283-302.

20.   Cheol Y. Kim, A. E. Clark and L. L. Hench, "Compositional Dependence of Calcium Phosphate Layer Formation in Fluoride Bioglasses®," *J. Biomed. Maters. Res.* **26** (1992) 1147-1161.

21.   L. L. Hench, C. G. Pantano, Jr., P. J. Buscemi and D. C. Greenspan, "Analysis of Bioglass Fixation of Hip Prostheses," *J. Biomed. Maters. Res.* **11[2]** (1977) 267-281.

22.   L. L. Hench and P. Buscemi, "A Method of Bonding a Bioglass® to Metal," U.S. Patent 4,159,358 (1979).

23.   J. K. West, G. Gonzales, G. LaTorre, and L. L. Hench, "Optimization of Coating 316L Stainless Steel with Bioglass®," *Bioceramics VI*, ed. P. Ducheyne, 1993.

24.   W. R. Lacefield and L. L. Hench, "The Bonding of Bioglass® to a Cobalt-Chromium Surgical Implant Alloy," *Biomaterials* **7[2]** 1986.

25.   J. K. West, A. E. Clark, M. Hall, and G. Turner, "In-Vivo Bone Bonding Study of Bioglass® Coated Titanium Alloy," in *CRC Handbook of Bioactive Ceramics*, eds. T. Yamamuro, L. L. Hench and J. Wilson, (CRC Press, Boca Raton, FL, 1990) pp. 161-166.

26.   J. D. Bobyn, R. M. Pilliar, H. V. Cameron and G. G. Weatherly, "The Optimum Pore Size for the Fixation of Porous Surfaced Metal Implants by Ingrowth of Bone," *Clin. Orthop. Relat. Res.* **150** (1980) 263.

Chapter 14

PYROLYTIC CARBON COATINGS

R. H. Dauskardt and R. O. Ritchie
Center for Advanced Materials, Materials Sciences Division,
Lawrence Berkeley Laboratory
and
Department of Materials Science and Mineral Engineering
University of California, Berkeley, CA 94720, U.S.A.

INTRODUCTION

Carbon is the most frequently found element in all organic molecules and compounds and as such performs a vital role in biological processes. As a crystalline material it can exist in a bewildering number of forms some of which offer the most outstanding biocompatibility, chemical inertness and thromboresistance of any of the ceramics (or "bio-ceramics") used in biomedical applications. These properties have, for example, made various forms of carbon the preferred material where interface to blood flow is required. Alternatively, they can be attached to both soft and hard tissue which is a prerequisite for a wide range of biomedical devices. Where device design and mechanical strength of the carbon form permit, components may be fabricated entirely of carbon. However, in the majority of biomedical applications carbon is used as a versatile coating, the principal mechanical properties often being derived from the underlying substrate material.

Three types of carbon are commonly used for biomedical devices: the low-temperature isotropic (LTI) form of pyrolytic carbon, glassy (vitreous) carbon and the ultralow-temperature isotropic (ULTI) form of vapor-deposited carbon. These three forms of carbon have a disorded lattice structure and are collectively referred to as turbostratic carbons. Although pyrolytic carbons were developed originally for elevated-temperature (e.g., as coatings for nuclear fuel particles) applications, LTI pyrolytic carbon has found wide appeal in the biomedical materials industry, in particular for mechanical cardiac-valve prosthetic devices, as it has been shown to be highly thromboresistant and to have inherent cellular biocompatibility with blood and soft tissue; moreover, it displays excellent durability, strength and resistance to wear, and had been thought to be immune to cyclic fatigue failure.

Indeed, the majority of modern mechanical heart-valve prostheses utilize components manufactured from silicon-alloyed, low-temperature isotropic (LTI)-pyrolytic carbon, either as a coating on a polycrystalline graphite substrate or as a monolithic material (where the substrate has been machined away). Alternatively, more complex shapes and even flexible materials may be obtained from coating fabricated metal components or polymeric sheets or fabrics with a thin, impermeable layer of (ULTI) carbon by vapor deposition. Larger components can be fabricated by the controlled heating of a preformed polymeric body to form glassy carbon. Due to its inherently low density and weakness, however, glassy carbons are typically used as a thick coating which is reinforced by the underlying substrate.

With the exception of the LTI carbons which are co-deposited with silicon, all the carbons in clinical use are pure elemental carbon. Up to 20 wt% silicon is most often added to LTI carbon to improve mechanical properties without significant changes in biocompatibility. Structurally the clinically important turbostratic carbons all from part of the disordered end of the wide range of crystalline states of carbon; from the perfect, three-dimensional graphite forms through the partially ordered graphite-like structures to the nearly amorphous state. Note, however, that by far the most important structural carbons for use in biomedical applications are the LTI pyrolytic carbons. For this reason more attention will be directed to these materials in our subsequent discussions.

In the following sections, we first discuss the composition, structure and processing procedures of the three clinically significant turbostratic carbons. These are unique compared to those of the more perfectly crystalline states of diamond or graphite. Then the important resulting biomedical and mechanical properties are considered. Finally, we will examine a number of biomedical applications and devices and conclude by discussing some important issues in device reliability and life prediction procedures.

CARBON STRUCTURES

While the microstructure of turbostratic carbon might seem very complicated due to its disordered nature, it is in fact quite closely related to the structure of graphite. In a graphite crystal, the carbon atoms are arranged in flat sheets or layer planes and the sheets stacked in a regular sequence to produce the three-dimensional graphite lattice as shown in Fig. 1a. Each carbon atom is strongly covalently bonded to six nearest neighbor atoms, forming a hexagonal array within each layer, while the layers are relatively weakly bonded by van der Waals attractions. The weak bonding between planes results in the highly anisotropic properties of single crystals of graphite, however, when a solid is composed of many small crystals having random orientations, the bulk material behaves in an isotropic fashion.

In graphite, the layer planes are stacked in a regular ABAB sequence. It is possible to disorder the sequence by disrupting the stacking through random rotations or displacements of the layers relative to eachother. Carbon materials having such disordered lattice plane stacking are known as *turbostratic* structures (Fig. 1b). In graphite the crystal size might be as large as 100 nm in diameter, the disordered regions of turbostratic carbon or "crystallites" are only around 10 nm in size. Randomly oriented turbostratic carbon crystallites are assembled to produce the bulk material as shown in Fig. 1c. As mentioned above, this random packing results in isotropic mechanical and physical properties at the macro scale.

Missing carbon atoms lead to many vacant lattice sites in each of the turbostratic layer planes. Imperfect matching of small segments of these planes also leads to wrinkles

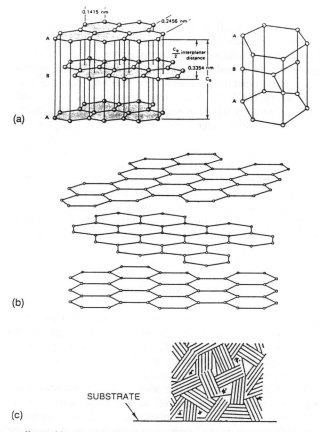

Fig. 1. The crystallographic arrangement of carbon atoms in (a) hexagonal graphite where parallel layer planes are in a regular sequence and (b) a turbostratic carbon where the layers are arranged without order. Randomly oriented turbostratic crystallites are assembled to produce a bulk material in (c).

or distortions within each plane. Furthermore, there may also be a substantial fraction of disorganized carbon between crystallites in the bulk material. As might be expected, such distortion and voiding in the structure may result in a marked effect on both the density and strength of turbostratic carbons. Densities, therefore, range from 1,400 kg.m^{-3} to a theoretical limiting value of 2,200 kg.m^{-3}. High density LTI carbons are the strongest bulk form of turbostratic carbon and their strength can be increased by adding silicon as discussed in the following section. ULTI carbon can also be produced with high densities and strengths, but is available only as a thin coating (0.1 - 1.0 μm) of pure carbon. Glassy carbon is inherently a low density material and as such is weak.

PROCESSING

Dense and high strength LTI pyrolytic carbon components are typically made by co-depositing carbon and silicon carbide on a polycrystalline graphite substrate via a chemical vapor-deposition, fluidized-bed process using a gas mixture of a silicon-containing carrier gas with a hydrocarbon (e.g., propane, methyltrichlorosilane and helium gas mixtures) at elevated temperatures. The resulting material contains typically 10 wt% silicon, often in the form of discrete sub-micron ß-SiC particles randomly dispersed in a matrix of roughly spherical micron-size subgrains of pyrolytic carbon; the carbon itself has a subcrystalline turbostratic structure, with a crystallite size typically less than 10 nm.

The system used for production of pyrolytic carbon coatings in shown schematically in Fig. 2. The fluidized bed coater is in principle a simple apparatus which consists of a vertical tube furnace (or reactor) containing a bed of granular particles, usually zirconium oxide. The reactant gas stream enters the reactor and fluidizes (i.e. supports and agitates) the bed of particles in which the components to be coated are suspended. Due to the high temperatures (typically in the range 1000 to 1500°C), the hydrocarbon gas "cracks" or pyrolyzes according to the reactions:

$$C_3H_8 \rightarrow 3C + 4H_2 \text{ , and} \tag{1}$$

$$CH_3Cl_3Si \rightarrow SiC + 3HCl. \tag{2}$$

The solid products of these pyrolysis reactions are carbon and silicon carbide which deposit as a coating on everything in the bed including the fluidizing particles and the graphite components to be coated. The fluidizing bed system is equipped with metering devices so that the quantity of carrier gas, hydrocarbon and silicon containing hydrocarbon can be controlled to give the proper concentrations for deposition.

The silicon carbide present in silicon alloyed pyrolytic carbon does not dissolve in the carbon matrix but instead forms discrete second phase particles of silicon carbide with cubic crystal symmetry as shown schematically in Fig. 3. The quantity of silicon present has a marked influence on the mechanical properties of pyrolytic carbon. As the silicon content increases, fracture strength, wear resistance, hardness and the elastic modulus increase significantly. Both the content and distribution of silicon carbide particles is quantitatively measured using X-ray diffraction (XRD) and scanning electron microscopy (SEM) techniques and must be carefully controlled to achieve optimum properties.

Glassy or vitreous carbons are formed by slowly heating a solid polymeric preform such as cellulosics or phenol-formaldehyde resin to drive off volatile constituents. The heating rate during formation must be carefully controlled to allow diffusion of the volatiles through the polymeric mass and prevent bubble formation within the part. This requirement typically limits the thickness of glassy carbon components to less than 7 mm. These carbons are composed of random crystallites of the order of 5 nm in size. While the glassy carbons only achieve a maximum density of ~ 1,500

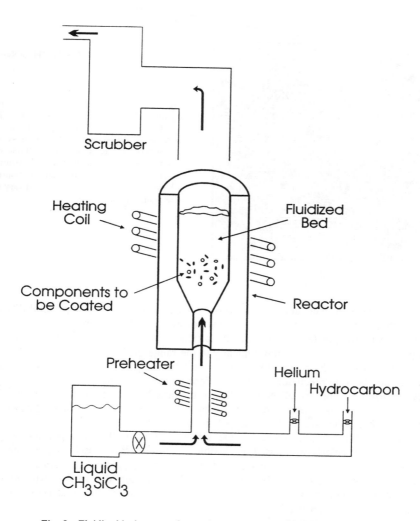

Fig. 2. Fluidized bed process for coating components with LTI pyrotytic carbon.

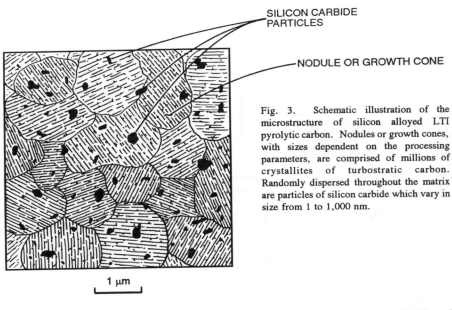

SILICON CARBIDE
PARTICLES

NODULE OR GROWTH CONE

Fig. 3.   Schematic illustration of the
microstructure of silicon alloyed LTI
pyrolytic carbon. Nodules or growth cones,
with sizes dependent on the processing
parameters, are comprised of millions of
crystallites of turbostratic carbon.
Randomly dispersed throughout the matrix
are particles of silicon carbide which vary in
size from 1 to 1,000 nm.

1 μm

$kg.m^{-3}$, they have, in fact, quite a low porosity. This leads to the low permeability of
glassy carbons to liquids and gasses since the porosity is not interconnected.

The ULTI vapor-deposited carbons can be produced with much higher densities and
strengths but only as a thin coating typically in the range of 0.1 - 1.0 mm in thickness.
The coating is formed by a hybrid vacuum process by using a catalyst to deposit carbon
from a carbon-bearing precursor. With variations in processing parameters, the density,
crystallite size and isotropy of the coating can be varied within quite wide ranges. In
addition, cooling of the substrate during coating allows the coating of low melting point
materials such as Dacron[*], Teflon[*] and polyurethane parts and sheets.

The adhesion of the ULTI coating to the substrate is naturally an important design
consideration particularly when used on flexible substrates. When used as a coating on
certain clinical stainless steels and titanium (Ti-6Al-4V) alloys, a bond strength exceeding
70 MPa is achieved. This excellent bond strength is, in part, due to the formation of
carbides at the carbon/metal interface. The bond strength with other materials that do
not form carbides is typically lower.

---

[*]Trademark, E. I. DuPont de Nemours, Inc., Wilmington, Delaware.

MECHANICAL AND BIOMEDICAL PROPERTIES

The mechanical properties of the various turbostratic carbons are closely related to their microstructures with a strong correlation of most of these properties with coating density. Other microstructural features such as crystallite size, structure and orientation, grain size and composition are also important in determining the resulting properties. Some of the more important mechanical properties of the clinically useful turbostratic carbons, together with a typical polycrystalline graphite substrate onto which coatings might be deposited, are presented in Table 1.

Table 1. Structural and Mechanical Properties of Graphite and Biomedical Turbostratic Carbons.

	Polycrystalline Graphite Substrate	Silicon Alloyed LTI Pyrolytic Carbon	Glassy (vitreous) Carbon	ULTI Vapor Deposited Carbon
Density (kg.m^{-3})	1,500-1,800	1,700-2,200	1,400-1,600	1,500-2,200
Crystallite Size (nm)	15-250	3-5	1-4	8-15
Expansion Coefficient (10^{-6} K^{-1})	0.5-5	5-6	2-6	-
Hardness (DPH)	50-120	230-370	150-200	150-250
Young's Modulus (GPa)	4-12	27-31	24-31	14-21
Flexural Strength (MPa)	65-300	350-530	69-206	345-690
Fracture Strain (%)	0.1-0.7	1.5-2.0	0.8-1.3	2.0-5.0
Fracture Toughness (MPa$\sqrt{m}$)	~1.5	0.9-1.1	-	-

Strength and Modulus

Due to the highly anisotropic properties of graphitic crystals, the macroscopic mechanical properties of polycrystalline carbons depends strongly on degree of crystal orientation. Also, the porosity of the material affects the strength by reducing the internal area over which the stress is distributed and by creating local regions of high stress. Accordingly, carbons with a high degree of preferred orientation and high density are extremely strong and stiff in directions parallel to the preferred layer planes. On the other hand, in turbostratic carbons where the crystallites are randomly arranged and the material is not fully dense, lower strengths and dramatically reduced moduli are measured. Therefore, while typical LTI pyrolytic and ULTI vapor deposited materials still retain a comparatively high strength, their moduli is uncharacteristically low. The greater porosity of the glassy carbons generally results in even lower strengths contributing to a weak structure albeit still stronger than polycrystalline graphite.

The moduli of the turbostratic carbons are close to 21 GPa and hence are close in magnitude to the modulus of bone (see Chapter 1). A carbon implant device under physiological loading and in contact with bone can therefore deform elastically with the bone and minimize stress concentrations which might otherwise develop. As a comparison, the modulus of surgical stainless steel or titanium alloys is almost an order of magnitude higher than that of bone (Chapter 1).

Hardness and Wear

The hardness or resistance of a material to indentation is an important property of turbostratic carbons since, as with most materials, it closely correlates with resistance to wear. Hardness is also important for LTI pyrolytic carbons since it correlates with the silicon carbide content, carbon crystal size and the deposition temperature and is therefore an indirect indicator of these parameters. In addition, wear rates typically decrease noticeably with increasing hardness. Silicon-alloyed LTI pyrolytic carbon has superior wear resistance compared to other forms of graphite or even unalloyed pyrolytic carbon. Excellent wear resistance is particularly important for components of artificial heart valves which articulate against each other and are exposed to significant wear, and possible cavitation erosion, during a patient's lifetime.

Fracture Resistance

The high strength and low moduli of turbostratic carbons results in large strains to failure when compared to other brittle ceramics. For example, while the total strain to rupture of an alumina ceramic specimen is $\geq 0.1\%$ and those for polycrystalline graphites are in the range 0.1 - 0.7%, the fracture strain for LTI pyrolytic carbons is 1.5 - 2.0% and can be as high as 5% for ULTI vapor deposited carbons. The high fracture strains are thought to result in part from the network of strong C - C covalent bonds in the graphitic layer planes which must be broken before failure by shear or cleavage can be accommodated. These large fracture strains are important properties for coatings of flexible polymeric substrates which must be able to sustain significant bending and flexing without cracking of the coating.

While the high fracture strains measured using smooth, uncracked specimens are high, they are sometimes incorrectly taken as an indication of material toughness. The catastrophic failure of many engineering structures occurs well below their strength or fracture strain. The toughness or "resistance to fracture" of turbostratic carbons must, therefore, be measured in the presence of cracks or flaws which may be present in the final device.

Using a fracture mechanics approach, the *fracture toughness* or critical stress intensity factor, K_c, can be measured by inserting an "atomically-sharp" crack of known length into a specimen and loading until fast fracture occurs. (In simple terms, the stress intensity, K, is given as $Q \sigma \sqrt{\pi a}$, where σ is the applied stress, a is the crack size, and Q is a geometry dependent factor of order unity.) Using this approach the toughness of silicon alloyed pyrolytic carbon has been measured in the range 0.9 - 1.1 MPa$\sqrt{m}$, only slightly higher than that of soda lime glass and less than half the fracture toughness of alumina or silicon nitride ceramics. Similar to other ceramics, these low values of fracture toughness remain a significant limitation for the reliable design of many pyrolytic carbon devices, particularly in the presence of tensile residual stresses in the coatings of composite components. Therefore, while the fracture strain does affect the fracture toughness depending on the specific fracture mechanism, the *magnitude* of the uniaxial fracture strain measured in a tensile specimen will be quite different to the highly constrained fracture strain in the triaxial stress field ahead of a sharp crack. For this reason, fracture strain and fracture toughness often show poor correlation.

In linear-elastic materials, it is also possible to express the toughness as a critical strain energy release rate, G_c, which is related to the fracture toughness by $G_c = K_c^2/E$ (where E is Young's modulus). It is interesting to note that because of the low moduli of the turbostratic carbons, in this formulation the converted toughness appears to be quite high, typically $0.04 - 0.3$ kJ.m^{-2}, compared to other ceramics which have much higher moduli and hence lower G_c values in the range $0.02 - 0.1$ kJ.m^{-2}. This apparent contradiction in toughness rankings is best rationalized by recognizing that the actual energy required to cause fracture in a device might be large due to the large deflections which result from the low moduli of turbostratic carbons. However, such fracture can occur at very low stresses depending on the device configuration. The onset of fracture is therefore nearly always calculated from a design stress approach using the fracture toughness, K_c; energy calculations based on G_c are rarely used in mechanical design against fracture.

Cyclic Fatigue

An important finding concerns the assumed insensitivity of turbostratic carbons and other brittle ceramic materials to cyclic fatigue degradation. Early studies claimed that the fatigue endurance strength of these materials was virtually identical to the single-cycle fracture stress, i.e., that cyclic stresses less than this stress do no microscopic damage. However, studies employing fracture-mechanics type testing procedures with pre-cracked samples demonstrate that fatigue cracks can grow under alternating loads in monolithic pyrolytic carbon and pyrolytic-carbon coated graphite composites in both ambient temperature air and blood-analog environments. The typical crack-growth rate per loading cycle, da/dN, for a pyrolytic carbon composite together with similar data for other brittle ceramic materials are shown in Fig. 4 as a function of the stress intensity range, $\Delta K = K_{max} - K_{min}$, where K_{max} and K_{min} are the maximum and minimum values of the applied cyclic stress intensity. Also included in the figure for comparison is the crack-growth behavior of two structural metallic alloys.

While the data in Fig. 4 show a clear (cyclic) fatigue effect, in view of past skepticism over fatigue in brittle materials, it is necessary to demonstrate unequivocally that the crack growth is cyclically induced and not simply a consequence of stress-corrosion (static fatigue) cracking at maximum load. To achieve this, crack extension was monitored with a) the stress intensity cyclically varied between K_{max} and K_{min} and b) the stress intensity held constant at the same value of K_{max}. This procedure was periodically repeated throughout the range of growth rates; a typical result, for $K_{max} =$

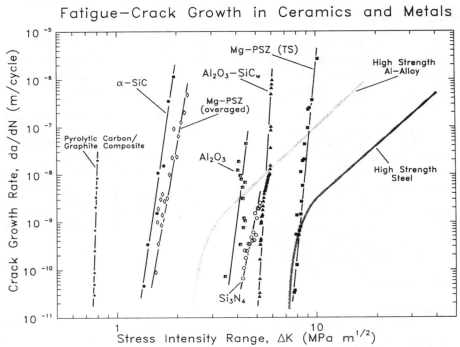

Fig. 4. Cyclic fatigue-crack growth behavior of a LTI carbon/graphite composite shown in comparison with a range of ceramics including alumina, silicon nitride, silicon carbide and zirconia, and two structural metallic alloys.

1.36 MPa$\sqrt{}$m and K_{min} = 0.14 MPa$\sqrt{}$m, is shown in Fig. 5. Whereas crack extension proceeds readily under cyclic loading conditions (region a), upon removal of the cyclic component by holding at the same K_{max} (region b), crack-growth rates are markedly reduced. Clearly, similar to behavior reported for other ceramics, a true cyclic fatigue effect is apparent; furthermore, subcritical fatigue crack-growth rates under cyclic loading appear to be far in excess of those under equivalent sustained loading.

    These observations of cyclic fatigue-crack growth in ceramics and pyrolytic carbons have important consequences for many prosthetic devices which are exposed to repeated physiological loading. This phenomenon is particularly important in heart-valve applications, as the human heart typically beats 38 million times each year, necessitating the structural design of prostheses for fatigue lifetimes in excess of patient lifetime, i.e., between 10^9 and 10^{10} cycles. Indeed, some structural failures of mechanical heart valves manufactured from pyrolytic carbon have been attributed to progressive fracture by fatigue. We discuss the importance of cyclic-fatigue effects in life prediction and device reliability in a later section.

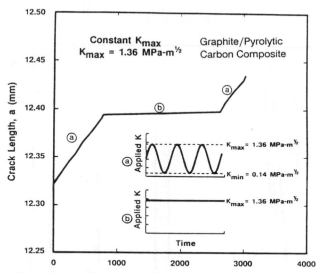

Fig. 5. The effect of sustained-load vs. cyclic loading conditions, at a constant Kmax, on the subcritical crack growth in pyrolytic carbon coated graphite tested in Ringer's solution at 37°C. Note how crack-growth rates under cyclic loading (region a) are far in excess of those measured under sustained loading (region b).

Stress-Corrosion Cracking

Similar to other ceramics, turobstratic carbons are prone to subcritical crack growth under the synergistic action of an applied load and a moist (or corrosive) environment. Such stress-corrosion (static fatigue) crack-growth behavior is plotted in terms of the crack velocity with respect to time (da/dt) as a function of the applied monotonic stress intensity in Fig. 6. The data include a LTI pyrolytic carbon/graphite composite in an environment of Ringer's lactate (blood analog) solution at a temperature of 37°C, glassy carbon in water and in air, and data for a pyrolytic graphite in air. Crack velocities span many orders of magnitude from 10^{-9} to 10^{-1} m/sec and similar to cyclic fatigue-crack growth, show a marked sensitivity to the applied stress intensity. Like cyclic fatigue, it is important to characterize the stress-corrosion cracking behavior of turbostratic carbons in order to permit reliable design.

Residual Stress

Biomedical turbostratic carbon coatings often exist in a state of residual stress. Like most coatings cooled from a higher processing temperature, the residual thermal stress state depends on the difference in thermal expansion mismatch between the coating and the underlying substrate material. Variation in the structure and grain size of the coating from the interface with the substrate to the coating surface may also affect the

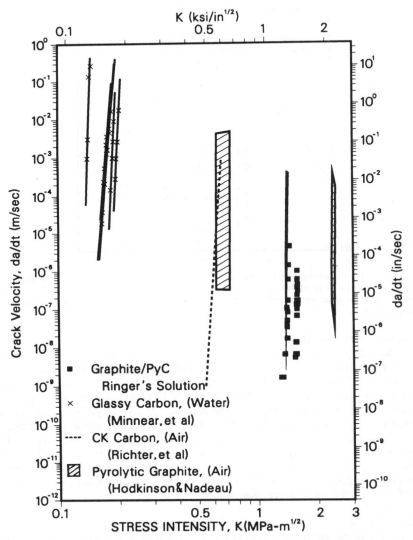

Fig. 6. Stress-corrosion crack-growth behavior of a graphite/LTI carbon coated composite tested in 37°C Ringer's solution shown in comparison with data for glassy carbon tested in water and air and for pyrolytic graphite in ambient temperature air.

residual stress state. Both effects depend on the coating thickness. For LTI pyrolytic carbon-coated graphite components, the greater thermal expansion coefficient of pyrolytic carbon compared to graphite results in a tensile residual stress in the coating. Depending on processing temperature and microstructural variation, measurements indicate a tensile stress of up to 60 MPa in the coating. Such residual stresses may be detrimental to the integrity of the coating. The residual stress must be added to the nominal applied stress when evaluating the likelihood for fracture in the coating.

Biocompatibility

From a biomaterials viewpoint, the greatest attribute of the three forms of turbostratic carbon is their excellent cellular biocompatibility and thromboresistance. This is particularly true of the high purity LTI pyrolytic and ULTI vapor deposited carbons. Depending on the polymeric precursor and the processing temperature, the purity of the glassy carbon forms may result in variability in biocompatibility. While the biocompatibility of the turbostratic carbons has been studied extensively, no theory explaining their excellent properties exists.

In cardiovascular applications, the compatibility of LTI pyrolytic carbons with blood has received the most attention. Unlike soft and hard tissue reactions to implanted materials which tend to be slow to develop, the rejective reactions in blood are dramatic and swift. Most materials in contact with blood quickly activate the tissue's clotting mechanism. LTI carbon has been shown to be equivalent to siliconized glass which causes little damage to blood. Theories to explain the excellent cellular biocompatibitily of LTI carbon with blood range from conditioning of the carbon surface with a passivating protein layer through selective adsorption to the complete inertness of the LTI carbon surface to proteins in general. A general observation, however, is that smooth or polished surfaces are better than rough surfaces, possibly because roughness may provide locations where cells can adhere and serve as thrombotic nuclei.

While ULTI vapor deposited coatings have also been shown to have excellent biocompatibility and thromboresistance with blood, glassy carbons have been studied primarily for attachment to both soft and hard tissue. When implanted, glassy carbons in general do not provoke an inflammatory response in adjacent tissue. Similar behavior has been reported for the LTI and ULTI carbons. Note, however, that although smooth surfaces promote better thromboresistance, rougher surfaces which allow tissue ingrowth and attachment provide stronger interfaces to soft or hard bone tissue.

APPLICATIONS

The excellent cellular biocompatibility of the three biomedical turbostratic carbons together with selected mechanical properties have resulted in a wide range of successful applications. Bokros has compiled a list of applications and these are reproduced in Table 2. We restrict our discussion to cardiovascular and dental applications in the following sections.

Table 2. Successful Applications of Glassy, LTI and Vapor-Deposited ULTI Carbons (after J. C. Bokros).

Application	Material	Material Characteristics Contributing to Success			
		Cellular Compatibility	Strength	Wear Resistance	Stiffness
Mitral and aortic heart valves	LTI	X	X	X	
Dacron and Teflon heart valves sewing rings	ULTI	X			
Blood access device	LTI/titanium	X			
Dacron and Teflon vacular grafts	ULTI	X			
Dacron, Teflon and polypropylene septum and aneurysm patches	ULTI	X			
Pacemaker electrodes	Porous glassy carbon-ULTI-coated porous titanium	X			
Blood oxygenator microporous membranes	ULTI	X			
Otologic vent tubes	LTI	X	X		
Subperiosteal dental implant frames	ULTI	X			
Dental endosseous root form and blade implants	LTI	X	X		X
Dacron-reinforced polyurethane aoplastic trays for alveolar ridge augmentation	ULTI	X			
Percutaneous electrical connectors	LTI	X	X		
Hand joints	LTI	X	X	X	X

Cardiovascular

Mechanical cardiac valve prostheses are designed to regulate blood flow continuously in hostile physiological environments for periods in excess of patient lifetimes. Valve components are exposed to cyclic loading, flexing and bending, wear at surfaces exposed to articulation and cavitation erosion on surfaces exposed to blood flow. These requirements represent one of the most demanding biomaterial applications. They also represent, however, one of the greatest successes of bioceramics in the form of silicon alloyed LTI pyrolytic carbon which is currently used in the majority of artificial heart valves.

Many modern heart valves are based on the tilting disk design which open and close as the heart beats, allowing blood flow under near-normal rheological conditions. Some designs involve two semi-circular leaflets (bileaflet valves) instead of a single disk or occluder. The disk or leaflets are contained in a circular housing which has a cloth sewing ring on the outer diameter to facilitate attachment to the heart (Fig. 7). The disk or leaflets and housing are manufactured from LTI carbon coated graphite, although some devices use a cobalt-chromium or titanium alloy housing. Unfortunately, the complex shapes of some of the metal housings require the use of casting and welding technologies in their manufacture. As a result, serious problems have arisen with some of these devices leading to catastrophic failures by fatigue of the valve housing.

Fig. 7. A typical bi-leaflet artificial heart valve with components made from silicon alloyed LTI pyrolytic carbon (courtesy Baxter Medical Inc.).

One of the primary restrictions on artificial heart valves is the uncertain lifetime under complex physiological loading due to degradation by cyclic fatigue or stress corrosion. While heart valves manufactured from pyrolytic carbon coated graphite are currently used extensively with more than 200,000 in use, several structural failures of LTI carbon coated components have been attributed to cyclic fatigue. These conclusions are questionable due to difficulty in distinguishing fractographic features produced under cyclic and monotonically loaded conditions.

The identical morphology of cyclic fatigue and overload (fast) fracture surfaces is not unique to pyrolytic carbon (Fig. 8a,b); similar behavior is shown by the graphite core of the composite material (Fig. 8c,d), and by most ceramic materials where cyclic fatigue fractures are documented. This is in marked contrast to metallic materials where the characteristic cyclic-fatigue fracture mode, i.e., striations (Fig. 8e), is quite distinct

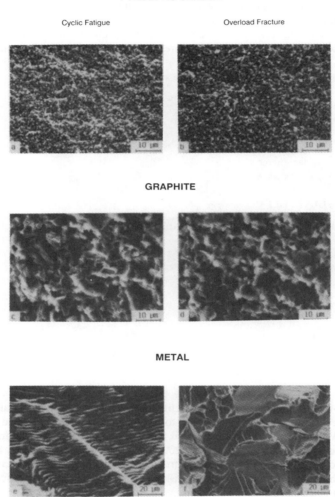

Fig. 8. Low and high magnification scanning electron micrographs of cyclic fatigue and overload (fast) fracture in a,b) the LTI pyrolytic-carbon cladding, c,d) the polycrystalline graphite substrate, and e,f) a metallic material (low-strength ferritic steel), showing e) fatigue striations and f) transgranular cleavage fracture.

to the transgranular cleavage (Fig. 8f), microvoid coalescence or intergranular fracture seen for failure under monotonic loads. The reason for this is not clear, due to current uncertainties in the micromechanisms of ceramic fatigue. The implication is that post-failure analysis of fracture surfaces in pyrolytic carbon can be very deceptive.

Dental

The close elastic modulus match between glassy and LTI carbons and bone make the carbons candidate materials for a number of load-bearing dental implant applications. Artificial tooth roots with sizes up to 11 mm in length and 5 mm in diameter have been manufactured from glassy carbon. LTI carbon with its superior strength has been used in more complex implant designs. Where the size and complexity of implants does not allow construction from carbon, components can be fabricated from metal alloys and coated with a thin impermeable layer of ULTI vapor deposited carbon. Subperiosteal dental implants and endosseous tooth root replicas have been fabricated using these techniques. Such designs combine the superior mechanical properties of the metal with the chemical properties of carbon.

DEVICE RELIABILITY

Careful mechanical design and accurate life prediction, which must accurately simulate realistic failure modes, are essential elements for the reliable use of ceramic implants exposed to complex physiological loading and environmentally-induced degradation in many clinical applications. The structural design of cardiovascular-assist devices places particularly demanding requirements on the pyrolytic carbon-coated components. To maintain structural integrity, prosthetic heart valves must therefore be designed to endure fatigue lifetimes greater than $\sim 10^9$ cycles in blood analog environments; current FDA requirements demand that this is achieved using so-called damage-tolerant design procedures. These procedures rely upon the fracture-mechanics based concept that a conservative (worst-case) estimate of the structural life can be defined in terms of the time, or number of loading cycles, for the largest undetected crack to grow to critical size, generally defined in terms of the material's fracture toughness, K_c. A similar analysis might be required for other forms of crack growth, for example, stress-corrosion cracking in components subject to sustained (non-cyclic) loads.

There are three principal materials-property inputs to such calculations; the initial flaw population or size of the largest (worst-case) crack pre-existing in the device (which must be defined by quality control), the critical crack size for catastrophic failure of the device (which in most cases is defined by the material's fracture toughness, K_c), and the rate at which the incipient cracks will grow subcritically between these two limits. Of these, the crucial input for fatigue life prediction is the rate of subcritical crack growth; specifically, relationships are required which define the crack-growth increment per cycle, da/dN, as a function of a crack-driving force such as the stress intensity K. In prostheses fabricated from metallic materials such as cobalt and titanium alloys, such cyclic fatigue-crack growth data are widely available in handbooks or can be measured using ASTM Standard techniques. In prostheses manufactured from pyrolytic carbons, fatigue data have only recently become available, in part because it had been incorrectly assumed that pyrolytic carbon was immune to cyclic fatigue failure. Damage-tolerant design and life-prediction procedures for both metallic and pyrolytic carbon implants are based on fatigue-crack propagation data derived from conventional ASTM-style (e.g., compact-tension) test pieces containing long (typically ~2 to 20 mm in length), pre-existing, through-thickness cracks.

Whereas all current analyses for pyrolytic-carbon/graphite implants utilize crack-propagation data conventionally measured on specimens with long, through-thickness cracks, the reality of actual flaw growth for valves in service is more likely primarily associated with small, part-through (half-penny shaped) cracks which initiate predominantly on the surface of the pyrolytic-carbon coating and propagate both along the surface of the pyrolytic carbon and depth-wise toward the graphite core. Such small-crack behavior is undoubtedly of special importance to most ceramics simply because ceramic components will in general not be able to tolerate the presence of physically long cracks. Although the collection of small-crack data is both complex and labor-intensive, life-prediction procedures should be based on such data.

Studies of small-crack growth behavior in other ceramic and ceramic-composite materials suggest that a reasonable correspondence can be obtained between the growth-rate behavior of long, through-thickness cracks and small, part-through surface cracks, *provided cyclic crack-growth rates are characterized in terms of the appropriate stress-intensity factor, incorporating both external (applied) and internal (residual) stresses.* Such internal stresses may result during fabrication of the pyrolytic-carbon/graphite composite and/or from surface damage such as scratches or gouges incurred during handling of the device. With the appropriate characterization, in principle either form of growth-rate data could be utilized for life prediction.

Despite the highly successful and expanding use of pyrocarbons in the biomedical industry, since their initial development during the mid-1970's, little fundamental materials research has been published on these materials. Little is known about the mechanistic role of the microstructure in influencing their fracture and fatigue properties, so critical for the structural integrity of heart-valve prostheses. For the continued use of turbostratic carbons as a structural material for such safety-critical applications as medical implant devices, the mechanistic and microstructural basis underlying critical mechanical properties, such as strength, ductility, toughness, and resistance to fatigue, wear and erosion, must be further investigated.

ACKNOWLEDGEMENT

This work was supported by the Director, Office of Energy Research, Office of Basic Energy Sciences, Materials Sciences Division of the U.S. Department of Energy under Contract No. DE-AC03-76SF00098.

READING LIST

W. V. Kotlensky, "Deformation in Pyrolytic Graphite," *Trans. Met. Soc. AIME* **223** (1965) 830-832.

W. V. Kotlensky and H. E. Martens, "Structural Changes Accompanying Deformation in Pyrolytic Graphite," *J. Am. Ceram. Soc.* **48** (1965) 135-138.

J. C. Bokros, L. D. LaGrange, and F. J. Schoen, "Control of Structure of Carbon for use in Bioengineering," in *Chemistry and Physics of Carbon*, ed. P. L. Walker (Dekker, New York, 1972) pp. 103-171.

J. C. Bokros, R. J. Akins, H. S. Shim, A. D. Haubold, and N. K. Agarural, "Carbon in Prosthetic Devices," in *Petroleum Derived Carbons*, eds. M. D. Deviney and T. M. O'Grady (American Chemical Society, Washington, D.C., 1976) pp. 237-265.

A. D. Haubold, R. A. Yapp, and J. C. Bokros, "Carbons for Biomedical Applications," in *Encyclopedia of Materials Science and Engineering*, Vol. 1, ed. M. B. Bever (Pergamon Press, Oxford, U.K./MIT Press, Cambridge, 1986) pp. 514-520.

R. B. More, A. D. Haubold, and L. A. Beavan, "Fracture Toughness of Pyrolite, Carbon," *Trans. Society for Biomaterials* **15** (1989) 180.

R. H. Dauskardt and R. O. Ritchie, "Cyclic Fatigue-Crack Growth Behavior in Ceramics," *Closed Loop* **17** (1989) 7-17.

R. O. Ritchie, R. H. Dauskardt, W. Yu, and A. M. Brendzel, "Cyclic Fatigue-Crack Propagation, Stress-Corrosion and Fracture-Toughness Behavior in Pyrolytic Carbon-Coated Graphite for Prosthetic Heart Valve Applications," *J. of Biomed. Maters. Res.* **24** (1990) 189-206.

R. O. Ritchie and R. H. Dauskardt, "Cyclic Fatigue in Ceramics: A Fracture Mechanics Approach to Subcritical Crack Growth and Life Prediction," *J. Ceramic Soc. Japan* **99** (1991) 1047-1062.

R. O. Ritchie, R. H. Dauskardt, and F. J. Pennisi, "On the Fractography of Overload, Stress Corrosion and Cyclic Fatigue Failures in Pyrolytic-Carbon Materials Used in Prosthetic Heart-Valve Devices," *J. Biomed. Mater. Res.* **26** (1992) 69-76.

Chapter 15

BIOCERAMIC COMPOSITES

P. Ducheyne, M. Marcolongo, E. Schepers*

Department of Bioengineering, University of Pennsylvania, Philadelphia, USA
*Department of Prosthetic Dentistry, University of Leuven, Leuven, Belgium

INTRODUCTION

Man-made materials have been utilized for reconstructive and corrective procedures in the body for several centuries, but it was not until the late nineteenth century, when Lister described the principle of aseptic intervention, that any reasonable chance of success was possible. This was the first milestone on the way to our present day, highly sophisticated use of materials in the body. Through the twentieth century, until the late sixties, metals were the primary types of materials considered for implantation. The main impetus for use of ceramics in augmenting body tissues followed from Hulbert's work. Hulbert reasoned that metals, when implanted into the body, were not in their highest oxidation state, and would undergo ionization in the body. Depending upon the extent of ionization and the characteristics of the resultant dissolution product, there may be some degree of effect. Most ceramics are stable and do not ionize, and bioinert ceramics, such as alumina and carbon, are used as implant materials today and are described in Chapters 2 and 14.

It was not long before another class of ceramics was proposed for use in the body, the bioactive ceramics. It was argued that no material is truly inert, not only because of chemically induced reactions, but also from reactions due to the physical presence of the implant. It was proposed that an implant material would achieve highest compatibility if it allowed normal tissue at its surface.

The early development of bioceramics was prompted by considerations of biocompatibility, but implant devices must posses an appropriate range of mechanical properties. It was realized that for a number of biomedical applications, ceramics alone, either bioactive or inert, could not meet the diverse requirements of safe and effective *in vivo* functioning. This turned attention toward composite materials which could take advantage of the desirable properties of each of the constituent materials, while mitigating the more limited characteristics of each component.

Bioceramic composites can be divided in to three categories: bioinert, bioactive, and biodegradable composites. In these categories, the ceramic phase can be the reinforcing material, the matrix material, or both. Synthesizing a successful composite, follows identification of desirable properties as well as inferior characteristics of each material.

The desirable properties of bioceramics are:

inert: minimal biological response, high wear resistance
bioactive: enhanced bone tissue response, bone bonding
resorbable: material is replaced by normal tissues, thereby excluding possible
                long term effects

However, these materials are limited for use in the body by:

inert: limited mechanical properties in tension
bioactive: limited tensile strength and fracture toughness
resorbable: rate of strength reduction due to resorption may be too rapid.

There are several bioceramic composites which have been fabricated and analyzed (Table 1). These include stainless steel fiber/bioactive glass (the first bioceramic composite) and titanium fiber/bioactive glass composites. These discontinuous metal fiber/ceramic composites maintain bioactivity, while increasing the fracture toughness and strength of the material in comparison with that of the ceramic alone. There are also ceramic reinforced/ bioactive ceramic composites (e.g., zirconia reinforced/AW glass and apatite/wollastonite (AW) two phase bioactive glass ceramic), and bioceramic augmented polymeric matrices (e.g., calcium phosphate reinforced polyethylene). It has been shown that these composites are bioactive and have higher mechanical strength than the bioceramics themselves.

A look at bioceramic composite fabrication methods, mechanical properties, including fracture mechanics data as far as known, and *in vivo* reactions for static and dynamic load conditions will follow. The specific examples in this chapter are the stainless steel/bioactive glass and titanium fiber/bioactive glass-ceramic composites. However, it must be understood that the principles of composite design, fabrication, and analysis are applicable to many different bioceramic composites.

Table 1. Some Bioceramic Composites.

Inert	Carbon fiber reinforced carbon
	Carbon fiber polymetric matrix materials (polysulfone, poly aryl ether ketone, poly ether ketone ketone, poly ether ether ketone)
	Carbon fiber reinforced bone cement
Bioactive	A-W Glass-Ceramic
	Stainless steel fiber reinforced Bioglass®
	Titanium fiber reinforced bioactive glass
	Zirconia reinforced A-W glass-ceramic
	Calcium phosphate particle reinforced polyethylene
	Calcium phosphate fiber and particle reinforced bone cement
Resorbable	Calcium phosphate fiber reinforced polylactic acid

FABRICATION METHODS

Stainless steel fiber/bioactive glass composites are made of 45S5 bioactive glass and AISI 316L stainless steel fibers. The nominal composition for the bioactive glass is: 45wt% SiO_2, 24.5% CaO, 24.5% Na_2O, and P_2O_5. AISI 316L stainless steel has the following compositional range for the more important alloying elements: 16-20wt % Cr, 10-14 % Ni, 2-4 % Mo, less than 0.03 % C.

The preparation of the stainless steel fiber/bioactive glass composite involves a number of steps which includes preparing the fiber preform, impregnating the preform with the glass matrix, and heat treating the composite. Initially, the required amount of fiber for a given geometry is weighed and compacted under pressure, the fibers are sintered at 1250°C and then the uniformity of the preform's density is inspected using radiography. The surface of the sintered metal fiber preform is oxidized for 10 min in air at 800°C. The total metal shrinkage is controlled by holding the porous preform at 400°C for 20 min, and then it is immersed into molten glass maintained at 1350 to 1380°C, and finally the glass-impregnated preform is annealed at 400 to 500°C for 4 hours and furnace cooled. The processing parameters have been optimized experimentally to achieve this successful procedure.

For an effective stress transfer between the glass matrix and the reinforcing metal fibers, there must be a good bond between the glass and the metal fibers. This is achieved through the oxidation of the metal fibers before immersion in the molten glass. For a good glass to metal bond, the oxide layer on the metal itself must be adherent. Subsequently, this oxide layer must not dissolve completely in the glass during the immersion process. The enrichment of the main alloying elements in the inner layer of the formed oxide is responsible for the good adherence of the oxide on the metal. At the oxide-glass interface, the oxide is only partly dissolved. As a consequence, there is a graded transition from the glass to the metal fibers. Figure 1 illustrates this principle. There is diffusion of the iron in the glass, and there is also diffusion of the silicon into the oxide. This graded glass-metal interface corresponds to chemical adhesion of the constituents, so stress transfer from the glass matrix to the reinforcing metal fibers is much more effective than through a purely frictional interaction at the interface.

There are several variables which affect glass impregnation: viscosity of the molten glass, thermal expansion coefficients of the metal substrate and glass, oxidation and roughness of the metal surface, metal temperature at time of immersion, duration of immersion, time and temperature of annealing, volume fraction of the metal fibers and size of porosity.

Many of these variables were optimized through research in coating of metal rods with a glass layer, including viscosity, temperature and time of immersion. Of the remaining variables, the most significant challenges to the composite fabrication were thermal expansion of the metal and glass, and oxidation of the metal surface.

The mismatch between the thermal expansion coefficients of the glass and the metal fibers is significant. The values for stainless steel AISI 316L in 100μm fiber compact are 20.0 x 10^{-6}°C^{-1} (up to 200°C) and 21.8 x 10^{-6} °C (up to 400°C) and for bioactive glass 45S5 is 18.0 x 10^{-6} °C (up to 450°C). By combining the two materials,

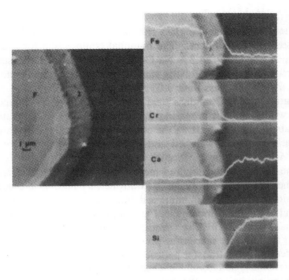

Fig. 1. Interfacial chemical analysis of the stainless steel fiber-to-glass-to-bone.

there is considerable potential for development of internal stresses during the cooling of the composite.

A variation on this general principle of impregnating a fiber preform can be found in the method used for fabricating a titanium fiber/bioactive glass composite. In using titanium, the difference in coefficients of thermal expansion is much greater. The thermal expansion coefficient of titanium is 8.7×10^{-6} $^\circ C^{-1}$ in the 25 to 500°C temperature range. As a result it is impossible to reinforce bioactive glass 45S5 with titanium fibers. However, a titanium/bioactive glass composite is desirable since the titanium fiber, having excellent biocompatibility, would eliminate adverse tissue effects arising from the susceptibility of stainless steel fibers to crevice corrosion.

The bioactive glass used in the stainless steel fiber/bioactive glass composite is one of the most highly reactive bioceramics known. This is not always the best type of bioactive material for a composite because of the large reaction layer formed between the glass and bone, which leads to a weak interfacial strength. In designing the titanium fiber/bioactive glass composite, the glass composition was designed to have a slower bioactivity rate.

The fabrication process begins with the selection of the glass composition. 8-12% by weight Na_2O, 4-8% P_2O_5, 50-54% SiO_2, 0.5 -5% CaF_2 and the balance CaO leads to a bioactive glass-ceramic with a thermal expansion coefficient of 9×10^{-6} $^\circ C^{-1}$ and a reaction layer about two orders of magnitude smaller than that of bioactive glass 45S5. The powder mix which will yield this composition, is melted, homogenized, and cast.

It is then cooled to obtain non-crystallized glass. The glass is annealed at 650°C for 4 hours, then ground finely to a mean particle size of 15 μm. The powder obtained has a homogeneous composition.

The titanium fibers, 4 mm long and 50 μm diameter, are pickled and cold compacted directly to the desired configuration. The titanium preform is placed in a copper or glass mold with the glass powder surrounding it, and vacuum closed. Hot isostatic pressing (HIP-ing) is carried out at a temperature in the range of 700 to 800°C at a maximum pressure of 1000 bars. This HIP-ing process results in glass impregnation of the titanium preform and partial crystallization of the glass, the extent of which depends of the HIP-ing cycle. Finishing includes polishing to ensure that the Ti-fibers are on the composite surface.

With this process, the surface distribution of the Ti is homogeneous, and the ratio of the area of visible fibers to the area of bioactive glass-ceramic is about 65:35.

MECHANICAL PROPERTIES

Tensile tests were done on a series of stainless steel/bioactive glass composites fabricated with 50, 100, and 200 μm diameter fibers, molded to different fiber volume fractions in the range of 0.4 to 0.6. The excess outer glass rim was removed by sand blasting and subsequent grinding on consecutive SiC paper on grits up to 320. Tensile test results are shown in Table 2, along with the properties of the parent glass and the porous fiber material. The results show that an increase in fiber volume percent of the 50 μm fibers caused a decrease of yield strength, while the ultimate strength remains unchanged; the modulus of elasticity also decreases. This observation apparently contradicts the theoretical assessment of the modulus of elasticity of composites, which

Table 2. Initial Tensile Tests of Stainless Steel Fiber/Bioglass® Composites.

Sample	Sample Type	$\sigma_{0.01}$(MPa)	SD	UTS (MPa)	S	E(GPa)	SD
50-S-45	Bioglass®-steel	59	5	80	17	112	13
50-S-60	composite	45	3	81	9	83	4
100-S-45		49	12	55	7	98	10
100-S-60		55	2	97	11	107	8
200-S-60		40	2	54	9	108	18
50-S-45	Porous fibre	20		70-110		35	
50-S-60	metal	42		140		65	
100-S-45		10		70		20	
100-S-60		30		135		65	
Bioglass® 45-S-5	Parent glass			42			

*Sample name is composed as follows; 1st figure represents the fiber diameter in μm, S represents stainless steel, and the second figure represents vol% of fibers.
®Registered trademark, University of Florida, Gainesville, Florida

projects a larger modulus for an increasing fraction of the higher modulus constituent (usually to a point of maximum stiffness, then any additional fibers cause a decrease in modulus, typically 60% is the maximum). In this case, the phenomenon may be caused by the fact that the distance between neighboring fibers has become too small to have a strengthening effect on the glass. In such a case, it is essentially the properties of the fiber phase of the composite which are measured. The compressive stress fields of neighboring fibers probably interfere with each other to such an extent that no appreciable strengthening can be achieved.

On the basis of these results, a second series of tensile and three-point bending tests were limited to composites made of either 45 vol% 50 μm (50-S-45) and 60 vol% 100 μm (60-S-100) fibers. Each of these specimen types had three different surface finishes, with final grinding grits of 80, 320, and 600. The results of this second series of tensile and bending testing are compiled in Table 3. Figure 2 shows a series of specimens after testing: this figure clearly documents the extensive plasticity that can be achieved by modifying bioactive glass into a glass-metal fiber composite. The surface finish had no appreciable effect on the mechanical properties of the different samples, and so the data can be considered *en masse*. Both the bending and tensile strengths of the composites are significantly higher than those of the parent glass or the fiber mesh. There seems to be little difference in the mechanical properties of the 50-S-45 and the 100-S-60 composites.

Table 3. Mean Strength Values of Two Stainless Steel Fiber/Bioglass® Composites, as Measured by Tensile and 3-point Bending Tests.

Sample	Mode of Testing	$\sigma 0.01$(MPa)	SD	$\sigma_{0.2}$(MPa)	SD	UTS(MPa)
50-S-45	Tension[*]	60.9	11.8	78.3	11.2	91.9
	Bending[†]	----		167.6	38.4	290.4
100-S-60	Tension[*]	57.0	8.6	73.3	7.1	97.9
	Bending[†]	----		162.9	31.4	339.9

[*]Twenty specimens
[†]Ten specimens

The strength values of the composites are close to the upper bound value which can be assumed by

$$\sigma_c = \sigma_{uf}\nu_f + \sigma_{um}\nu_m \tag{1}$$

where σ_c is the composite strength, σ_{uf} is the ultimate fiber strength, ν_f is the fiber volume fraction, and ν_{um} is the matrix volume fraction. This equation is valid within the assumptions of continuous reinforcement of aligned fibers, equal strains in both components and absence of internal stresses. With a σ_{uf}=530MPa and σ_{um}=42MPa, the composite strengths given in Table 4 are obtained. However, since internal compressive stresses are present in the glass fraction, and since the matrix stress, σ_{um}, is probably larger as the glass is strained as small volumetric units, the strength by just combining

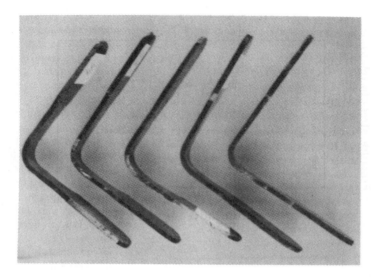

Fig. 2. Series of mechanically tested specimens: notice the extensive deformability of this glass-metal composite.

Table 4. Calculated and Measured Strengths of Two Optimized Microstructures of Stainless Steel Fiber/Bioglass® Composites.

Composite	Composite Strength (MPa)	
	Calculated	Measured
50-S-45	262	290
100-S-60	335	335

These calculated values are very close to the measured values.

two materials should be lower than the calculated numbers. There is still room for increasing the strength of this particular fiber reinforced glass composite, probably by full optimization of the fiber diameter, the respective volume fraction of each constituent and the fiber/glass bonding.

In another series of tests, the four point bend fracture strengths were measured as a function of strain rates. These tests were performed on the 50-S-45 and 100S-60 composites. As seen in Fig. 3, there is a linear relationship between the logarithms of fracture strength and strain rate, which indicates that the properties of the composite are still markedly dependent on the properties of the ceramic phase, since fracture mechanics theory of ceramics, unlike the theory for metals, prescribes this linear relationship.

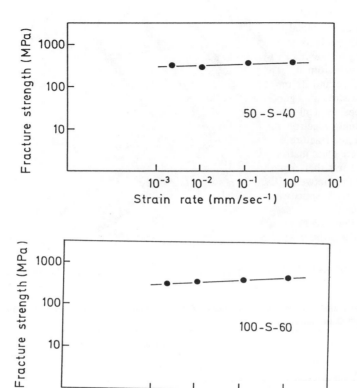

Fig. 3. Fracture strength as a function of strain rate for two stainless steel fiber/Bioglass® composites.

There is a marked difference in fracture between the two types of composites is bent heavily with local cracks at the surface. A fractographic analysis demonstrates the significance of these results.

DUCTILITY AND FRACTOGRAPHIC ANALYSIS

The stainless steel-bioactive glass composites display a marked ductility. In tension testing, up to 10% deformation to failure is observed. In three-point bend testing plastic deformation until a bend angle of over 90 degrees is invariably obtained; this was the maximum attainable with the experimental set-up used.

The onset of plastic deformation of the composite is probably caused by plastic deformation of the metal fibers. Subsequently, multiple cracking of the matrix occurs. Fractographic analysis shows cracks in the glass, perpendicular to the tensile direction, with blunting of the cracks by fibers. Cracking is extensive at the fracture surface, but is limited away from it. There is no debonding between fiber and glass except at the fracture surface and its immediate vicinity. This indicates that stress transfer between fiber and glass remains good until structural damage is close to yielding complete failure of the composite. Only then does debonding and fiber pull-out take place.

Scanning electron microscopical (SEM) examination reveals that the extensive elongation at the fracture surface is almost solely due to the stretching and pull-out of the fibers from the glass matrix; there is little glass between the fibers and it is the fibers that are torn and not their sinter-bonds. Figure 4 are SEM micrographs of this finding. After debonding between glass and fiber occurred, the stretched fibers elongated without the glass. Only at spots where the cracking was perpendicular to the fiber does the glass remain on the fiber.

The reinforcement of the glass caused by the fibers is threefold:

1. *The cohibitive effect:* if the bond between the fibers and the glass is good, stress transfer between both is effective. The main portion of the stress then acts upon the constituent with the greatest modulus of elasticity, in the present case, the metal fibers. As the stress in the glass matrix is decreased, subcritical crack growth is hindered.

2. *The cooperative effect:* if cracks grow in the glass matrix, they are blunted by the presence of the fibers. The stress concentration at the tip of the crack is reduced initially by the plastic deformation of the fibers. A subsequent mechanism which can be operative when substantial gross deformation has occurred, is the decohesion between the glass and fibers, followed by pull-out of the fibers.

For both effects, the cohibitive and cooperative effect, the glass-to-metal interface plays an important role. There must be a compromise between the contradictory requirements for strength and toughness. Best strength is reached for a high interfacial bond between the glass and the fibers, but the fiber pull-out which adds considerably to toughness, requires decohesion between the glass matrix and the metal fibers. In the present composite, initially good interfacial bonding is achieved for high strength, and fiber pull-out occurs with substantial gross deformation of the fibers.

3. Crack initiation and propagation is also hindered by residual compressive stresses in the glass matrix. This is due to the difference in thermal expansion between fibers and glass, which puts the glass in compression. Here also, a good interfacial bond is essential to carry the induced stresses at the glass/metal interface.

TISSUE RESPONSE

Static Load Conditions

In view of potential clinical applications of bioactive glass composites the biological tissue response must be evaluated, examining the difference in material's and host's tissue response to bulk glass and bioactive glass metal composites, the effect of

(a) Overview of the fracture surface (fiber diameter: 50μm).

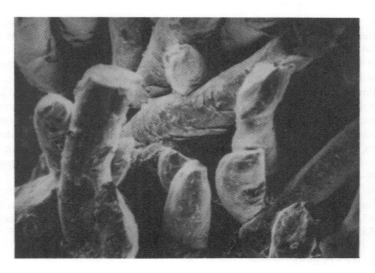

(b) Fiber pull-out and fiber-rupture.

Fig. 4. Scanning electron micrographs documenting some of the reinforcement mechanism of the metal fiber-bioactive glass composites.

Fig. 4. Continued.

(c)   Oxide scale spalling off at ultimate failure indicating excellent adhesion between matrix and fiber.

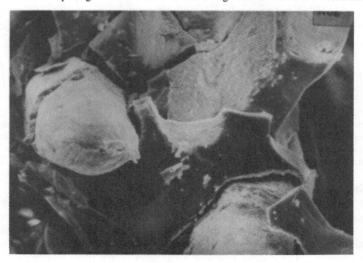

(d)   Glass remnant in the center of the photograph, indicating continuous matrix-to-fiber stress transfer
           till final fracture.

a possible exposure of the metal fibers, the subsequent interaction of released metal and glass ions, the influence of time on the interfacial reactions and the thickness of the reacted glass layers.

Implantation experiments were first done under static load conditions to evaluate the bonding behavior of the stainless steel/bioactive glass composite. Bioactive glass composite plugs with a diameter of 5mm and a length of 8mm were implanted for 8 weeks in the femora and tibiae of dogs, together with control bulk bioactive glass and control bulk stainless steel specimens. The specimens were implanted under statically loaded conditions. The shear strength of the bond-plug was measured by a push-out test.

The results of the push-out test are presented in Fig. 5. Stainless steel specimens were encapsulated by fibrous tissue. The plugs required only minimal force to be pushed out of the capsule. In contrast, the bioactive glass and the bioactive glass composite plugs bonded to bone. Comparing the measured shear strengths with respect to position of each plug and thus the intrinsic strength of the surrounding bone, the experiments do not show a significant bonding difference between bioactive glass and bioactive glass composites.

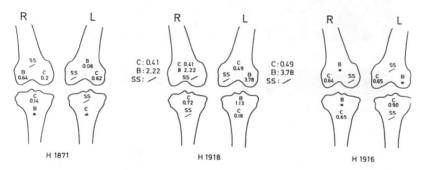

Fig. 5. Mechanical test results of stainless steel fiber/bioactive glass composite push-out testing. Shear stress units are in MPa. C represents composite, B for bioactive glass, SS for stainless steel. An asterisk means that the bone surrounding the specimen was damaged inadvertently during dissection. (From Gheysen et al. *Biomaterials*, V4, (1983) with permission.)

Stainless Steel Fiber Reinforced Bioactive Glass as Tooth Root Implant

Beagle dogs were used as experimental animals in a dental root implant study. The premolars were extracted at least one year prior to implantation. With appropriate anesthesia cavities were prepared in the edentulous areas. Low-speed drills (approximately 300 RPM) with abundant cooling by physiological saline were used for all the preparations. Three dogs received subgingivally six fiber-reinforced bioactive glass implants with a pin of bioactive glass on the surface and one bulk bioactive glass, as a control. They were killed 4 months after implantation. In one animal five bulk bioactive glass implants were installed in the lower jaw and two in the upper jaw. This animal was killed after 16 months.

Following the usual preparation for light and SEM no difference in the reaction pattern between bulk bioactive glass and fiber reinforced composite glass was observed at 4 months (Fig. 6). The various reaction layers present on the composite glass were similar to those reported previously. Energy dispersive X-ray (EDX) analysis clearly demonstrated the formation of a Si-rich layer covered by a CaP-rich layer at the outer surface.

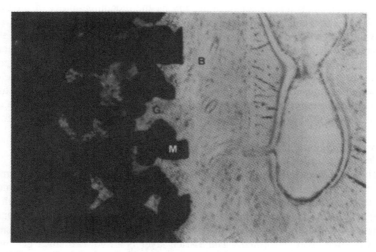

Fig. 6. Micrograph of the interfacial zone between stainless steel-bioactive glass composite and bone tissue.

Histologically, connection between the implant surface and bone is readily apparent: it is preferentially present at the cortical bone border. No interposed fibrous tissue can be seen between the implant and the bone. The osteocytes at the interface are regularly distributed and some cell processes in the canaliculi of these cells appear to be in a very close relationship with the reacted implant surface. EDX analysis of the Ca-P rich layer in such an interface shows an increasing Ca and P concentration at the outer glass surface and a smooth transition of the Ca and P peak intensities towards the bone tissue. This results in a compositional gradient between bioactive glass and bone tissue. In contrast, when the implant is surrounded by fibrous tissue the Ca and P profiles increase toward the outer surface but drop immediately to zero when the neighboring tissue is scanned. This fibrous encapsulation of the implant is preferentially seen at the apical part of the implant, near the infra-alveolar nerve where no bone tissue is present at the time of installation. The fibrous tissue capsule consists of dense packed collagen fibers and contains few cells. These observations confirm that bioactive glass is not osteoinductive since bone connection is only formed when bone tissue is present in the immediately vicinity of the glass surface at the time of installation. This lack of osteoinductivity also is substantiated by the absence of bone tissue at the glass surface around a perforation to the maxillary sinus. Close adaptation of the epithelial tissue to the glass surface is observed here. Whereas no osteoinduction can be ascribed to the glass, it is osteoconductive. An outgrowth of bone can be observed in the apical and the cervical directions, where it starts from the initial contact area with cortical bone.

When the implant is intact, EDX analysis does not reveal any ion diffusion from the metal fibers into the outer glass rim of the implant or in the surrounding tissues. Thus, the outer glass rim is effective in preserving the biocompatibility of the bioactive glass itself, and prevents any influence of the stainless steel fiber. However, if metal fibers are directly exposed to the surrounding tissue fluids, Fe ions from the stainless steel fibers are detected by EDX point analysis in tissues of these areas. Other metal ions are not detected, but this can be due to the detection limit of this technique (approximately 0.01 - 0.1%). These metal ions inhibit the interfacial osteogenesis in these areas and can even elicit an inflammatory cell reaction. Since these reactions are not consistently observed, it is suggested that the tissue response in areas with a cracked glass rim is due to the synergistic action of dissolved glass ions and metal fiber ions.

Stainless Steel Fiber Reinforced Bioactive Glass
Dental Implants Under Dynamic Load Conditions

The mechanical properties of the bioactive glass are substantially improved by reinforcing the glass with stainless steel fibers and it may appear that a structurally reliable material is available. However, one still must consider the strength of the reacted glass layer at the outer surface. The bonding of the bioactive glass to bone tissue is associated with a series of chemical reactions, resulting is a Si-rich layer on top of which a CaP-rich layer is formed. The Si-rich layer is a hydrated silica gel, with low mechanical properties. In a study in which chewing forces acted on the stainless steel fiber-reinforced bioactive glass implants with an outer rim of pure glass, it was shown that even in the presence of an established glass-bone bond, this reaction zone is susceptible to shearing upon loading. Upon a topical shear failure of the reacted glass layer, the glass leaching reactions are perpetuated and eventually the underlying metal fibers are exposed. Since it has already been shown that the stainless steel fibers must remain covered by a rim of glass, the exposure to the surrounding tissue fluids of the stainless steel fibers is the last step in the failure of stainless steel-glass composites to achieve long term bonding under functional load conditions.

Titanium Fiber Reinforced Bioactive Glass Ceramic
Dental Implants under Dynamic Load Conditions

The outcome of clinical studies with stainless steel/45S5 composites is drastically altered by modifying the composite in two ways, replacing stainless steel by titanium, and modifying the bioactive glass to be less reactive. A study aimed at determining the integrity of the bond under functional load conditions for time periods of up to two years is nearly completed.

Implants were installed in the partial edentulous lower jaws of beagle dogs, four in each half. After three months of subgingival installation of the "root" parts of the implants, the permucosal parts were connected. On top of the implants, fixed crowns or bridges were installed. This resulted in the application of chewing forces to the bone-glass composite interface. Clinically these implants have been functioning very well during the experimental period. This must be contrasted to the glass-stainless steel fiber implants which all failed clinically within two weeks of transmittal of occlusal forces.

The formulation of this new composite, comprising titanium fibers and bioactive glass-ceramic, both intentionally exposed at the surface, was based on two animal studies. One of these studies illustrates that bioceramic composite development can be based on a fundamental understanding of the properties to be achieved.

Tissue Response to a Bioactive Glass-metal Interface

In order to study whether the exposure of any metal in contact with glass to surrounding tissues is precluded, a comparative analysis of bioactive glass in contact with titanium, wrought Co-Cr alloy and stainless steel was done. A series of 24 implants was installed in the femora and tibia of two beagle dogs and resected after 3 months. These implants were made from a bioactive glass cylinder jacket and a cylindrical bolt and nut with the same diameter as the glass jacket and made of either stainless steel AISI 316L, commercial pure titanium, or wrought Co-Cr alloy.

In this study bioactive cylinders showed an intensive bone bonding and osteoconductivity, resulting in a complete bony encapsulation of the glass. The metal bolts and nuts showed a direct contact with bone tissue, but only in areas of presumed bone contact at the time of implantation. Osteoconduction was not seen on these metal surfaces. The extent of bone contact decreased from titanium to wrought Co-Cr alloy to stainless steel.

Osteogenesis was frequently disturbed at the stainless steel or cobalt chromium to glass interface, shown by a focal absence of bone formation in this area. This ring-shaped non-calcified tissue contained non-inflammatory cells. Occasionally inflammatory cells like macrophages and histiocytes were seen but only with stainless steel which evoked a much more pronounced reaction than the Co-Cr alloy, which in its turn was more intensely reactive than titanium. In the case of titanium, bone formation did not appear to be hindered. When small crevices occurred between the bioactive glass cylinder jackets and the metal bolts or nuts, the superiority of titanium was evident. These crevices were mostly invaded by fibrous tissue, except for titanium where bone grew into the entire crevice (Fig. 7).

The results of this study together with findings on reduced glass reactivity prompted the synthesis of the titanium-glass ceramic composite which currently has undergone extensive evaluation under non-functional load conditions. Once histological evidence of bonding under long term functional load conditions is seen, human clinical trials can begin.

The experiments showed that bone bonding is established at the bioactive glass islands in between the titanium fibers at the implant surface. Bone tissue jumps from one bioactive glass island to another, and, with only a small titanium fiber separating the glass, close bone apposition to this titanium surface is achieved. Figure 8 shows bone

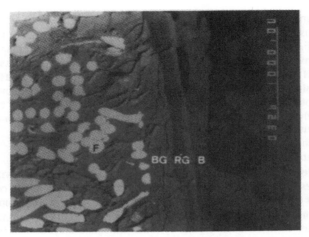

Fig. 7. This back-scattered image shows the good penetration of the glass between the sintered stainless-steel fibers and the good wettability of the metal fibers. The reacted glass layer is bonded to the bone tissue. BG = unreacted bioactive glass, RG = reacted glass layer, B = bone, F = metal fiber.

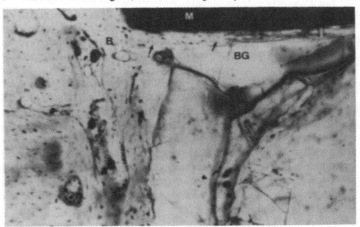

Fig. 8. Bone grows (arrows) into the crevice between the bioactive glass and the metal, adheres to the glass surfaces and contacts the titanium surface. Oxidized titanium sample. B = bone, BG = bioactive glass, M = metal. Giemsa stain, original magnification X100.

tissue in direct contact with the glass ceramic-titanium composite. The optimal distribution ratio of titanium fibers and glass varies between 60/40 and 70/30. The glass islands are slightly dissolved after three months of implantation. This creates undercut areas at the implant surface by which the initial exclusively chemical bonding of the implant in the jawbone is enhanced by mechanical interlock.

CONCLUSION

Bioceramic composite synthesis serves the purpose of enhancing mechanical behavior of the parent ceramic, while maintaining its excellent biocompatibility and, if applicable, bioactivity. The approach of reinforcing bioactive glass with metal fibers imparts to the bioactive glass an unusual fracture toughness. If the fiber selected is titanium, this fiber can be exposed to body fluids and a well-bonding glass composite is obtained.

The titanium fiber bioactive glass composite represents a fundamental advance in biomaterials, which derives from consideration of the failure mode of current implants with bioactive coatings as bonding vehicles. Dental implants or hip prostheses with plasma sprayed calcium phosphate coatings are known to fail at the ceramic to metal interface. The advent of the titanium fiber-glass composite represents a new era in that a continuity between metal implant and bioactive coating may be achieved, using fibers that are sintered onto the implant substrate and then run throughout the coating up the surface.

READING LIST

P. Ducheyne, and L. L. Hench, "The Processing and Static Mechanical Properties and Metal Fibre Reinforced Bioglass®," *J. Materials Science* **17** (1982) 595-606.
A. Evans, *Fracture in Ceramic Materials* (Noyes Publications, Park Ridge, NJ, 1984).
E. Schepers, M. De Clercq and P. Ducheyne, "Tissue Response to a Bioactive Glass-metal Interface," in *Implant Materials in Biofunction, Advances in Biomaterials, Volume 8*, ed. C. de Putter, G. de Lange, K. de Groot, and A. Lee (Elsevier Science Publishers B.V., Amsterdam, 1988) pp. 79-85.
S. Tsai, and H. Hahn in *Introduction to Composite Materials* (Technomic Publishing Co, Inc., Lancaster, PA, 1980).

Chapter 16

DESIGN OF BIOACTIVE
CERAMIC-POLYMER COMPOSITES

W. Bonfield
IRC in Biomedical Materials
Queen Mary and Westfield College
London, E1 4NS, UK

INTRODUCTION

The basis for tailor-making bioactive hydroxyapatite-polymer composites as skeletal implants to mimic hard tissue is reviewed. Hydroxyapatite (HA) reinforced polyethylene composite was originally conceived as a biomaterial for bone replacement on the basis of producing appropriate mechanical compatibility, as well as the necessary biocompatibility.[1] It was demonstrated that an increase in the volume fraction of particulate HA from 0 to 0.5 (50 volume percentage) produced an increase in the Young's Modulus of this composite, so as to approach the lower band of the range of values associated with bone itself. Cortical bone at the ultra-structural level is a hydroxyapatite reinforced collagen composite.[2] Thus, the equivalence of microstructure and deformation behavior give hydroxyapatite-polyethylene composites a special property as a bone analogue material. In contrast to current orthopaedic materials, HA-reinforced polyethylene composites offer the potential of a stable implant-tissue interface during physiological loading. The fracture characteristics of the composite depend on the hydroxyapatite volume fraction. There is a transition from ductile to brittle fracture as the volume fraction increases above approximately 0.45.[3] Hence it is possible to tailor-make the composition of the composite, so as to achieve any particular combination of Young's Modulus combined with the appropriate fracture characteristics. With respect to bone substitution, an optimum combination appears to be approximately 0.4 volume fraction of HA, when the fracture toughness of the composite is still significantly greater than that of cortical bone. *In vivo* studies in a loaded animal model demonstrated that, with a volume fraction of HA greater than approximately 0.2, the composite demonstrated bone apposition around the implant.[4,5] Recent high resolution electron microscopy[6] has demonstrated continuous hydroxyapatite lattice images across the composite-bone interface, indicating continuity at the ultrastructural level. This result was in contrast to the control polyethylene implants, which demonstrated fibrous encapsulation. Hydroxyapatite reinforced polyethylene composite, with a volume fraction of 0.4 of HA, has now been used clinically as an implant for reconstruction of the orbital floor.[7] The particular properties of the composite allowed the implant to be shaped by the clinician during the operation and to be implanted without any fixation. Results, now extending into the fourth year, have demonstrated successful osteointegration of the implants and a very good clinical result in this lightly-loaded application.

Following from the work on HA-reinforced polyethylene composite have been studies on other particulate filled composites with potential as biomaterials. These include work on HA-reinforced polyhydroxybuterate[8] and HA-reinforced polylactide.[9] In both cases the use of degradable polymer matrices offers interesting potential for skeletal scaffolding materials. However, it is still essential in the formulation of these composites to obtain the required mechanical properties necessary for a loaded application. With respect to thermosetting materials, HA additions have been made to polymethylmethacrylate bone cement in an attempt to introduce some bioactivity to this material.[10] This route presents some difficulty, due to the brittle nature of the matrix material, and a more promising alternative appears to be PEMA bone cement,[11] which, with considerable ductility, approaches more closely the condition observed for polyethylene. It should be emphasized that a _ductile_ polymer matrix is an essential precursor for effective stiffening and strengthening with HA additions.

To achieve the desired mechanical properties, it is necessary experimentally to have precise control of the processing conditions, so as to ensure a homogeneous distribution of HA.[12] In addition, the mean particle size and the particle size distribution, as well as the surface area, of the HA can be varied, so as to produce different mechanical properties. However, the technology to tailor-make bioactive composites of any particular combination, so as to deliver a range of prescribed mechanical and biological behavior, is now well established for the HA-polyethylene system[13,14] as well as having considerable potential for other bioceramic-polymer combinations.

ANALOGUE COMPOSITE DESIGN

The key data[15] for the HA-polyethylene system are shown graphically in Fig. 1,[16] which plots volume fraction of hydroxyapatite against Young's modulus and strain to fracture. It can be seen that from a starting value of 1.3 GPa for zero volume fraction (i.e., 100% polyethylene), the Young's modulus increases with volume fraction to exceed the lower bound of values associated with cortical bone, at the 0.5 volume fraction (50 volume percentage) HA composition. It should be noted that 0.5 volume fraction of HA corresponds to ~0.75 weight fraction (75 weight percentage), i.e. HA is the major constituent of the composite. The stiffening produced is accompanied by a decrease in the strain to failure, but the composite retains appreciable ductility until a HA volume fraction of ~0.4. Hence composites with volume fractions <0.4 deliver the ceramic in a fracture tough condition.

Biological activity is also indicated in Fig. 1. The essential result established _in vivo_ is that fibrous encapsulation is noted for polyethylene controls and for composites with less than 0.2 volume fraction of HA, but bone apposition is achieved for composites with HA volume fraction greater than 0.2. In terms of the interfacial shear strength of the implant-tissue interface, the transition from fibrous encapsulation to bone apposition for the HA-polyethylene system produces approximately a ten fold increase in value

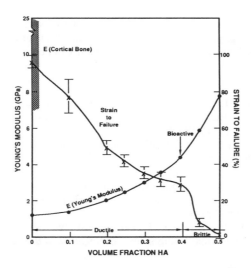

Fig. 1. Effect of volume fraction of hydroxyapatite (HA) on Young's modulus (E) and strain to failure of hydroxyapatite reinforced polyethylene composite, in comparison to cortical bone, as represented by Hench[16] based on data by Bonfield.[15] (Reproduced by courtesy of the *Journal of the American Ceramic Society*.)

(from ~1 to ~10 MPa). Hence formulation of a HA-polyethylene composite, with HA volume fraction between 0.2 and 0.4, gives a bioactive, fracture tough, modulus matching bone implant.

The medical application of such analogue composites is in prostheses requiring fixation to bone and a stable implant-tissue interface. The matching of deformation behavior across the implant-tissue interface means that polyethylene wear debris is not a problem. However the composite is not intended for an articulating application, such as for a joint surface. With that restriction, there remain a myriad of potential skeletal implant applications for appropriately designed HA reinforced polyethylene composites.

The experimental data shown in Fig. 1 can not be modelled on the basis of simple rule-of-mixture calculations combining polyethylene (E=1.3 GPa) with hydroxyapatite (E=80 GPa), an approach which gives theoretical values much higher than the measured results. A statistical finite element approach, based on the concept of Varonoi cells,[16] has recently been applied to the results of Fig. 1 and gives satisfactory agreement between experiment and theory. The details of this approach are beyond the scope of this short chapter, but give confidence that the curves shown in Fig. 1 may be used as first order-predictors for the experimental behavior of other HA-polymer composites, as well as a guide to the formulation of alternative bioceramic-polyethylene combinations.

ACKNOWLEDGEMENTS

The author acknowledges the considerable input of various colleagues over the past decade, to the program of research on bone-analogue materials and the continuing support of the UK Science and Engineering Research Council.

REFERENCES

1. W. Bonfield, M. D. Grynpas, A. E. Tully, J. Bowman, and J. Abram, "Hydroxyapatite Reinforced Polyethylene - A Mechanically Compatible Implant," *Biomaterials* 2 (1981) 185-186.

2. W. Bonfield and C. H. Li, "Anisotropy of Non-elastic Flow in Bone," *J. Appl. Phys.* 38 (1967) 2450-55.

3. W. Bonfield, J. C. Behiri, C. Doyle, J. Bowman and J. Abram, "Hydroxyapatite Reinforced Polyethylene Composites for Bone Replacement," in *Biomaterials and Biomechanics* eds. P. Ducheyne, G. VanderPerre and A. E. Aubert (Elsevier Publ. Amsterdam 1984) pp. 421-426.

4. W. Bonfield, C. Doyle, and K. E. Tanner, "In vivo Evaluation of Hydroxyapatite Reinforced Polyethylene Composites," *Biological and Biomechanical Performance of Biomaterials*, eds. P. Christel, A. Meunier, and A.J.C. Lee (Elsevier Publ. Amsterdam, 1986) pp. 153-158.

5. C. Doyle, Z. B. Luklinska, K. E. Tanner, and W. Bonfield, "An Ultrastructural Study of the Interface Between Hydroxyapatite Polymer Composites and Bone," *Clinical Implant Materials*, eds. G. Heimke, U. Soltsz, and A.J.C. Lee, (Elsevier Pub. Amsterdam, 1990) pp. 339-344.

6. W. Bonfield and Z. B. Luklinska, "High Resolution Electron Microscopy of a Bone-Implant Interface," in *The Bone-Biomaterials Interface*, ed. J. E. Davies (University of Toronto Press, 1991) pp. 89-94.

7. R. N. Downes, S. Vardy, T. E. Tanner, and W. Bonfield, "Hydroxyapatite-Polyethylene Composite in Orbital Surgery," *Bioceramics Volume 4*, eds. W. Bonfield, G. W. Hastings, and K. E. Tanner (Butterworth Heinmann, Oxford, 1991) pp. 239-246.

8. C. Doyle, K. E. Tanner, and W. Bonfield, "*In Vitro* and *In Vivo* Evaluation of Polyhydroxybutyrate and of Polyhydroxybutyrate Reinforced with Hydroxyapatite," *Biomaterials* 12 (1991) 841-847.

9. A. Van Sliedregt, *Hydroxyapatite-Polylactide Composites for Reconstructive Surgery* (Doctoral dissertation, University of Leiden, 1993).

10. A. Sogal and S. F. Hulbert, "Mechanical Properties of a Composite Bone Cement-Polymethylmethacrylate and Hydroxyapatite," in *Bioceramics Volume 5*, eds. T. Yamamuro, T. Kokubo, and T. Nakamura (Kobunshi Kankokai, Kyoto, 1992) pp. 213-224.

11. S. N. Khorasani, S. Deb, J. C. Behiri, M. Braden, and W. Bonfield, "Modified Hydroxyapatite Reinforced PEMA Bone Cement," *Bioceramics Volume 5*, eds.

T. Yamamuro, T. Kokubo, and T. Nakamuro (Kobunshi Kankokai, Kyoto, 1992) pp. 225-232.

12.   J. Abram, J. Bowman, J. C. Behiri, and W. Bonfield, "The Influence of Compounding Route on the Mechanical Properties of Highly Loaded Particulate Filled Polyethylene Composites," *Plastics and Rubber Processing and Application* **4** (1984) 261-269.

13.   W. Bonfield, J. Bowman, and M. D. Grynpas, "Prosthesis Comprising Composite Materials," UK Patent Application 8032647 (1980), UK Patent GB2085461B, The Patent Office, London (1984).

14.   W. Bonfield, J. A. Bowman, and M. D. Grynpas, M.D., "Composite Material for Use in Orthopaedics," US Patent 5, 1991, 017, 627.

15.   W. Bonfield, "Hydroxyapatite Reinforced Polyethylene as an Analogous Material for Bone Replacement," *Bioceramics: Materials Characteristics Versus In Vivo Behavior*, eds. P. Ducheyne and J. E. Lemons (Annals of the New York Academy of Sciences, 1988) Vol. 523, pp. 173-177.

16.   L. L. Hench, Bioceramics: From Concept to Clinic," *J. Am Ceram Society* **74** (1991) 1487-1510.

17.   F. J. Guild and R. J. Young, "A Predictive Model for Particulate-Filled Composite Materials, Part 1 Hard Particles," *J. Mater. Sci.* **24** (1989) 298-306.

18.   F. J. Guild and W. Bonfield, "Predictive Modelling of Bioactive Hydroxyapatite-Polyethylene Composites," Submitted to *Biomaterials*.

Chapter 17

RADIOTHERAPY GLASSES

Delbert E. Day and Thomas E. Day
Graduate Center for Materials Research
University of Missouri-Rolla
Rolla, MO 65401

INTRODUCTION

Radiotherapy glasses are defined as radioactive glasses used for *in situ* irradiation, beta or gamma radiation, of targeted organs inside the body. Glasses used for this purpose must not only be biocompatible, but also chemically insoluble in the body during the time that the glass is radioactive to prevent the unwanted release of the radioisotope from the targeted site. The development of radiotherapy glasses was motivated by the need to deliver large (> 10,000 rads), localized doses of beta radiation to diseased organs in the body in such a way as to minimize, and ideally avoid, damage to adjacent healthy tissue. Irradiating malignant tumors inside the body by external beam radiation is limited in several important ways. A major limitation is that the maximum dose which can be safely delivered is constrained by the need to protect surrounding healthy tissue and is usually too small (≤ 3,000 rads) to be therapeutic. Furthermore, external irradiation with energy radiation such as gamma, which is needed (often causes damage to healthy tissue) . The lower energy beta radiation is not well suited for delivery by external means because its smaller range in tissue may be too small for it to reach the target site. Since beta radiation is preferred in many cases, a means of *in situ* radiation would be advantageous.

For use as *in vivo* radiation delivery vehicles, radiotherapy glasses should be (1) biocompatible and nontoxic to the body, (2) chemically insoluble during the time the glass is radioactive, and (3) have high chemical purity. Aluminosilicate glasses containing yttrium and rare earth (RE) cations such as Sm, Ho, Re, and Dy satisfy these criteria. Furthermore, they have the advantage that radioisotopes such as Y-90, Sm-153, and Ho-166, can be made by neutron activation as the last step in the manufacturing process so that the glass can be manufactured in the normal way avoiding the handling of radioactive materials.

Yttrium aluminosilicate (YAS) glasses have been successfully used in clinical trials for more than five years. This is the only commercial use at this time, and uses YAS glass microspheres, containing Y-90, to irradiate malignant tumors in the liver. Depending upon the size of the liver, the desired dose, and the diameter of the microspheres, from 1 to 15 million microspheres of radioactive YAS glass, 15 to 35 μm in diameter, are injected into the hepatic artery which is the primary blood supply for the target tumors. The microspheres are sized so that the blood carries them into the capillary bed of the liver, but they are too large to pass completely through the liver and enter the circulatory system. Since the distribution of the radioactive microspheres follows the blood flow, the microspheres will concentrate in the tumor which has a greater than normal blood supply, and irradiate the tumor with β rays. In one case, 80% of the dose reached the tumor. Since Y-90 has a half life of 64.1 hrs, the radioactivity decays to a negligible level in about 21 days.

Radiotherapy glasses can be made in a variety of shapes, such as irregular particles, fibers, or spheres. Microspheres are the present shape of choice since the diameter can be carefully controlled and the smooth spherical surface helps with easy delivery of the particles to the target.

The RE aluminosilicate glasses are currently being evaluated for applications such as the irradiation of diseased kidneys prior to surgical removal, radiation synovectomy of arthritic joints, and the irradiation of malignant tumors in the liver. This chapter focuses on YAS glass microspheres which have been in commercial use in Canada since 1991.

PROCESSING

A typical manufacturing sequence for preparing radiotherapy glass microspheres is given in Fig. 1. The first step is the melting of a homogeneous mixture of high purity powders, such as Y_2O_3, Al_2O_3, and SiO_2, in a platinum crucible. Melting typically occurs at 1550 to 1650°C for the RE aluminosilicate compositions inside the glass forming areas depicted in Fig. 2. After melting, the chemically homogeneous melt is quenched to room temperature and crushed to a powder of the desired size. This powder is spheriodized by passing the particles through a gas/oxygen flame where each particle is melted, forms a sphere by surface tension forces, and becomes solid during cooling. The microspheres are then screened to obtain microspheres of the desired size. An example of the uniform and highly spherical microspheres made in this way is shown in Fig. 3. The final step is the irradiation of the glass microspheres with neutrons so as to form the desired quantity of radioisotope. YAS glasses are easily irradiated, forming Y-90, to a specific activity up to 5 mCi/mg of glass. After irradiation the microspheres are ready for packaging and shipment. Naturally, it is important that high purity raw materials, free of neutron activatable impurities, be used and that care be taken during the various manufacturing steps to avoid chemical contamination of the glass.

In addition to preparing glasses by conventional melting as just described, YAS glasses have been made by sol-gel processing. Property measurements made on a sol-gel derived YAS glass indicate that it should be acceptable for human use.

COMPOSITIONS

As evident from Fig. 2, glasses can be obtained from a wide range of Y, Sm, and Ho aluminosilicate compositions which melt below 1600°C. At this time, these are the only RE aluminosilicate systems where the boundaries for glass formation have been determined. However, glasses have been prepared from isolated aluminosilicate

1. GLASS MELTING

a. Select chemically pure raw materials (oxides) which do not contain any impurities that would form undesirable radioisotopes during neutron irradiation.

b. Mix raw materials to form a homogeneous mixture of powders.

c. Melt raw materials to form homogeneous glass.

2. SPHERIODIZATION (MICROSPHERE FORMATION)

a. Crush glass to particles of desired size.

b. Inject particles into gas-oxygen flame to melt each particle and form solid glass sphere (flame spray powder).

c. Collect microspheres in suitable container.

3. SIZING -- screen or separate microspheres into desired size range.

4. NEUTRON ACTIVATION -- irradiate microspheres in nuclear reactor (several days) until desired level of radioactivity is achieved. Package microspheres for delivery to physician.

Fig. 1.   Steps in manufacturing radiotherapy glass microspheres.

compositions which contain ReO_2, Dy_2O_3, or Er_2O_3. Since a large range of β-emitting RE radioisotopes can be incorporated into aluminosilicate glasses, it is possible to select one which is best suited to the particular type and size of the target organ. This compositional flexibility is an inherent advantage of radiotherapy glasses. In cases where some amount of gamma radiation is desired, neutron activatable gamma emitting radioisotopes, such as Na-24, K-42, or P-32, can also be incorporated into the aluminosilicate glass matrix.

An aluminosilicate glass is well suited for radiotherapy use since (a) no unwanted radioisotopes are formed by the neutron activation of Al, Si, or O, (b) these glasses have a high chemical durability, being essentially insoluble in the body, (c) microspheres with a high specific activity can be easily obtained because of the large amount (40 to 70 wt%) of RE oxide which can be present in the glass, (d) homogeneous melts can be prepared at reasonable temperatures ($< 1600°C$), and (e) particles of the glass are easily spheriodized in a flame because of the viscosity characteristics of the glass.

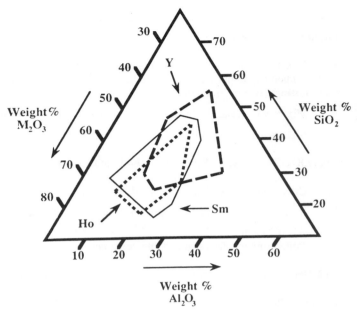

Fig. 2. Glass formation region ($< 1600°C$) for Y_2O_3, Ho_2O_3, or Sm_2O_3 aluminosilicate glasses.

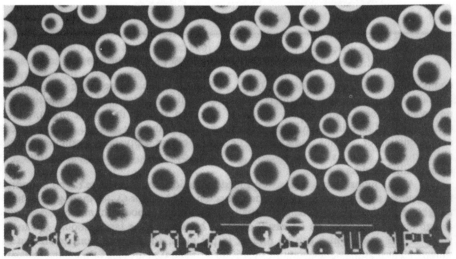

Fig. 3. Appearance of "typical" glass microspheres made according to the procedure described in Fig. 1. White bar in lower right hand corner is 100 μm.

PROPERTIES

Chemical Durability

Glasses used for radiotherapy purposes need to be highly durable during the time they are radioactive, since the means of confining the radioisotope to the target organ is to keep it inside a chemically insoluble microsphere. *In vitro* and clinical tests on radioactive YAS glasses have demonstrated their superior chemical durability. More than 100 patients have been injected with radioactive YAS glass microspheres over the past five years, with no reports of any premature or unwanted release of radioactive Y-90 in the body.

In vitro tests on a wide range of YAS glasses, containing from 9 to 30 Y_2O_3, 11 to 35 Al_2O_3, and 48 to 72 SiO_2, mol%, have shown that these glasses have an excellent chemical durability in deionized water and in saline at 37°C; this durability varies only slightly with chemical composition. An example of the small amount of yttrium released from a typical YAS glass, is shown in Fig. 4. The only data of practical interest is that for the first three weeks since the glass is no longer radioactive after that time. Slightly more yttrium is leached from YAS glasses at higher temperatures, 50°C, or in the HCl solution (both of which are used for accelerated testing), but the amount present in deionized water at three weeks, <5 ppm/cm^2 of glass, is too small to be of concern.

A comparison of the small amount of yttrium released from a glass in either bulk form or as glass microspheres or powder is shown in Table 1. The results for microspheres, which are relevant to the use of such glasses in the body, show that little

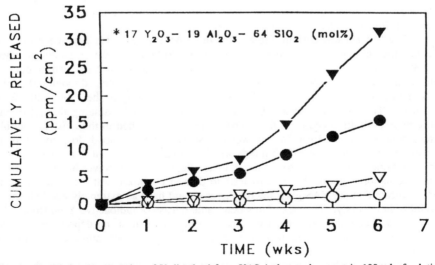

Fig. 4. Cumulative concentration of Y dissolved from YAS-4 glass and present in 100 ml of solution. Open circles and triangles are for DI water at 37 and 50°C, respectively, while closed circles and triangles are for 12M HCl (pH = 2) at 37 and 50°C, respectively. (Ref. Erbe, 1991).

Table 1.   Weight Percent Yttrium Released Per gm of $17Y_2O_3$-$19Al_2O_3$-$64SiO_2$, Mol%, Glass. (Ref. Erbe 1991).

	% Y Related/gm of Glass	
Conditions	3 wks	4 wks
DI Water at 37°C		
CM* Bulk Glass	0.02	0.04
CM Microspheres (25 to 35 μm)	0.06	0.09
CM Powder (20 to 38 μm)	0.20	0.27
SG* Powder (20 to 38 μm)	0.11	0.20
Saline at 37°C		
CM Bulk Glass	0.03	0.07**
CM Microspheres	0.04	0.11**
SG Powder	0.81	1.19

*CM means conventionally melting glass while SG means a glass prepared by sol-gel techniques.
**Measured at 6 weeks.

yttrium is released from the microspheres or powder even though the surface area of these samples is 300 times larger than that of the bulk sample.   There is no significant difference in the amount of yttrium released from microspheres tested in either deionized water or saline at 37°C.

The *in vitro* test data in Table 1 for the YAS microspheres have been used to calculate the amount of radioactive   Y-90 that would be released in a patient injected with 300 mCi of Y-90.   The solid data points in Fig. 5 show the calculated amount of radiation released to the body due to the very slight chemical attack (dissolution) of the YAS glass beads.   The calculation takes into account the decay of the Y-90 (half life of 64.1 hrs).   The solid line labeled C in Fig. 5 was calculated assuming that all of the radioactive Y-90 dissolved from the microspheres was absorbed in the most susceptible tissue, bone marrow.   Even in this worse case scenario, the total dose to the bone marrow is estimated at less than 5 mrads which is roughly equivalent to a chest x-ray or about the same dose that a person living in Leadville, Colorado receives in one year from cosmic radiation.   All of the *in vitro* tests to date in deionized water and saline up to 50°C indicate that the YAS glass microspheres should have an extremely good chemical durability in the body.   The lack of any detectable release of radioactive Y-90 from YAS glass microspheres that have been injected into humans that is, no depression of bone marrow activity, substantiates the *in vitro* test results and demonstrates the suitability of these glasses for use in humans.

Overall, the RE aluminosilicate glasses have excellent durability in deionized water and saline, but their durability should be expected to vary somewhat with temperature and with the RE concentration and the specific RE cation in the glass.   In

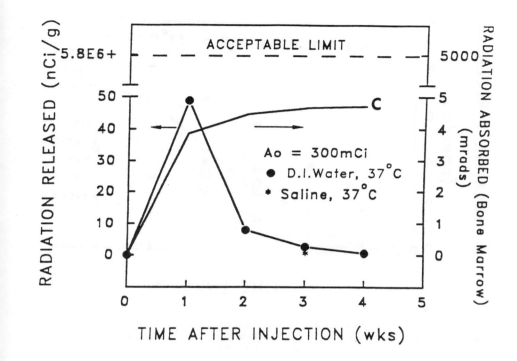

Fig. 5. Calculated amount of Y-90 radiation released (nCi/g) from YAS-4 microspheres (25 to 35 μm) immersed in (•) DI water (pH 6.9) at 37°C for up to 4 wks or in (*) isotonic saline (pH 6.2) at 37°C for 3 wks. Calculated from data (Table 1), assuming an initial injected dose of 300 mCi and taking into account the decay of the radioactive Y-90. Curve C represents cumulative absorbed dose (mrads) assuming that all radiation from released Y-90 is absorbed by bone marrow. (Ref. Erbe, 1991).

general, the durability in deionized water or saline tends to decrease slightly with increasing concentration of the RE cation in the glass as shown in Fig. 6 where slightly more yttrium is dissolved from YAS glasses of higher yttrium content, YAS-9 and -11 contain 27.4 and 30 mol% Y_2O_3, respectively, while the YAS-4 glass contains 17 mol% Y_2O_3. Samarium aluminosilicate (SmAS) glasses are also highly durable in deionized water at 37°C, their dissolution rate ranging from about 30 x 10^{-9} gm/cm²/min to 2 x 10^{-9} gm/cm²/min, which is quite similar to the dissolution rate for YAS glasses. While SmAS glass microspheres have not been used in humans at this time, the chemical durability of these glasses is considered acceptable for human use and there has been no reported release of Sm-153 from SmAS glass microspheres injected into the kidneys of rabbits.

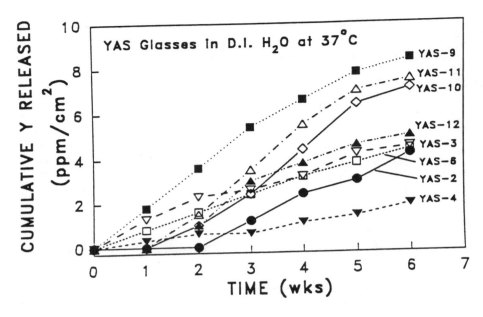

Fig. 6. Cumulative concentration (ppm) of Y released into 100 ml of DI water at 37°C per surface area (cm²) of bulk glass. The YAS-4, YAS-12, and YAS-9 glasses contain 17.0, 30.0, and 27.4 mol% Y_2O_3, respectively. Experimental error ± 2.0 ppm. (Ref. Erbe, 1991).

The excellent chemical durability of RE aluminosilicate glasses is attributed to the absence of alkali and alkaline earth oxides in these glasses, which typically lower the chemical durability of silicate glasses, and to the presence of small highly charged cations which can form strong chemical bonds with oxygen. The RE aluminosilicate glasses have a strongly bonded, three dimensional network structure which is not easily attacked by aqueous solutions having a pH between 6 and 8. *In vitro* measurements of the chemical durability of Y, Sm, Ho, and a few Re aluminosilicate glasses indicate that most RE aluminosilicate glasses should have a chemical durability satisfactory for *in vivo* use.

Density and Refractive Index

Since the molecular weight of the rare earth oxides is much higher than that of Al_2O_3 and SiO_2, the density of the RE aluminosilicate glasses depends primarily on the concentration of the RE oxide. As shown in Fig. 7, the density of YAS glasses ranges from about 2.8 gm/cm³ at 10 mol% Y_2O_3 to about 4.0 gm/cm³ for glasses containing 30 mol% Y_2O_3. Comparable SmAS glasses have a higher density, ranging from about 3.4 to 4.6 gm/cm³, which is consistent with the higher molecular weight of Sm_2O_3.

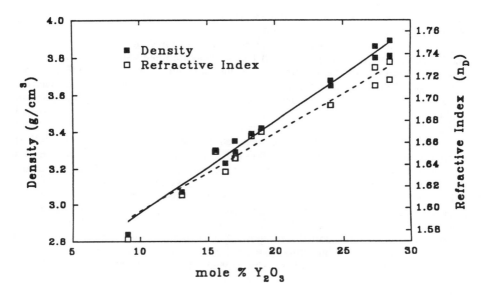

Fig. 7. Density (solid line) and refractive index (dashed line) of YAS glasses. Lines are least squares fit to data points.

The density of the RE aluminosilicate glasses is obviously considerably higher than that of blood, but this has not caused any problems in the injection of these glasses into humans or test animals. During injection, precautions are necessary to insure that the microspheres do not settle out of solution, but simple agitation is adequate to keep the microspheres suspended.

The refractive index of the glasses is not important to their use for radiotherapy purposes, but this is another property which depends primarily upon the concentration of the RE oxide, see Fig. 7. The relative amounts of alumina and silica in these glasses is of lesser importance to properties such as refractive index, density, and thermal expansion coefficient.

CLINICAL RESULTS

The tissue response to radiotherapy glasses varies according to the tissue being irradiated and clinical data relating tissue response to radiotherapy glasses exists only for the liver and kidney. This chapter discusses effects only in the liver.

The liver is referred to as a reverting post-mitotic cell type. This means that the liver does not normally divide or renew itself, as does the skin or lining of the digestive tract, but has the capability to do so. If the capacity of the liver to function is decreased by some type of injury (chemical, trauma, etc.), it will be stimulated to renew itself in order to maintain normal body function.

Liver tumors, like other tumor types, undergo rapid mitotic division and are highly sensitive to ionizing radiation. All tissues that are rapidly dividing cell types are extremely sensitive to the effects of ionizing radiation. This is due to the large percentage of the time that the cells' genetic material is condensed in the nucleus. Ionizing events reaching the condensed genetic material in the nucleus of the cell will lead to cell death. Since the normal liver is not usually in a dividing state, it is resistant to the effects of low to moderate levels of ionizing radiation.

The "average" doses that have been delivered with radiotherapy glasses to liver tumors have ranged from 5,000 to 15,000 rads. This would normally be considered a large human dose and it is difficult to predict the exact dose delivered to just the tumor. As previously mentioned, the distribution of the glass microspheres in the liver is believed to depend on the blood flow. Many hepatic tumors are classified as hypervascularized, which means that the blood flow to the tumors exceeds that to the normal surrounding tissue, and, consequently, a larger than normal fraction of microspheres will be transported and deposited in the tumor. This increases the radiation dose to the tumor while minimizing the exposure of the surrounding normal tissue. This localization of the radiation explains why patients, treated with radioactive Y-90 glass microspheres can tolerate much higher doses than those treated by whole liver radiation methods.

One study used dogs as a model for hepatic arterial injection of both non and radioactive YAS glass microspheres. Doses exceeding 30,000 rads were delivered to the livers of these dogs. Delivered doses of non and radioactive microspheres (143 to 562 mg) were measured in units of mCi per gm of liver tissue and ranged from one to twelve times the anticipated human dose. The dogs were grouped by varying dose levels and a control group that received nonradioactive microspheres was used to determine the physical impact of the microspheres alone on liver function.

Doses of nonradioactive microspheres delivered were up to six times that of the anticipated human dose. Minimal changes were detected, such as changes within the walls of the central veins, in the appearance of the hepatocytes, and in the tissue architecture. Hepatocellular function and damage were within normal limits. There were no signs of portal fibrosis or cirrhosis.

Changes seen in the liver injected with radioactive YAS microspheres, were similar to the findings of other irradiation studies in dog liver. These changes included histologic changes in the portal areas of the liver. Doses as high as 35,000 rads were delivered, but did not cause total necrosis and were judged by clinical standards as compatible with survival. Doses up to 15,000 rads were well tolerated and showed little change in liver function. No microspheres were found in the bone marrow. Even in dogs receiving more than 15,000 rads, no bone marrow suppression occurred. At doses above 25,000 rads, the consequences of hepatic cirrhosis would probably pose significant problems.

In a preliminary study, a transient increase in body temperature has been noted, but this lasted only a few days. In some patients with a history of previous liver disease (chronic alcoholism), ulcerations of the lower stomach and upper small intestine have occurred. When treated, these conditions were self-limiting. In almost all patients receiving YAS radiotherapy glasses, liver enzymes were mildly elevated. This effect was not dose-related and lasted from a few days to weeks.

Clinical applications of radiotherapy glasses have been on liver and kidney tumors. Most of this work has been with liver tumors, since patients with liver cancer can enter the terminal phase within four to six months of diagnosis. This has sparked a major effort in investigating ways of delivering ionizing radiation *in vivo* to treat these malignant tumors.

Work is currently underway to discover whether very large doses delivered to the kidneys will reduce the shedding of malignant cells, which can spread the tumor, during surgical removal of a diseased kidney.

In summary, any tissue that is relatively insensitive to low or moderate amounts of ionizing radiation in which these unique microspheres can be deposited, by either the blood flow or surgical implantation, is a potential candidate for this new form of radiotherapy.

SUMMARY

Rare earth aluminosilicate glass microspheres have proved to be well suited for radio-therapeutic use in humans. These microspheres provide a new and unique method of irradiating diseased internal organs with beta radiation, in amounts which exceed those that can be delivered by other means. YAS glass microspheres have been safely used for more than five years to irradiate, up to 15,000 rads, malignant tumors in the liver in more than 100 patients. Since the glass microspheres tend to distribute themselves in the liver in proportion to the blood flow, the actual dose to the tumors is believed to be much larger than the average dose to the entire liver since the microspheres tend to concentrate in the tumors, because of their vascularity. In one case, 80% of the YAS microspheres were estimated to lodge in the tumor vascular bed, giving and estimated dose of 32,000 rads. Samarium aluminosilicate glass microspheres have been used to irradiate, up to 15,000 rads, the kidneys in rabbits without any harmful side effects or detectable damage to adjacent tissue.

A major advantage of the RE aluminosilicate glass microspheres is excellent chemical durability in the body. Since the glasses are insoluble in body fluids, the radioactive RE isotope is confined to the target organ and prevented from entering the circulation. The maximum time which these glass microspheres will remain in the body is currently unknown, but YAS microspheres have been in one patient for more than four years with no reported problems. If needed, it should be possible to develop RE containing glasses which will gradually degrade in the body when they are no longer radioactive.

The use of RE aluminosilicate glass microspheres is still at an early stage in treating liver cancer, but results are promising. The currently recommended dose range is 8,000 to 15,000 rads. In one group of 39 adenocarcinoma patients treated with 5,000 to 11,000 rads in a phase I-II study, the average survival time was 9.7 months from the date of treatment with the microspheres. This compares with a median survival time of 12 months from the time of diagnosis for patients treated with conventional chemotherapy. All of the patients treated with the YAS glass microspheres had undergone one chemotherapy treatment and diagnosis may have occurred several months prior to injection with the YAS microspheres. Treatment with glass microspheres takes about one hour, followed by a few hours in the hospital for observation; treatment as "day-patient". Chemotherapy requires several repeated treatments over several weeks. Increased liver enzymes are a common side effect following treatment with the glass microspheres, and transient fever, increased pain, and nausea and vomiting are less common side effects.

Radiotherapy glass microspheres are being considered for treating other diseases and other types of cancer. Using radiotherapy glasses to kill cancer cells in diseased kidneys prior to surgical removal has already been mentioned. Ideally, it should be possible to use radiotherapy glass microspheres to irradiate any diseased organ having a capillary bed. The *in situ* irradiation of arthritic joints with beta emitting RE aluminosilicate glass microspheres is also under study. The stifle joints in rabbits have been injected with usable quantities of glass microspheres without any noticeable physical damage to the joint for periods up to one year. In rabbits the glass microspheres were found imbedded in a layer of the synovial tissue. In this application, the radioactive glass microspheres were used to perform a radiation synovectomy of the diseased joint.

READING LIST

G. J. Ehrhardt and D. E. Day, "Therapeutic Use of Yttrium-90 Microspheres," *Nucl. Med. Biol. Intl. J. Radiat. Appl. Instr., Part B*, **14(3)** (1987) 233-42.

R. F. Brown, L. Lindesmith, and D. E. Day, "Ho-166 Containing Glass for Internal Radiotherapy," *Nucl. Med. Biol. Intl. J. Radiat. Appl. Instr., Part B*, **18(7)** (1991) 783-90.

M. J. Hyatt and D. E. Day, "Glass Formation and Properties in the Yttria-Alumina-Silica System," *J. Am. Ceram. Soc. - Comm.* **70(10)** (1987) C-283-87.

E. M. Erbe and D. E. Day, "Properties of Sm_2O_3-Al_2O_3-SiO_2 Glasses for *In Vivo* Applicationsm," *J. Am. Ceram. Soc.* **73(9** (1990) 2708-13.

E. M. Erbe, *Structure and Properties of Y_2O_3-Al_2O_3-SiO_2 Glasses* (Ph.D. Thesis, University of Missouri-Rolla, May, 1991).

M. J. Herba, F. F. Illescas, M. P. Thirlwell, et al., "Hepatic Malignancies: Improved Treatment with Intraarterial Y-90," *Radiology*, **169(1)** (1988) 311-14.

S. Houle, T. K. Yip, F. Shepherd, et al., "Hepatocellular Carcinoma: Pilot Trial of Treatment with Y-90 Microspheres," *Radiology* **172(2)** (1989) 857-60.

I. Wollner, C. Knutsen, P. Smith, et al., "Effects of Hepatic Arterial Yttrium 90 Glass Microspheres in Dogs," *Cancer* **61(7)** (1988) 1336-44.

D. E. Day and G. J. Ehrhardt, U.S. Patents 4,789,501; 5,011,677 and 5,011,797.

Biomedical Materials, *MRS Bulletin* **XVI (9)** (1991) 26-81.

J. C. Harbert and H. A. Ziessman, "Therapy With Intraarterial Microspheres," in *Nuclear Medicine Annual*, ed. L. M. Freeman and H. S. Weissmann (Raven Press, NY, 1987) pp. 295-319.

Chapter 18

CHARACTERIZATION OF BIOCERAMICS

Larry L. Hench
Advanced Materials Research Center
University of Florida
Gainesville, Florida 32611

INTRODUCTION

The final form and properties required of a bioceramic depend upon the function served by the material as an implant (Chapter 1). The forms used include: powders, coatings and bulk shapes. The properties of interest for most implant applications are mechanical performance and surface chemical behavior. Control of properties requires control of each of the processing steps illustrated in the top half of Fig. 1. (See Chapter 1 for a review of these processing steps.) In order to ensure that the required final properties are achieved prior to implantation, it is essential to characterize the bioceramic. This chapter summarizes the concepts and instrumental methods involved in characterization of a bioceramic.

CERAMIC PROCESSING STEPS

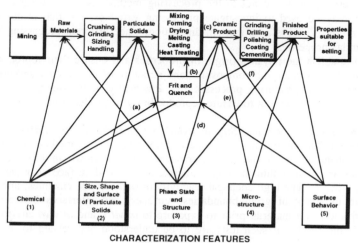

CHARACTERIZATION FEATURES

Fig. 1. The relationship between characterization features and products of ceramic processing steps.

Characterization has been defined by the Materials Advisory Board of the National Research Council in the United States as:

"Characterization describes those features of the composition and structure (including defects) of a material that are significant for a particular preparation, study of properties, or use, and suffice for the reproduction of the material."

Thus, to characterize a bioceramic it is necessary to evaluate its composition, structure, and surface sufficiently that the material and its properties can be reproduced.

The Goal of Characterization is the Assurance of Reproducible Properties
The bottom half of Fig. 1 illustrates the five major classes or features of characterization. They are:
1) Chemical composition,
2) Size, shape, and surface of particulates,
3) Phase State and Structure,
4) Microstructure,
5) Surface.
Information is required on all five features to achieve reproducibility of properties for a material and device.

The arrows between the bottom and top half of Fig. 1 relate the characterization features, one of the five classes, to the appropriate output in the ceramic processing sequence. Two points are emphasized. First, each characterization feature indicated can be quantitatively evaluated. Second, characterization is directed towards the output of the processing steps and not to the processing method itself. The same analytical techniques may be used to investigate processing variables, but that is not materials characterization. This is an important distinction, since different processing steps can lead to similar properties.

To establish the physical origin of properties and ensure their reproducibility it is necessary to determine the composition and structure of the product and not just the specifics of the manufacturing process used. Practically, it is essential to control the reproducibility of processing to ensure reproducibility of product properties.

Characterization is the Critical Step in Relating Processing to Properties
The mechanical and surface properties important for biomedical applications of ceramics depend upon all five characterization features shown in Fig. 1. Because bioceramics are used within the body, microstructural and surface characteristics are especially critical. The high chemical reactivity of body fluids, enzymes, and cells can easily lead to attack of grain boundaries, surfaces, and interfaces between phases. Characterization of a material prior to exposure to *in vitro* and *in vivo* environments is essential if the results of such tests are to be understood and related to the potential long-term performance of the ceramic as an implant.

STRATEGY OF A CHARACTERIZATION PROGRAM

Many biomaterials have been tested in animals and even in people without adequate characterization. This is often because the developers of new biomaterials are unaware of the importance of characterization. It may also be because of expense. A strategy is necessary to minimize costs but still ensure that sufficient tests are done.

The steps in establishing a characterization strategy are:

1) Determine the analytical costs of various instrumental methods available for each of the five classes of characterization features shown in Fig. 1.

2) Determine the level of accuracy required of the tests for each feature.

3) Decide the statistical confidence level desired for the tests.

4) Establish the number of samples required to achieve the desired confidence in reproducibility.

5) Determine the relative importance of the five characterization features on the final properties desired.

6) Determine the relationships between properties and each of the characterization features.

7) Based upon the results of steps 1-6, compute the economics of alternative analytical tests to achieve reproducibility of properties within the desired confidence levels.

8) Select the least expensive test or combination of tests.

EXAMPLE OF A CHARACTERIZATION PROGRAM

Let us consider as an example the characterization of a bioactive glass dental implant, 45S5 Bioglass® Endosseous Ridge Maintenance Implant (ERMI)®. The ERMI is an FDA approved Class III device (Chapter 19) manufactured and marketed by U.S. Biomaterials Inc. under license from the University of Florida. The clinical use of the ERMI is described in Chapter 4. The characterization program is described in terms of the five characterization features shown in Fig. 1. The tolerances given in the example are used only to illustrate the steps involved in establishing a characterization program. The actual tolerances imposed commercially by U.S. Biomaterials Inc. are company confidential.

The property of most concern for the final product is its **bioactivity**; i.e., ability to bond to tissues. This is a surface chemical property. The mechanical requirements for the ERMI are minimal since it is a buried implant (Chapter 4) and is subjected only to small compressive loads. The characterization strategy is therefore designed to ensure reproducible bioactivity of the ERMI.

Feature 1: Chemical composition. The nominal composition of the 45S5 glass is 45% SiO_2, 24.5% Na_2O, 24.5% CaO, and 6% P_2O_5, all in weight percent. Extensive studies of composition versus implant behavior, discussed in Chapter 3, show that SiO_2 content is the primary compositional variable that affects the rate of bonding and bioactivity of the implant.

Animal experiments show that with compositions which vary from 42 to 52% SiO_2 the implant will bond to both bone and soft tissue. There is very little effect on bioactivity with variations in Na_2O or CaO content. P_2O_5 content can also vary $\pm 2\%$ with little effect on bioactivity. Consequently, analytical limits on the bulk composition of the implant can be set with tolerances of $\pm 1\%$ for all four major chemical constituents. This level of compositional control is easy to achieve in glass processing. The tolerance of chemical analysis can be set at any value up to $\pm 0.5\%$ and detect variations within the $\pm 1\%$ control range. Automated analytical methods such as X-ray fluorescence, routinely used in the glass industry, are inexpensive and achieve this level of tolerance easily for Si, Na, Ca and P.

In vivo studies, reviewed in Chapter 3, show that cation impurities with $3+$, $4+$, $5+$ charges, such as Al^{3+} or Ta^{5+}, can inhibit formation of a hydroxy-carbonate apatite (HCA) layer on a glass and destroy bioactivity. The compositional limit for the impurities varies for each element but the effect becomes significant at concentrations $> 1\%$. Thus, a limit for total impurity content of multi-valent cations is set at 0.5%, which allows for a margin of analytical error. X-ray fluorescence analysis is also suitable for this measurement and it can be obtained at the same time as the bulk compositional analysis of a glass batch, thereby minimizing costs.

Feature 2: Size, shape and surface of particulate solids. The ERMI can be made from either a raw glass batch or from glass frit. (A raw glass batch is a mixture of the constituent oxide powders which are melted in a crucible, refined, homogenized and cast into the final implant shape, following a schedule similar to that illustrated in Chapter 1. A glass frit is an intermediate processing step which produces large glass particles by quenching a molten batch of glass into water or onto a steel plate or rollers. The particles of a glass frit are either remelted and cast into a shape in a forming process or ground into a specific powder size for use in applications such as periodontal repair or bone augmentation, see Chapter 3.)

If a glass frit is used in processing it results in an additional, intermediate processing step, as indicated in the top half of Fig. 1. The frit can be analyzed for chemical composition (path A), remelted (path B) and cast into an implant (path C). Since no additional finishing is required, the implant can then be characterized for phase state and structure (path D), microstructure (path E), and surface behavior (path F).

This sequence of processing-characterization steps has the advantage that chemical compositional analysis of a large batch of glass frit is less expensive than analyses of many small raw batches of glass. A disadvantage is the potential for pick-up of impurities during fritting-remelting operations. There is also the economic advantage that a composition that does not meet specifications will be detected before the costs of fabricating implants are incurred.

The decision as to which combination of processing and characterization steps to follow is based upon relative economics and depends very much upon scale of operations and ability to control the cleanliness of each process step. The goal of characterization is to ensure that, whichever sequence is chosen, the results are the same and within the tolerance levels required for Good Manufacturing Practices by regulatory agencies (Chapter 19).

An advantage of the intermediate glass fritting step is that mixtures of compositions can be made and then formed into shapes and densified by sintering or hot-pressing. This is the sequence of processing followed in the manufacture of A/W glass-ceramic (Cerabone®), as described in Chapters 5 and 6. The resulting glass-ceramic has a composite structure with superior mechanical properties but still retains bioactivity.

It must be remembered that a decision on which processing steps to use in making implants must be accompanied by a decision on characterization steps. The characterization strategy must be compatible with processing and *vice-versa*. The costs of characterization are additional to processing costs and must not be excessive.

Feature 3: Phase state and structure. The objective of this characterization step is to determine whether the implant is crystalline or amorphous. ERMI implants are amorphous, glassy materials and crystals are undesirable because they can lead to heterogeneous surface reactions. Improper melting, casting, or annealing can lead to crystallization (Chapter 1). Therefore, a characterization step is necessary to ensure that crystallization has not occurred. Several methods are available, including X-ray diffraction (XRD). The XRD spectrum (A) in Fig. 2 shows that a 45S5 Bioglass®

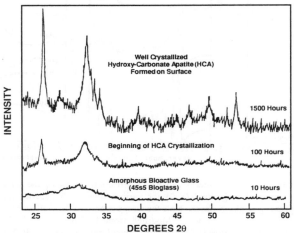

Fig. 2. X-ray diffraction pattern of the surface of 45 wt% bioactive silica glass containing 6 wt% P_2O_5 after exposure to a simulated physiological solution for 10, 100, and 1500 hours.

implant is amorphous, no diffraction peaks are present.   There is only a broad diffraction region, characteristic of the short range ordering of the silicate structure in glasses (Chapter 1).

Crystals grown in glass due to processing problems are usually visible to the eye and the least expensive method of characterization is visual inspection.

Complex multi-phase, polycrystalline materials such as bioactive glass-ceramics, require characterization of the type of each crystal phase present and the volume fraction of the crystal phases.   A mixture of crystal phases and glassy matrix, such as in A/W glass-ceramic, requires an analysis capable of determining the volume fraction of the amorphous matrix.   This is difficult, except by subtraction; i.e, one must determine the crystal phase percentages by XRD or optical microscopy, then subtract from 100% to obtain the percent of glass present.   Figure 3 shows a XRD spectrum of A/W glass-ceramic with a mixture of apatite and wollastonite crystal phases.

Feature 4: Microstructure.   This is one of the most important characterization features for bioceramics.   Microstructure usually controls mechanical properties and can influence surface behavior due to the presence of phase boundaries, missing grains or

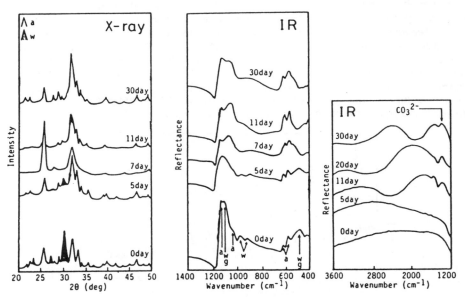

Fig. 3.   Thin film x-ray diffraction patterns and FT-IR reflection spectra of the surface of A/W glass-ceramic exposed to simulated body fluid (K-9) for various times; a = apatite crystal phase, w = wollastonite crystal phase, g = glassy phase.   Note the disappearance of the wollastonite phase as a surface layer of hydroxy carbonate apatite grows.   (Modified from T. Kokubo, S. Sakka, T. Kitsugi, T. Yamamuro, M. Takagi, and T. Shibuya in *Ceramics in Clinical Applications*, ed. P. Vincenzini (Elsevier, Amsterdam, 1987) p. 175.

inclusions. For example, the ASTM and ISO standards for alumina implants (Chapter 2) establish a limit of <7 μm on mean grain size. Quantitative microscopy, using either optical or scanning electron microscopy (SEM), is the method usually used in analyzing microstructure. Characterization involves the determination of size distribution of grains, size distribution and volume fraction of porosity, phase boundary area, and connectivity of phases in a multi-phase microstructure. For details of use of quantitative microscopy (also called stereology) to obtain these features consult DeHoff in the Reading List.

For a glass implant, such as an ERMI, microstructure is not so important. Visual inspection is sufficient to ensure that inclusions or bubbles are absent. However, microstructural characterization of a multi-phase glass-ceramic requires analysis of the size, volume fraction and distribution of each of the phases. This can be a big effort and is usually done with an image analyzing computer. In Chapters 5-10 are shown micrographs of the complex microstructures present in multi-phase bioceramics. A characterization program must establish which of the many microstructural features is most important in controlling the mechanical and surface properties of the material and concentrate the analysis on one or two features, otherwise costs become prohibitive.

Microstructural characterization of a two-phase bioactive composite, such as stainless steel-reinforced Bioglass® (Chapter 15) or polyethylene-HA (Chapter 16) is particularly important, since the volume fraction of the high modulus phase controls the mechanical properties. The second phase must be dispersed homogeneously to be effective. A critical volume fraction of the bioactive phase is necessary for the composite to be bioactive. Thus, analyses of volume fraction and distribution of the two phases are critical in a characterization program for composites.

Feature 5: Surface behavior. The surface of a biomaterial becomes an interface with tissues upon implantation. Consequently, characterization of surface features are of vital concern if an implant is to achieve reliable long term behavior. There are many different surface analytical methods available. The most important problem is often the determination of which surface feature is relevant to the biological and biomechanical performance of an implant. Experiments are required to establish the relationships between surface analyses and *in vivo* and *in vitro* behavior. Only after these relationships are known is it possible to decide on the characterization method(s) to use and the tolerances required to ensure reproducibility.

The characterization of Bioglass® ERMI dental implants is used to illustrate surface analytical methods and how to decide on a characterization criterion. There are two approaches for analyzing the mechanisms and reactions at an implant surface. First, the constituents released into a test environment or surrounding tissues can be analyzed. The traditional wet chemical methods, such as atomic emission or absorption spectroscopy, or inductively coupled plasma (ICP) chemical analysis, can be used to do this. See Adair, in the Reading List, for a summary of analytical methods available and their levels of tolerance.

When bioactive glass implants are exposed *in vivo* or to *in vitro* solutions Na, Ca, Si, and P ions are released from the glass surface. The loss of cations is due to ion exchange with H^+ and H_3O^+ ions from the solution. These reactions correspond to

Surface Reaction Stages 1 and 2 discussed in Chapter 3. The decrease of P in the
solution is due to the formation of an amorphous calcium-phosphate layer on the glass
surface (Stage 4). Solution analysis, however, provides no understanding of the changes
occurring on the surface of the implant. For example, Stage 3-formation of the silica gel
layer or Stage 5-crystallization of the HCA layer, cannot be identified from solution
analysis alone.

A second approach, surface analysis of the material, is also required to determine
compositional gradients or phase changes taking place. Figure 4 summarizes the range
of instrumental methods available for surface characterization. They can be classified
into two groups: 1) those sampling deep (up to 1.5 μm) into the surface, and 2) those
that examine only the "outer surface" (0.5-5.0 nm) of the material.

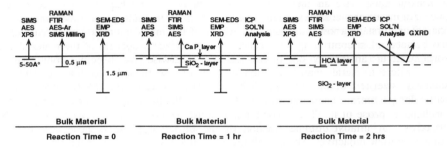

Fig. 4. Alternative methods for surface characterization of a bioactive implant. (A) Surface before
reaction, Reaction Time = 0, (B) After one hour reaction, (C) After two hours reaction. Note the effect
of instrumental sampling depth on the surface analysis.

Results from several of these methods applied to the surface characterization of
bioactive glasses, ceramics and glass-ceramics follow.

Figure 4 also summarizes the problem faced in characterizing the surface and
interface of a bioactive ceramic. The composition and phases of the material change
with time. Thus, analysis of the surface prior to implantation, Fig. 4A, may or may not
be predictive of the interface of the material after implantation, illustrated in Figs. 4B
and 4C. The analytical methods chosen must be capable of following these kinetic
changes in a reproducible manner. For a bioactive glass-ceramic or bioactive composite
with multiple phases, the time dependent changes of each phase need to be understood
before a characterization program can be established.

Infrared Reflection Spectroscopy (IRRS). This is one of the most versatile, rapid
and cost-effective means of surface analysis, especially with the use of a Fourier
transform IR (FTIR) spectrometer. As shown in Fig. 4, IRRS examines the surface of
a glass or ceramic to a depth of approximately 0.5 μm, depending upon the index of
refraction and density of the surface layer. It is non destructive, requires no vacuum or
sample preparation and is applicable to samples of any dimension, even those with
curvature. An implant can be analyzed before and after *in vitro* or *in vivo* testing.

See p. 47

An FTIR can be used in either specular reflection or diffuse reflection mode, as discussed by LaTorre and Hench, in the book edited by Adair, in the Reading List. The diffuse mode is particularly useful when there is considerable scattering from surface reaction layers. Analysis of powders requires use of a diffuse scattering stage.

The IRRS method gives quantitative information of the chemical composition of the surface since it is sensitive to the vibrational modes which are characteristic for each molecular constituent of the material. Since Si-O-Si bonds differ in energy from Si-O-Na or Si-O-Ca bonds their vibrational frequencies are different. Changes in these vibrations can be detected when the surface composition is altered by chemical reaction with a solution, as illustrated in Fig. 3 for A/W bioactive glass-ceramic. Thus, the kinetics of surface compositional change can be followed with IRRS. Details are discussed in Chapters 3 and 5.

An FTIR surface analysis can also detect phase changes occurring within a surface layer. The crystallization of a hydroxyl carbonate apatite (HCA) layer on a bioactive glass surface is apparent in Fig. 5. The single P-O mode of the amorphous calcium phosphate layer (Surface Reaction Stage 4, discussed in Chapter 3) is transformed into two separate P-O modes when the apatite crystals are formed. Eventually three P-O modes characteristic of well-developed apatite crystals are observed, as shown in Fig. 5, after longer times. The crystalline HCA layer is also detected with XRD, as shown in Fig. 2. Because of its versatility and speed, the diffuse reflection FTIR method can be used as a quantitative quality assurance test method with either bulk samples, such as the ERMI dental implant, or powders.

Infrared microscopy combines optical microscopy with IR spectroscopy which makes it useful for analysis of surface features, surface profiles, and microstructural features. This method is described in the LaTorre and Hench article in the Reading List and applied to the analysis of a bioactive glass-bone interface.

Fourier Transform Raman Spectroscopy can also be used for analysis of surfaces before and after reaction. It is a method with very high resolution and is especially sensitive to the extent of crystallization of the surface. The Raman modes arise from molecular vibrations of the surface but electron transitions involved are different than IR modes which yields complementary information on the composition and phase state of the surface layers.

Scanning Electron Microscopy (SEM) with Energy Dispersive X-ray Spectroscopy (EDS). SEM-EDS is a rapid characterization method for surfaces and interfaces and is probably the most widely used instrument for this purpose. It must be remembered that there is a deep sampling depth of several μm (Fig. 4) for this method. The important advantage of the SEM method is that microstructural and surface features can be observed and their dimensions and area fraction measured. The composition of the surface and microstructural features can then be analyzed chemically with EDS. The size of the electron beam can be modified and the area analyzed varied. A characterization program must establish experimentally a standard sampling area and standards for the analytical results, with appropriate calibrations, in order for the data to be quantitative.

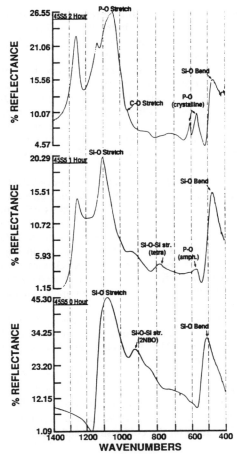

Fig. 5. FTIR diffusion reflection spectra of a 45S5 bioactive glass before reaction (0 hrs) and after 1 hr and 2 hrs reaction in simulated body fluid. Note formation of amorphous calcium phosphate at one hour and hydroxy carbonate apatite at two hours.

Figure 6 shows SEM-EDS results from the interface between a 45S5 Bioglass® implant and a rat tibia, bonded for 30 days. Areas 1 and 2 show a large Si signal from the silica-rich layer formed on the glass. Areas 3 and 4 are from the calcium-phosphate-rich layer and bone, respectively.

Electron Microprobe Analysis (EMP). This electron beam method is excellent for determining the thickness and compositional gradient of reaction layers formed on bioceramic implants. Figure 7 illustrates the type of data obtained for a 45S5 Bioglass®-bone interface. The intensity of the X-ray signal for each element (which is proportional to composition) is plotted as a function of distance across the interface. Moving from the surface of the implant to the bone, there is a large increase in concentration of Si due to the formation of a repolymerized SiO_2 layer on the glass. The thickness of this

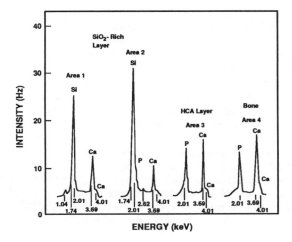

Fig. 6. Scanning electron microscopy-energy-dispersive X-ray analysis of the interface between a 45S5 Bioglass® implant and a rat tibia, bonded 1 month. Areas 1 and 2 show a large signal from the silica-rich layer formed on the glass. Areas 3 and 4 are from the calcium phosphate-rich film and bone, respectively.

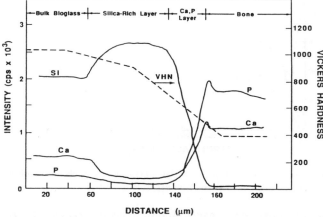

Fig. 7. Compositional profile across a rat tibia-bioactive glass (45S5) interface after 1 yr. Obtained with electron beam microprobe. Dashed line is Vickers Hardness gradient across the interface.

layer is about 80 μm. Next, there is a gradual decrease in the Si signal accompanying a progressive increase in Ca and P intensity as the electron beam traverses the HCA layer grown on the glass. The thickness of the HCA layer is about 20 μm. The Ca and P signals then drop to a constant value characteristic of bone mineral. The dashed line, in Fig. 7, is the Vickers hardness gradient across the interface.

Determining interfacial compositional profiles such as shown in Fig. 7 is essential in understanding the behavior and *in vivo* performance of a bioactive implant, but is not desired for characterization of an implant. The objective of characterization of the surface is to ensure that a profile such as in Fig. 7 will be routinely obtained without having to perform an additional 30-180 day test in animals.

Auger Electron Spectroscopy (AES). As shown in Fig. 4, AES examines the outermost surface of a material with a resolution on the scale of 0.1 nm. AES combined with AR ion milling, which removes 1 to 50 μm slices of a surface between AES analyses, is one of the most accurate methods for determining compositional gradients within the outermost layers of a bioceramic. Figure 8 illustrates a compositional profile of a 45S5 Bioglass® implant removed from a rat bone after only one hour of implantation. The SiO_2-rich layer and Ca-P-rich layer have already developed within this short reaction time. C and N signals from adsorbed biological constituents are also detected within the interfacial reaction layers. This result is especially important from the standpoint of surface characterization because it means that the kinetics of formation of the surface reaction layers on the implant is the most important feature of the material.

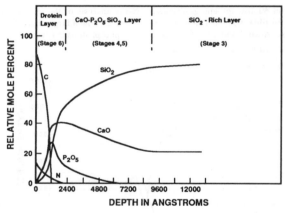

Fig. 8. Surface compositional profile of bioactive glass (45S5 Bioglass®) obtained with Auger electron spectroscopy and Ar-ion beam milling after one hour exposure to rat bone.

Secondary Ion Mass Spectroscopy (SIMS). This method also involves removal of surface layers of the material, atom by atom, and measurement of the concentration of the atoms removed, with a mass spectrometer. It yields a very accurate compositional profile including many elements. The instrument is expensive and requires considerable expertise in its operation but produces the best overall results for complex interfaces.

Surface Charge Analysis. The surface charge of a bioceramic influences the adsorption of proteins and other biological constituents and affects the response of tissues in contact with it. The sign and magnitude of surface charge are determined by the composition of the material, the crystal phases in the material, the defects in crystalline phases, and the pH of the solution in contact with the material. A material will have a zero point of charge (ZPC) at a particular pH where surface positive charges balance surface negative charges. At pH values above this point, also known as the isoelectric point (IEP), the material will exhibit a positive charge. At pH values below the IEP the material will exhibit a negative charge. The pH of body fluids is acidic, <7, during wound healing, slightly alkaline, 7.4, in normal conditions, and alkaline, >7.4, during bone mineralization. Thus, it is important that bioceramics used in bone repair have surface charges compatible with the range of solution pH occurring during the healing and mineralization of bone.

One of the most useful methods for determining surface charges of bioceramics is a zeta potential measurement. Ducheyne and co-workers (see the Reading List) discuss this method and its use in characterizing the behavior of various calcium phosphate ceramics. They have shown that the cationic defect structure of stoichiometric and Ca-deficient hydroxyapatite has a large effect on surface charge of the materials. They also show that the magnitude and duration of the changes in zeta potential of the apatites are related to an ion exchange between the hydrated layer around the ceramic surface and a net precipitation of new material. The zeta potential method should be considered as a characterization method for calcium phosphate ceramics to determine the relative effect of defect structures on surface properties.

The Decision. Deciding on a specific test or combination of tests is difficult but must be done. As discussed in Chapter 3, experiments established that the critical step in the reaction stages of a bioactive glass implant is formation of the HCA layer. The tissue response correlates directly with the time of HCA formation. Thus, the critical characterization test is the determination of the time of HCA formation. As described above, many different analytical methods can be used to show development of an HCA layer on the glass: XRD, FTIR, FTRaman, SEM-EDS, EMP, AES, and SIMS. Which one should be chosen?

Step 7 in "Development of A Characterization Strategy" states that the decision on test(s) to be used should be based upon relative economics, without sacrifice of accuracy. Economic analysis of the alternative methods listed above requires taking into account a number of factors: time of analysis, time and cost of sample preparation, cost of analysis, cost of instrument, maintenance costs, operator costs, operator skill, reproducibility with different operators, availability of instrument.

When these factors are taken into consideration, the favored characterization method for bioactive glasses is FTIR analysis, before and after an *in vitro* test simulation of the physiological environment. The characterization requirement is that the glass develops a fully crystallized hydroxyl-carbonate apatite (HCA) reaction layer within a specified period of time, for example 20 hours, when exposed to a simulated body fluid (SBF) at 37°C.

The SBF chosen for the characterization program was based upon experiment. Kokubo and co-workers (see the Reading List) showed that the *in vitro* behavior of A/W glass-ceramic most closely matched *in vivo* behavior when the material was tested in the SBF composition listed in Table 1. This composition was based upon analyses of human body fluids, previously reported, and includes important constituents such as soluble calcium, phosphate ions and carbonate ions. The chemicals used in making up the SBF solution are given in Table 2, courtesy of Professor Kokubo.

Comparative studies of various bioactive glass compositions and SBF solutions, summarized in Chapter 3, showed that the rate of HCA formation on many bioactive glasses exposed to the solutions given in Table 1 correlates well with *in vivo* results.

Thus, the critical surface characterization test for a bioactive glass is the time taken for HCA formation when exposed to SBF K-9 at 37°C.

Table 1. Ion Concentration (mM) of SBF and Human Blood Plasma.

Ion	Simulated Fluid	Blood Plasma
Na^+	142.0	142.0
K^+	5.0	5.0
Mg^{2+}	1.5	1.5
Ca^{2+}	2.5	2.5
Cl^-	147.8	103.0
HCO_3^-	4.2	27.0
HPO_4^{2-}	1.0	1.0
SO_4^{2-}	0.5	0.5

Table 2. Reagents for Preparing SBF#9.

Order	Reagent	Purity	Amount
1	NaCl	For Biological work	7.996g
2	$NaHCO_3$	Certified A.C.S.	0.350g
3	KCl	Certified A.C.S.	0.224g
4	$K_2HPO_4 \cdot 3H_2O$	99+%	0.228g
5	$MgCl_2 \cdot 6H_2O$	Assay 99.7%	0.305g
6	1N-HCl		40 ml (about 90% of total amount of HCl to be added)
7	$CaCl_2$	Assay 99.6%	0.278g
8	Na_2SO_4	Certified A.C.S.	0.071g
9	$NH_2C(CH_2OH)_3$	Assay 100.1%	6.057g

SUMMARY

This chapter has shown that it is necessary to characterize many different features of an implant in order to determine which are the critical ones to achieve reproducibility. After the critical characterization feature(s) have been determined it is necessary to determine the tolerances required for reproducible performance. Then it is possible to compare the relative economics of the variety of methods suitable for characterization. A systematic effort is required in order to develop a cost effective set of tests that guarantee reproducibility. It is time consuming to follow such a systematic test program but the eventual savings and reliability that result are worth it in the end.

READING LIST

L. L. Hench and R. W. Gould, eds., *Characterization of Ceramics* (Marcel Dekker, N.Y., 1971).

C. G. Pantano, Jr., D. E. Clark, and L. L. Hench, *Corrosion of Glass* (The Glass Industry, N.Y., 1979).

T. Yamamuro, J. Wilson, and L. L. Hench, eds., *Handbook of Bioactive Ceramics: Vols I and II* (CRC Press, Boca Raton, FL, 1990)

T. Kokubo, H. Kushitani, S. Sakka, T. Kitsugi, and T. Yamamuro, "Solutions Able to Reproduce *In Vivo* Surface-Structure Changes in Bioactive Glass-Ceramic A-W," *J. Biomedical Mater. Res.* **24** (1990) 721-734.

P. Ducheyne, C. S. Kim, and S.R. Pollack, "The Effect of Phase Differences on the Time Dependent Variation of the Zeta Potential of Hydroxyapatite," *J. Biomedical Mater. Res.* **26** (1992) 147-168.

G. LaTorre and L. L. Hench, "Analysis of Bioactive Glass Interfacial Reactions Using Fourier Transform Infrared Reflection Spectroscopy," in *Characterization Methods for the Solid-Solution Interface in Ceramic Systems*, eds., J. H. Adair and J. A. Casey (American Ceramic Soc., Westerville, OH, 1993).

J. H. Adair and J. A. Casey, eds., *Characterization Methods for the Solid-Solution Interface in Ceramic Systems* (American Ceramic Society, Westerville, OH, 1993).

R. T. DeHoff in *Quantitative Microscopy*, eds. R. T. DeHoff and F. N. Rhines (McGraw-Hill, New York, 1968) p. 291.

L. L. Hench, R. J. Splinter, T. K. Greenlee and W. C. Allen, "Bonding Mechanisms at the Interface of Ceramic Prosthetic Materials," *J. Biomed. Maters Res.* **2[1]** (1972) 117-141.

U. Gross, R. Kinne, H. J. Schmitz, V. Strunz, "The Response of Bone to Surface Active Glass/Glass-Ceramics," *CRC Critical Reviews in Biocompatibility*, **4** (1988) 2.

T. Kokubo, "Surface Chemistry of Bioactive Glass-Ceramics," *J. Non-Cryst. Solids* **120** (1990) 138-151.

L. L. Hench, "Bioactive Ceramics," in *Bioceramics: Materials Characteristics Versus In Vivo Behavior*, Vol. 523, eds. P. Ducheyne and J. E. Lemons (Annuals of the New York Academy of Sciences, 1988) pp. 54-71.

J. Gamble, *Chemical Anatomy, Physiology and Pathology of Extracellular Fluid, 6 ed.* (Harvard University Press, Cambridge, 1967) pp. 1-17.

C. Y. Kim, A. E. Clark and L. L. Hench, "Early Stages of Calcium-Phosphate Layer Formation in Bioglass," *J. Non-Crystalline Solids* **113** (1989) 195-202.

O. H. Andersson, G. LaTorre, and L. L. Hench, "The Kinetics of Bioactive Ceramics, Part II: Surface Reactions of Three Bioactive Glasses," in *Bioceramics, Volume 3*, eds. J. E. Hulbert and S. F. Hulbert (Rose-Hulman Institute of Technology, Terre Haute, Indiana, 1992) pp. 46-53.

L. L. Hench, O. A. Andersson and G. P. LaTorre, "The Kinetics of Bioactive Ceramics, Part III: Surface Reactions for Bioactive Glasses Compared with an Inactive Glass," in *Bioceramics, Volume 4*, eds. W. Bonfield, G. W. Hastings and K. E. Tanner, (Butterworth-Heinemann Ltd., Guildford, England, 1991) pp. 155-162.

B. O. Fowler, "Infrared Studies of Apatites I Vibrational Assignments for Calcium Strontium and Barium Hydroxyapatite Utilizing Isotopic Substitution," *Inorg. Chem.* **13[1]** (1974) 194.

R. F. Le Geros, G. Bone, and R. Le Geros, "Type of H_2O in Human Enamel and in Precipitated Apatites," *Calcif. Tissue Res.* **26** (1978) 111.

F.C.M. Driessens, "Formation and Stability of Calcium Phosphates in Relation to the Phase Composition of the Mineral in Calcified Tissues," in *Bioceramics of Calcium Phosphate*, ed. K. de Groot (CRC Press, Inc., Boca Raton, FL, 1983) p. 7.

Chapter 19

REGULATION OF MEDICAL DEVICES

Emanuel Horowitz* and Edward Mueller**
*Johns Hopkins University
Materials Science and Engineering Department
Baltimore MD 21218
**Food and Drug Administration
Center for Devices and Radiological Health
Rockville, MD 20852

INTRODUCTION

The government regulation of medical devices and biomaterials is a complex and difficult task. Regulations should be effective in protecting patients from undue risk without discouraging technical innovations and development. A renowned surgeon, Dr. Dwight Harken, often stated, "A device is safe when it is safer than the disease it corrects and is the best available." Dr. Harken's words illustrate what must be at the heart of a good regulatory system, i.e. a capability of assessing risk-to-benefit ratios. It is important to characterize unambiguously product performance and to be able to assess risks and benefits in a comparable and defensible fashion. To do this requires a strong interplay between all areas of science, engineering, and medicine and most importantly an understanding, by the public, that regulations do not create new knowledge but instead coordinate existing knowledge. Because a product has been assessed in a regulatory sense does not mean it will never fail to perform.

In this chapter, a brief review of the Food and Drug Administration's (FDA) regulation of medical devices will be provided. This review will discuss the revision to the Medical Devices Act and will discuss some new initiatives the Food and Drug Administration Center for Devices and Radiological Health (FDA/CDRH) is considering to facilitate and improve the review process. In addition, a section has been devoted to the role of standards and shows how they playing an increasingly important role.

HISTORICAL PERSPECTIVE

Pre-Food, Drug and Cosmetics Act (1938)

Prior to 1938 and the passage of the Food, Drug, and Cosmetic Act, the federal government had limited power to regulate unsafe practices and protect the health and safety of the public. The Food and Drug Administration, which was established in 1930, had evolved from the Food, Drug and Insecticide Administration which was formed in 1927. In 1937 an elixir of sulfanilamide containing a poisonous solvent was marketed resulting in the death of 107 people, mostly children. This incident dramatized the need for new legislation designed to protect the public. The following year, Congress passed the Federal Food, Drug and Cosmetics Act of 1938.

The Food, Drug and Cosmetics Act of 1938

The Act of 1938 authorized the Food and Drug Administration (FDA) to regulate medical devices by establishing requirements for their safety. In addition, it defined a medical device as *any instrument, apparatus, or contrivance, including any and all components or parts, that were intended for use in the diagnosis, cure, treatment or prevention of disease in man or in other animals.*

Regulations were established to control misbranded or adulterated medical devices which were involved in interstate commerce. Misbranding meant that the labeling was incorrect or misleading or it did not properly identify the manufacturer, packager or distributor or provide information on the quantity of the contents of the package. Adulteration included devices which were dirty and/or were prepared or packaged under unsanitary conditions.

The 1938 legislation lacked true effectiveness because, unlike drugs, it did not require premarket clearance for medical devices which were involved in interstate commerce. This meant that medical devices could be marketed before being cleared by the federal government in accordance with accepted clearance procedures. During the period that this law was in effect the FDA was limited to enforcing labeling provisions and removing fraudulent medical devices from the marketplace.

The medical advances in science and engineering during the 1940's and 1950's impacted strongly on medical technology. New materials made of polymers, metals and alloys, and ceramics were introduced and found application in newly designed medical devices while advances in electronics led to the development of entirely new and more complicated medical systems. As the number of medical devices and implants being used on patients increased, particularly in applications which involved life-saving or life-sustaining treatment, the number of adverse reactions and failures grew. It soon became clear that the provisions of the Act of 1938, drafted to protect the public from unsafe or ineffective devices was inadequate. This led to the confusing situation where some courts of law ruled that certain medical devices could be treated as drugs and regulated accordingly. It was clear that new and more effective legislation was necessary.

The Medical Device Amendments of 1976

In 1969 president Richard Nixon appointed Dr. Theodore Cooper, the Director of the National Institutes of Health's National Heart Institute, to chair a presidential commission, later known as the Cooper Committee, to study the need for increased federal regulation of medical devices. The Cooper Committee found that problems traceable to medical devices had caused about 10,000 injuries and more than 700 deaths over a ten year period. Their findings and recommendations, spurred the passage of the Medical Devices Amendments of 1976,[2] which became law on May 28, 1976. The FDA now had the legal authority to regulate medical devices, especially with regard to their labeling, marketing, manufacture, processing, distribution and use. A sharp distinction was drawn between medical devices and drugs. Those products that are not metabolized or are not dependent on being metabolized in order to achieve their intended purpose were to be regulated as devices and not drugs. Furthermore, the 1976 definition of a device was somewhat expanded and encompassed instruments, apparatus, implements, machines, contrivances, implants, *in vitro* reagents, or other similar or related articles. It included any component or part which was used in the diagnosis of disease or other conditions, or in the cure, mitigation, treatment, or prevention of disease in man or other animals. Actions which were prohibited under the law included adulteration and misbranding of devices destined for interstate commerce.

The medical device types covered by the legislation of 1976 numbered about 1700 and were categorized in the following way:

Over-the-counter devices which did not require a prescription.

Prescription devices.

Investigational devices which were in the development phase and could be used on humans only to obtain safety and effectiveness data.

Custom devices which met the special needs and requirements of an individual patient.

Also included were critical devices which were defined in the Good Manufacturering Practice Regulation (GMP) as devices which were life supporting or life sustaining.

The Safe Medical Devices Act (SMDA) of 1990

The Act of 1990[3,4] provided new enforcement authority for the FDA, to strengthen the ability of the agency to regulate medical devices and prevent the use of unsafe and ineffective products. The law also modified the classification and reclassification of medical devices and, in turn, affected the premarket notification and premarket approval procedures. It also changed the Good Manufacturing Practices requirements. Most significantly, the new law provided FDA with a much broader authority to collect data and monitor devices after they have been cleared for marketing. Some of these new authorities, briefly discussed below, include user reporting, tracking, post marketing surveillance, special controls and standards, and revisions to Premarket Notification 510(k) and Premarket Approval (PMA) requirements.

User Reporting

When a medical device fails in service and causes death or serious injury, user facilities, which include hospitals, nursing homes, outpatient treatment and surgical facilities are required to report such cases to the manufacturer and, in cases of death, to the FDA. User facilities must submit a semiannual report to the FDA which summarizes their reports of such incidents. This User Reporting provision is an extension of the existing Mandatory Device Reporting regulation which required manufacturers to report death and serious injury information to the FDA.

Tracking

Manufacturers of certain high risk medical devices are required to establish a method for tracking their products whose failure would jeopardize the heath of the patient. For life sustaining medical devices, used outside a user facility and some kinds of permanent implants, the new regulations require that the device manufacturer establishes a system for identifying and locating a device at any given time.

All device manufacturers must report any removals or corrective actions taken to reduce the risk to the patient's health or remedy a violation of a medical device requirement. This requirement is intended to facilitate recall and notification provisions of the medical device law.

Post-Market Surveillance

The Post-Market Surveillance Study provisions of the Act are probably the most far reaching and immediate for devices manufactured from new materials. Two types of studies are outlined in the new law: mandatory and discretionary. Manufacturers of permanent implant devices and life sustaining and life supporting devices which are brought to market after January 1, 1991, will be required to conduct post market studies of the performance of their products. In addition, FDA may require the manufacturer of any device to initiate such a study of its performance, regardless of when the device was first marketed. The Congress' intention for these studies is expansion of the information available on the performance of the device over a larger population and for a longer period of time, beyond that gathered during the premarket testing. FDA will probably define the aspects of the device that are to be studied, e.g. specific populations may be targeted for study as "high risk" with the use of a new material, the qualifications of the principal investigator, and the characteristics of an acceptable study. The study protocol and the qualifications of the principal investigator require FDA approval prior to initiation of the study.

Special Controls

The 1990 Act redefines Class II to include any device which, by application of special controls, can be considered to be safe and effective. The classification and explanation of medical devices in Class I, Class II and Class III is discussed later in this chapter. Special controls may include performance standards, postmarket surveillance and patient registries. Furthermore, the procedures for establishing performance standards are simplified.

510(k)/PMA Revisions

The reclassification of Class III provisions for preamendment devices requires that manufacturers submit to the FDA, upon request, summary of information known to them about their devices, including any adverse safety and effectiveness data. Then FDA will decide whether such devices will remain in Class III or be down-classified into Class II or Class I.

With regard to 510(k) devices, the manufacturer is obliged to submit to FDA with the premarket notification, or make available, a summary statement containing safety and effectiveness information and data upon which an evaluation of the device may be performed. Some new provisions have been introduced on information required with regard to premarket approval applications which support the safety and effectiveness claim. The requirements and time constraints are quite involved and those interested in these requirements should refer to the PMA paragraphs in the law or consult the FDA

Future Directions

An important focus of the FDA's future activities in medical devices is risk assessment. There is a great need in the device area to weigh a product's risks against its benefits. In some cases the calculation is reasonably straightforward e.g. comparing the morbidity and mortality of a damaged heart valve with the use of a prosthetic valve. For many other devices, the relationship is more convoluted. How does one measure the benefits of a breast implant? How does one measure the risks of a material like silicone gel or oil, when its basic toxicology and pharmacokinetics are still uncertain. Straightforward or convoluted, the process is difficult.

To confound the issue, societal messages associated with risk assessment are mixed and inconsistent. Many people have been led to believe that no amount of risk from a medical product is acceptable, and that the government's role is to guarantee that the products it regulates are absolutely safe. That view is considered unrealistic, and greatly hampers the ability to communicate effectively about risk-benefit issues.

With regards to biomaterials and risk/benefit decision making, FDA needs a systematic means for evaluating/comparing clinical performance against *in-vitro* assessments of material and device characteristics and properties. As in medicine, biomaterials biocompatibility assessments are not absolute measurements but are very much comparative evaluations. Imperative in such an approach, is the use of consistent (standardized) methods of evaluation and a common knowledge base.

A direction FDA is currently exploring involves development of a compendium of knowledge about the performance and potential risks of specific biomaterials in specific environments - a BIOMATERIALS DATABASE. Such a compendium could be a resource for both manufacturers and government, particularly to streamline and improve the quality of the product approval process. Conceptually, this compendium could be compared to the drug master file used by drug manufacturers to develop "me-too" pharmaceuticals with reduced testing requirements.

For example, if a new product were manufactured from a certain formulation of a material, using a manufacturing process comparable with that listed in the compendium, requirements for preclinical testing may be reduced or waived. Such a system could eventually become an integral part of internationally harmonized procedures for premarket approval. As currently envisioned, this compendium would be a series of networked databases, as opposed to a single comprehensive database.

A second project proposed by FDA, as part of the U.S. government's Advanced Materials and Processing Program (AMPP), involves development of a systematic way to retrieve and study explanted devices on a national level. The proposal, planned to begin in 1994, involves close interactions with university medical centers, professional societies, industry and standards organizations to establish research protocols and a uniform system for collecting, storing and analyzing information about the fate of implanted devices in the body. This kind of information, if widely disseminated and shared, would be valuable to new materials and device research, development and sound regulation. The explant/retrieval database would be one of the networked databases in the above mentioned compendium.

A third area FDA has been working on is the definition of types of research needed to evaluate effectively product risks and benefits. In 1990, FDA published a document entitled "Research Agenda for the 90s"[5] which attempted to address these issues. This document which is currently being updated has been circulated widely in an attempt to stimulate needed research.

THE FDA REGULATORY SYSTEM

System Overview

The FDA has the authority to regulate devices during most phases of their development, testing, production, distribution and use. FDA's approach to this regulation focuses heavily on both the pre- and post-market phases of a product's lifetime. During the pre-market phase, FDA concentrates on providing a reasonable assurance that new products are adequately evaluated for safety and effectiveness. Implicit in this assessment is the concept of risk to benefit. This means that if the benefits significantly outweigh the risks for the intended application, the product would probably be approved for marketing in the United States. Since risk/benefit assessments for new technology involve considerable clinical judgment, FDA consults panels of clinicians, engineers, toxicologists and other experts familiar with the devices. These panels review the data provided by the manufacturer to support the claims for product safety and effectiveness. If the advisory panel believes the data supports the manufacturer's claims, an "approval-for-marketing" recommendation is made to FDA.

Once a product has been approved for use in the United States, the FDA's role shifts to the post-market monitoring of product reviewed performance and manufacturing practices. This activity ensures that the product design which was reviewed during the premarket phase is, in fact, the product which is manufactured and sold. To do this, FDA periodically inspects the medical device manufacturers to ensure compliance with Good Manufacturing Practices (GMPs). Further, the FDA evaluates failures reported under a variety of systems to determine whether appropriate actions have been taken by the manufacturer to reduce the risk to the users.

This discussion has focused on medical devices, not biomaterials. The regulation of biomaterials is an often confused area for developers of new medical devices.[6] Many believe that the FDA regulates all materials used in medical devices, in addition to regulating the device itself. Hence, FDA is regularly confronted with the following types of questions:

1. How do I get my biomaterial approved by the FDA?
2. How do I get a list of the approved materials for use in medical devices?

The answers depend entirely on the intended application of the materials. FDA regulates the end product (medical device), not the materials from which it is made. FDA does not approve biomaterials. Therefore, FDA does not maintain a list of approved biomaterials nor does it provide guidance on how to get a biomaterial approved for general application. There are, however, instances where a biomaterial is the end product, i.e., PPMA bone cement, hydroxylapatite, injectable collagen, etc. In these cases, the claims made for the material by the manufacturer are for specific applications and the manufacturer has to provide adequate *in-vitro* and *in-vivo* test data to demonstrate safety and efficacy in those applications.

Another question frequently asked is, "How can I satisfy the requirements of the FDA and market my product?" The FDA has published a flowchart shown in Appendix 1 which helps to clarify this procedure. Appendix 2 contains a list of special FDA publications which provide additional information on other topics related to medical devices.

Classification of Medical Devices

The key to FDA's regulatory approach for medical devices is the classification system. The level of regulation or control is governed by the "Class" in which the device is placed by the agency; Class I, II or III.

Classification recommendations by FDA are first published as proposals in the Federal Register. After receipt and consideration of comments, a final regulation classifying each device is published. Most devices have already been classified.

The Safe Medical Devices Act of 1990 significantly changed certain aspects of the regulatory system governing surgical and medical devices. The Food and Drug Administration is involved in an on-going process of developing guidance documents to explain and clarify the new law in terms of its impact on the medical community and the manufacturers of medical devices. Portions of one such document, "An Introduction to Medical Device Regulations[7]" have been used in sections 2 and 9 to provide the most current official view on some key aspects of the revised regulatory system.

The three levels of control based on device class are:

Class I devices, those needing the lowest level of regulation, are subject to the "General Controls" requirements. These include establishment (manufacturing site) registration, device listing, Premarket Notification and Good Manufacturing Practices (GMP).

Class II devices are subject to "Special Controls" as well as "General Controls" requirements. Special Controls may include labeling, a mandatory performance standard, etc.

Class III devices cannot be marketed until they:

-   have an approved Premarket Approval Application (PMA), or

-   as a result of Premarket Notification [510 (k)] submissions, have been found by FDA to be substantially equivalent to devices marketed before May 28, 1976 (preamendment devices). If a PMA has been "called" for that type of preamendment device, then, a PMA must be submitted. A 510 (k) can not be used in this case. (Premarket Notification is discussed in more detail later).

Thus, the class of a device must be known to see what regulations apply to it. Note, however, that all devices, regardless of class, are subject to General Controls. Some Class I devices are exempt from the Premarket Notification and/or the Good Manufacturing Practices requirements. Exemptions are listed in the final classification regulation for the specific device.

General Controls
Device firms must meet the following General Controls requirements:

-   Register each manufacturing location (establishment).

-   List marketed medical devices.

-   Submit a "Premarket Notification" [510(k)] before marketing a device that is new to the firm or that has been significantly modified. If a contract manufacturer previously manufactured a device for a manufacturer, and the relationship has dissolved, the contract manufacturer must submit a 510(k) if it wishes to continue to manufacture and to distribute the device under its own labeling.

-   Manufacture devices in accordance with the Good Manufacturing Practices (GMP) regulation.

These controls are explained below.

Establishment Registration
Unless exempt under 510(g) of the FD&C (1938) Act, the owner or operator of an establishment must register with FDA within 30 days after beginning any of these

activities: manufacture, preparation, propagation, compounding, assembly, or processing of a device intended for human use. Activities requiring registration include repackaging, relabeling, importing of foreign devices and specifications development.

Initial registration is made on the "Initial Registration Form" (FDA 2891). Thereafter, firms will receive and "Annual Registration Form" (FDA) 2891a) from FDA each year. If changes in registration status occur at other than the annual registration time, they must be submitted in writing to FDA within 30 days. Under SMDA, distributors are also required to register with FDA.

Listing Devices

Devices are to be listed on the "Device Listing Form" (FDA 2892). Unlike registration, listing is not updated yearly.

Only when a significant change occurs in one or more of the data elements on the form, does the firm submit a new listing form containing the changes.

Premarket Notification [510(k)][8]

At least 90 days before it intends to market a device for the first time, a firm must submit to FDA's Document Mail Center a "Premarket Notification," also called a "510(k) submission. The 510(k) submission must contain sufficient information to show that a device is substantially equivalent to a legally marketed device. A premarket Notification also is required for a product marketed by a firm when there is a significant change, including a different use that may significantly affect its safety and effectiveness.

There is no form for 510(k) submissions. However, a format can be found in Section 807.87 of 21 CFR (Code of Federal Regulations). Each submission must contain:

- The name of the device (both trade, common or usual and classification name).

- The establishment registration number. If the firm is not yet registered, a statement to this effect is sufficient.

- The class of the device (if known), or a statement that the class is not known, indicating the appropriate panel.

- Action taken to conform to any applicable FDA Special Controls for a Class II device, if a special control has been issued.

- Labels and labeling for the device, including any advertisements sufficient to describe the intended use of the product. Proposed labeling is sufficient.

- Persons submitting a premarket notification [510(k)] submission must provide to the FDA, as part of the submission, an adequate summary of information on safety and effectiveness or a statement that it will be made available if anyone asks for it.

- Persons submitting a 510(k) for a class III device must include a certification that they have looked at all available information on the device (and others like it) and must provide a summary of, and citation to, all adverse safety and effectiveness data respecting the device (and others like it).

- A statement (with accompanying data) indicating how the device is similar to and/or different from other comparable commercially available products.

- For a submission made because the device has undergone a change or modification that could significantly affect safety and effectiveness, sufficient data to show that consideration has been given to this change and its effect on the safety and effectiveness of the device.

- Any additional information requested by FDA to determine if the device is substantially equivalent. Such added information must be submitted within 30 days, or an extension of time to respond must be requested.

Ely[9] has reported that since 1976 more than 15,000 510(k)'s have been submitted to the FDA and only about 2 percent did not qualify.

Good Manufacturing Practices[10-12]

The Good Manufacturing Practices (GMP) regulation covers the methods, facilities, and controls used in pre-production design validation, manufacturing, packaging, storing, and installing medical devices. The GMP regulation identifies the essential elements required of the quality assurance program. FDA monitors compliance with the GMP regulation during inspection of the firm's manufacturing facilities.

To address the variety and complexity of devices, the GMP regulation designates two device categories: "noncritical" and "critical". General requirements apply to all devices, but critical devices must meet additional GMP requirements.

Investigational Device Exemption[13]

To allow manufacturers of devices intended solely for investigational use to ship these devices for use on human subjects, the FD&C Act authorizes FDA to exempt these firms from certain requirements. This exemption is known as an Investigational Device Exemption (IDE) and applies only to investigational studies gathering safety and effectiveness data for a medical device when using human subjects. If a device is considered to present "significant risk," IDE applicants submit information to FDA demonstrating that testing will be supervised by an Institutional Review Board (IRB), that appropriate informed consent will be obtained, and that certain records and reports will be maintained. For a "nonsignificant" risk device, submission to FDA is not necessary but IRB approval is still required.

A "significant risk" investigational device is one that:

-   Is intended as an implant and presents a potential for serious risk to the health, safety, or welfare of a subject.

-   Is purported or represented to be for use in supporting or sustaining human life, and presents a potential for serious risk to the health, safety, or welfare of a subject.

-   Is for a use of substantial importance in diagnosing, curing, mitigating, or treating disease (or otherwise preventing impairment of human health) and presents a potential for serious risk to the health, safety, or welfare of a subject.

-   Otherwise presents a potential for serious risk to the health, safety, or welfare of a subject.

Certain types of devices are exempted from the IDE regulation. These include custom devices, certain *in vitro* diagnostic devices, devices solely for veterinary use, nd devices that are substantially equivalent to preamendment devices used for the same intended purpose. Preamendment devices are those commercially marketed before May 28, 1976.

Premarket Approval[14,15]

An approved Premarket Approval (PMA) application is somewhat like a "private license" granted to the applicant to market a particular medical device. Other firms seeking to market the same type of device for the same use must also have an approved PMA. Class III devices may require PMA application approval, Premarket Notification, or both.

Premarket approval requirements differ between "preamendment" and "postamendment" devices.

Preamendment devices are those in commercial distribution before May 28, 1976. The agency is calling for such PMAs, as required in the 1990 Safe Medical Devices Act.

Before requiring a firm to have an approved PMA application in order to continue marketing a preamendment medical device, FDA must wait 30 months from the effective data of a final classification regulation for the device or 90 days after publication of a final regulation requiring the submission of a PMA--whichever is later.

Postamendment devices are those first commercially distributed after May 28, 1976. Manufacturers of Class III postamendment devices that are not substantially equivalent to preamendment Class III devices are required to obtain PMA application approval before marketing their device.

If a firm plans to market a device that is similar to a preamendment class III device for which PMA are not required, a Premarket Notification should be submitted. If FDA finds the new device substantially equivalent to the preamendment device, it will then be subject to the same requirements as the preamendment device. If the device is not substantially equivalent to the preamendment Class III device, then by statue a PMA is required or a firm may choose to petition to reclassify the device into Class I or Class II. SMDA also provides for limited situations when information provided in previous four-of-a-kind PMA devices may be used by FDA in reviewing new Class III devices. In this context four-of-a-kind means four PMA devices which have been previously approved and are equivalent. Also, FDA may temporarily suspend a PMA while determining if the PMA approval should be withdrawn.

Inspections

It is unlawful to refuse to permit inspection, as allowed by the provisions of the law, of any factory or establishment in which medical devices are manufactured, processed, packed or stored for introduction into interstate commerce.

The FDA is required to inspect manufacturers of Class II and Class III medical devices at least once every two years. A history of problems with a company or its products will influence the frequency of inspection visits. Where complaints have been received by the FDA, particularly on critical devices being used for life saving or life sustaining applications, the manufacturers of such products will be inspected more often.

The FDA compliance program covering the inspection of medical device manufacturers is designed to find out how well they are complying with the Good Manufacturing Practice requirements. The inspection reviews and evaluates records and data, labeling procedure, equipment and facilities and examines operations and personnel. Under the government-wide Quality Assurance Program the FDA sometimes asks other government agencies to inspect medical device manufacturing establishments. In cases where the inspection has revealed deficiencies the FDA sends a "Warning Letter," to the company pointing out the short-comings. A response within ten days is required from the manufacturer explaining how the deficiencies or violations will be corrected. If the response is judged to be too little or too late the FDA can initiate legal action.

Master Files for Medical Devices (MAF's)

Applications and forms submitted to the FDA often contain proprietary data or confidential commercial and financial information. In some cases an applicant for a Premarket Approval (PMA) or an Investigational Device Exemption (IDE) may need to use a subcontractor for some of the tasks related to the production and marketing of his medical device. The manufacturer may submit quality control procedures, materials composition and/or test data and manufacturing processes in an FDA master file instead of making it available to another company. The subcontractor may not object to the FDA reviewing his company's information but may oppose having the PMA or IDE applicant gaining access to his special files. The system was established by the FDA to protect trade secrets of medical device companies and their subcontractors while allowing scientific review and evaluation of these devices. There are various types of master files (e.g. biologic, drug, food and veterinary) in addition to medical master files for medical devices. Master files are designed for manufacturing procedures and controls, synthesis and specifications for chemicals and materials used in producing devices; packaging materials and data from non-clinical and clinical studies. There is a specified format and arrangement for the file to make it easy to use, store, and retrieve. The contents of the master file may be modified to include new test results and information or to explain changes in the manufacturing procedure or in the product.

Criteria for acceptance of an MAF includes the following:

- The MAF must provide factual information useful to the evaluation of the device.

- The MAF is not to be used by manufacturers of medical devices to protect their proprietary processes and data, but only when services or materials are to be made available to other device producers and would otherwise have to be duplicated.

The use of the proprietary information in the master file can only be approved by the owner (holder) of the file or by a designated agent.

FDA Guidance

The FDA has generated numerous guidance documents to help manufacturers and developers of new products with the regulatory requirements. For premarket approval requirements, there are over 35 device-specific guidance documents detailing useful studies. Examples include prosthetic heart valves, PTCA catheters, cochlear implants and ventricular assist devices. In addition other useful documents such as a survey of medical device standards, FDA's medical devices research agenda and CDRH office's annual reports are readily available. To facilitate dissemination of this information, FDA/CDRH has a Division of Small Manufacturers Assistance (DSMA) and has established a Personal Computer Bulletin Board System (BBS)* for immediate access to the documents. The BBS directory of available documents covers general device evaluation guidance, information on specific medical devices (i.e. cardiovascular, neurological, ophthalmic), tracking and reporting, manufacturers assistance and standards. the bulletin board is still under development and further information is available about the system by contacting the Center for Devices and Radiological Health (CDRH).

---

*The Bulletin Board System is based on RBBS PC Software and operates with the following communications parameters: 8 databits, 1 stopbit, no parity, 300-9600 baud. The phone number is 301-443-7496.

STANDARDIZATION

Vannevar Bush, a famous U.S. scientist who was an advisor to presidents during and after World War II, said "If men are to accomplish together anything useful whatever they must, above all, be able to understand one another". Dr. Bush, in one sentence, defined the need for standards and the standardization process. Standards clarify the meaning of materials, devices and systems in terms of their properties and performance characteristics. They promote understanding among buyers, sellers and users of materials and products; provide a systematic way of establishing and maintaining quality; assure interchangeability of parts and components; minimize unnecessary redundancy and reduce inventory. For medical devices they provide specifications for chemical, physical and mechanical properties of the materials used as well as standard methods for testing, analyzing and characterizing both the materials and the finished products. Although there are both mandatory and voluntary standards this discussion deals solely with voluntary standards which are principally developed by the private sector rather than mandatory standards which come most frequently from government agencies. Voluntary standards for materials usually deal with composition specifications and property characterization methods, whereas those for products tend to focus on defining product performance measures[16] and methods as assessment. Mandatory standards include the aspects identified in the voluntary standard and further specify product performance minimums. The voluntary standard often serves as a precursor to the development of a mandatory standard. In some instances where the voluntary standard is written with a specific application in mind, performance floors or minimums are defined and such standards tend to be more readily adopted to mandatory use.

National Standards

There are several hundred private organizations in the United States which engage in developing various types of voluntary standards. A limited number of these devote their full effort to this activity and play a dominant role in producing most of the voluntary standards in use. Two of these organizations are the American Society for Testing and Materials (ASTM)*and the Association for the Advancement of Medical Instrumentation.** ASTM Technical Committee F-4 on Medical and Surgical Materials and Devices was formed in 1962 and since that time has prepared and published more than 135 standards, specifications, practices and guidelines related to medical materials and devices. These standards deal with the materials and devices used by the various medical disciplines, including orthopedics, cardiovascular, plastic and reconstruc-tive surgery, and neurosurgery. A list of selected ASTM F-4 standards is presented in Table 1.

---

* American Society for Testing and Materials (ASTM), 1916 Race Street, Philadelphia, PA 19103

** Association for the Advancement of Medical Instrumentation (AAMI), 3330 Washington Boulevard, Suite 400, Arlington, VA 22201-4598.

Table 1. Examples of ASTM F-4 Standards.

1.	F451	Acrylic Bone Cement
2.	F1185	Composition of Ceramic Hydroxylapatite for Surgical Implant Application
3.	F603	High Purity Dense Aluminum Oxide for Surgical Implant Application
4.	F703	Implantable Breast Prostheses
5.	F639	Polyethylene Plastics for Medical Application
6.	F648	Ultra-High-Molecular Weight Polyethylene Powder and Fabricated Form for Surgical Implants
7.	F136	Stainless Steel Bar and Wire for Surgical Implants
8.	F621	Stainless Steel Forgings for Surgical Implants
9.	F620	Titanium-6A1-4V ELI Alloy Forgings for Surgical Implants
10.	F1027	Assessment of Tissue and Cell compatibility of Oro-Facial Prosthetic Materials and Devices
11.	F700	Care and Handling of Intracranial Aneurysm Clips and Instruments
12.	F641	Implantable Epoxy Electronic Encapsulants

The relationship between the voluntary standards system and the regulation of medical devices has been discussed by McNeill and Altman.[17] Marlowe and Mueller[18] have described this relationship in the context of the FDA and ASTM. In 1991 the FDA was participating Organization for Standardization (ISO) Technical Advisory Groups (TAG's). This provides important in-put from FDA into the voluntary consensus standards process. The FDA, in turn, becomes more aware of medical device technology developments, improved test methods and evaluation procedures, and chemical, physical, mechanical and performance requirements, which helps its medical device review, approval and enforcement activities.

The ASTM standards for bioceramics are discussed by Lemons and Greenspan in the appendix to this book. After a rather long quiescent period, the work on bioceramics in ASTM F-4 has markedly expanded. At present there are at least seven task groups actively developing standards for hydroxylapatite, calcium phosphate and ceramic coatings. These standards will define the properties and performance characteristics of the calcium phosphate type powders and coatings which are used to modify the surface of medical and dental devices intended for implantation in the body.

International Standards

The development and promulgation of international standards for medical devices is largely concentrated in two International Organization for Standardization (ISO) Technical Committees; ISO TC-150, Implants for Surgery and ISO TC-194, Biological Evaluation of Medical Devices.

TC-150 Implants for Surgery (DEVICE STANDARDS)

Twenty countries are currently participating (P) members of ISO TC-150, including the United States. The Committee members meet annually to introduce, draft and develop standards which gain international recognition and acceptance. These standards describe the materials and devices, define their property and performance requirements and provide test methods for evaluation and characterization. The actual work is carried out in various subcommittees and working groups. These include materials, cardiovascular implants, neurosurgical implants, bone and joint replacements, osteosynthesis, terminology, certification, and retrieval and analysis of implants.

The ISO TC-150 has already published a wide variety of surgical device standards which include standards on specific devices such as orthopedic and cardiovascular devices; surgical instruments; and test methods and procedures.

TC-194 Biological Evaluation of Medical and Dental Materials and Devices. (BIOCOMPATIBILITY STANDARDS)

The ISO TC-194 was established to address biological screening tests for medical devices and related materials. Their initial efforts were directed at consolidating existing national methods for performing such tests as opposed to developing new techniques. Fourteen countries participated on twelve working groups to develop a series of standards in the following areas:

WG 1 - Systematic approach to biological evaluation and terminology
WG 2 - Degradation aspects related to biological testing
WG 3 - Animal protection aspects
WG 4 - Clinical investigations in humans
WG 5 - Cytotoxicity
WG 6 - Mutagenicity, carcinogenicity, reproductive toxicity
WG 7 - Systemic toxicity
WG 8 - Irritation, sensitization
WG 9 - Selection of tests for interactions with blood
WG 10 - Implantation
WG 11 - Ethylene oxide and other sterilization process residues
WG 12 - Sample preparation and reference materials

Presently, the FDA relies on the Tripartite Biocompatibility Guidance[19] for assessing medical device/material toxicity. This guidance was established at the September 1984 meeting of the "Tripartite Subcommittee on Medical Devices". The group, composed of officials from the health departments of the United States Canada and the United Kingdom agreed that there was a need for such a guidance document which identified the types of biocompatibility data required to evaluate the toxicological effects of the final product and possible leachable chemicals or degradation moieties. The guidance was limited to the effect of the material on the host body and did not consider the effect of the host on the material response. The Tripartite Guidance distinguished the physical nature of the contact of the materials with the body from the duration of the contact. The document which was developed provided a rational framework for the application of the toxicity evaluation of medical devices. The guidelines were largely based on the standards developed by the ASTM, F-4 Committee on Medical and Surgical Materials and Devices (FT48)*. These standards identified more than a dozen biological tests for assaying the biocompatibility and toxicity of medical devices. These include irritation tests, cytotoxicity, acute systemic toxicity, hemocompatibility, mutagenicity and carcinogenesis bioassay.

Because of the similarity between the Tripartite Guidance and the standard developed in WG-1 of ISO-TC-194[20], and the other standards generated in TC-194, harmonization of the documents will probably occur in the future.

Activities in international standards are of interest to the medical device community because these standards will promote understanding across national boundaries, establish requirements, facilitate foreign trade, and set acceptable levels of quality and performance for medical devices.

REFERENCES

1.   Medical Devices: A Legislative Plan, April 1970. U.S. Department of Health, Education and Welfare, A Study Group on Medical Devices. (The Cooper Committee Report).

2.   Medical Device Amendment of 1976. Federal Food, Drug and Cosmetic Act as Amended and Related Laws, May 28, 1976 U.S. Government Printing Office, Washington, D.C. (1990) 0-248-576QL3.

3.   Highlights of Safe Medical Devices Act of 1990, FDA Booklet: FDA 91-4243. U.S. Department of Health and Human Services, Center for Devices and Radiological Health, Rockville, Maryland. August 1991.

4.   Conference Report, House of Representatives, Report No. 101-959 (1990).

5.   Research Agenda for the 1990's, FDA/CDRH - Special Publication, September 18, 1989.

---

* F748 Standard Practice for Selecting Generic Biological Test Methods for Materials and Devices, ASTM Vol. 13.01.

6. E. Mueller, R. Kammula and D. Marlowe, "Regulation of Biomaterials and Medical Devices", *MRS Bulletin*, Vol. XVI, **9** (September 1991) 39-41.

7. "An Introduction to Medical Device Regulation, Food and Drug Administration (HFZ-220), Center for Devices and Radiological Health, Division of Small Manufacturers Assistance, 5600 Fishers Lane, Rockville, Maryland 20857. (To be published).

8. Premarket Notification: 510(k) - Regulatory Requirements for Medical Devices. FDA 90-4158 U.S. Department of Health and Human Services, Public Health Service, Food and Drug Administration, Center for Devices and Radiological Health, Rockville, Maryland 20857.

9. J.L. Ely, "FDA Regulations and Policy Regarding New Materials," in *Contemporary Biomaterials* eds. John W. Boretos and Murray Eden, (Noyes Publications, Park Ridge, N.J.) Chapter 22, (1984) 626-644.

10. Medical Device Good Manufacturing Practices Manual, FDA 91-41 (1991).

11. Regulations Establishing Good Manufacturing Practices for the Manufacture, Packing, Storage, and Installation of Medical Devices, Federal Register (43FR 31508) July 21, 1978.

12. J.J. Riordan and W. Cotliar, "Complying with FDA Good Manufacturing Practice Requirements" AAMI, 3330 Washington Blvd. Suite 400, Arlington, VA 22201-4598 (1991).

13. Investigational Device Exemptions - Regulatory Requirements for Medical Devices. FDA 89-4159.

14. Premarket Approval (PMA) Manual. FDA 87-4214. October 1986.

15. Premarket Approval (PMA) Manual Supplement, FDA 91-4245.

16. E. Horowitz, "Performance Standards: A Role for ISR?" *Institute for Stndards Research, (ISR) Newsletter, Vol.2 Issue 3*, December 1991.

17. C. MacNeill and M. Altman, "The Government Role in Developing Voluntary Medical Devices Standards" *ASTM, Standardization News*, August 1987.

18. D.E. Marlowe and E.P. Mueller, "The FDA and ASTM", *ASTM Standardization News* (1991) 28-31.

19. Tripartite Biocompatibility Guidance for Medical Devices (1986). Toxicology Subgroup, Tripartite Subcommittee on Medical Devices. Department of Health and Human Services, Washington, D.C.

20. Biological Testing of Medical and Dental Materials and Devices, Part 1: Guidance and Selection of Tests. ISO 10993-1.

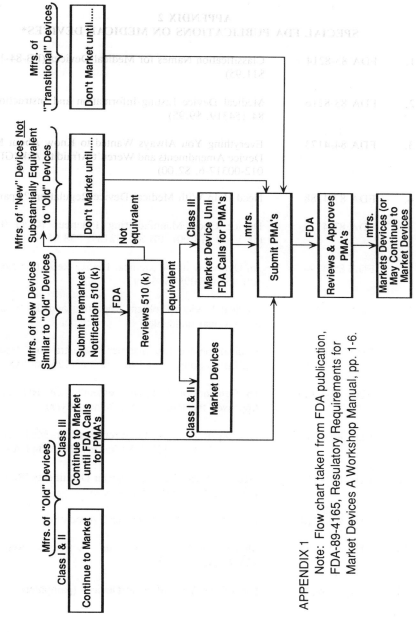

Bringing Devices to Market

APPENDIX 1

Note: Flow chart taken from FDA publication, FDA-89-4165, Resulatory Requirements for Market Devices A Workshop Manual, pp. 1-6.

APPENDIX 2
SPECIAL FDA PUBLICATIONS ON MEDICAL DEVICES*

1. FDA 83-8214   Classification Names for Medical Devices (PB-84-154863, $11.95)

2. FDA 83-8216   Medical Device Listing-Information and Instructions (PB 84-154319, $9.95)

3. FDA 84-4173   Everything You Always Wanted to Know About Medical Device Amendments and Weren't Afraid to Ask (GPO 017-012-00317-6, $2.00)

4. FDA 87-4188   Need Help With Medical Device Regulations? (pamphlet)

5. FDA 87-4179   Device Good Manufacturing Practices (GPO 017-012-00330-3, $18.00) (PB 88-132139, $38.95)

6. FDA 85-4194   An Overview of the Medical Device Reporting Regulation (PB 86-109709/AS, $9.95)

7. FDA 85-4196   Medical Devices Problem Reporting and the Health Care Professional (pamphlet)

8. FDA 87-4199   Medical Device Establishment Registra-tion: Information and Instructions, May 1987, (PB 88-123-666/AS, $12.95)

9. FDA 86-4202   To Cement Or Not To Cement?   Or Has The FDA Approved the Use Of This Device (flyer)

10. FDA 86-4203   Labeling-Regulatory Requirements for Medical Devices (GPO 017-012-00327-3, $2.75) PB 86-184348/AS, $11.95)

11. FDA 86-4208   Medical Device Federal Register Documents (Revised June 1986) PB 87-115481/AS, $13.95)

12. FDA 87-4218   Have A New Medical Device? (brochure)

13. FDA 87-4219   Medical Devices Standards Activities Report (PB 123641/AS, $19.95)

14. FDA 87-4223   Classifying Your Medical Devices (pamphlet)

15.   FDA 87-4222   An Introduction to Medical Device Regulations (pamphlet)

16.   FDA 88-4226   Medical Device Reporting: Questions and Answers, February 1988 (PB 88-192737/AS, $14.95)

17.   FDA 88-4227   Export of Medical Devices: A Workshop Manual, September 1988, (GPO 017-012-00338-9, $10.00) PB 89-119663/AS, $28.95)

18.   FDA 88-4228   Import of Medical Devices: A Workshop Manual, September 1988, (GPO 017-012-0037-1, $8.50) PB 89-11671/AS, $21.95)

19.   FDA 89-4165   Regulatory Requirements for Medical Devices May 1989[1]

*AVAILABLE FROM:

1)   The Superintendent of Documents, U.S. Government Printing Office (GPO) Washington, D.C. 20402

2)   National Technical Information Service (NTIS), Springfield, VA

3)   Center For Devices and Radiological Health, Food and Drug Administration (HFZ-265), 5600 Fishers Lane, Rockville, MD 20857

Chapter 20

ETHICAL ISSUES

Larry L. Hench
University of Florida
Advanced Materials Research Center
Alachua, FL 32615

INTRODUCTION

Millions of people currently have implants and their number is increasing rapidly.[1] It is important to examine the ethical issues associated with the increased use of artificial, non-living materials to repair, restore, or augment living tissues.

In part, the large increase in use of implants is because ethical concerns are less for artificial materials than use of living transplants.[2] Implants do not require donors. Thus, donor consent is not needed for an implant. However, many moral and social issues are as important for implants as for transplants. These issues include: informed patient consent, patient risk/benefit ratios, cost/benefit ratios, availability, reliability, and incidence of revision surgery. The number of implants far exceeds the number of transplants annually and it is important that these issues be addressed and boundaries for use of implants be established that take into consideration complex ethical concerns.

THE THEORETICAL PROBLEM

The science of biomaterials and implant design and performance is based upon well-established scientific principles of physics, chemistry, biology and physiology. The theoretical foundations of these historic disciplines are well developed and have been proven through centuries of experimental trial and error. There is little uncertainty that if the stem of a femoral head replacement is too small or the thickness of the femoral bone is too thin that fracture will result, i.e. the mechanics of the system can be calculated and the result predicted with a high level of confidence. Devices are designed using these biomechanical principles and reasonably high reliability has been achieved.

In contrast, there is conflict and uncertainty in the theoretical foundation for analyzing ethical issues.[3] Two main schools of ethics exist with major differences in their approach to achieving a moral judgement. They are: 1) The utilitarianism view developed by the English philosopher, John Stuart Mill and his successors. Their position is that an action is "right" if it leads to the greatest possible good or least possible bad consequences, i.e., "right is relative." 2) In contrast, the deontological (binding obligation) approach of the German philosopher, Immanuel Kant and successors such as W. D. Ross, argues that moral standards exist independent of utilitarian ends. Their position is that a moral life should not be conceived in terms of means and ends but an act is "right" because it satisfies the demands of some overriding principle of obligation. We should act as if our action was about to become a law of nature and all men were to hereafter act in the same way, i.e., "right is revealed."

Many modified versions of both Kant and Mill's philosophies have been developed; but the fundamental differences remain, at present, unreconcilable.[3] This is in part because the principles underlying human moral behavior cannot ethically be tested by using designed experiments, as can physical principles. Tests would violate the ethical principles being tested. Thus, there appears to be an Ethical Uncertainty Principle which limits the ability to describe moral behavior for individual humans. This is a parallel to the Heisenberg Uncertainty Principle which establishes bounds upon the knowledge obtainable for discrete particles in physical systems. The uncertainty in ethical theory make it difficult to reconcile differences of opinions as to the relative importance of three general principles for making ethical decisions.

THREE GENERAL PRINCIPLES FOR MAKING ETHICAL DECISIONS

Moral philosophers generally agree that three general principles exist for making ethical decisions (Table 1).[3] The first principle is Respect for Autonomy. This is the concept that each person has the right to decide what is best for themself. The Respect for Autonomy generally ranks highest in any hierarchy of ethical principles. This principle usually takes precedence in a medical situation and should seldom be violated in decision-making. However, this principle assumes that an individual is both capable of making a rational decision and desires to make such as decision. Difficulties arise when it is unclear whether an individual is capable of making rational decisions, as may be the case with the very young, the very old or in medical emergencies involving trauma or coma, or when an individual does not want to make a decision. Often, individuals do not have the ability to understand the consequences of their decision and prefer to abdicate this right to their physician. The moral dilemma of when "to pull the plug" for a terminally ill patient in an intensive care unit is a consequence of the conflict of this principle and the principle of beneficence.

Table 1. Three General Principles for Making Ethical Decisions.

-RESPECT FOR AUTONOMY (The concept of personal self-governance. The principle of a person's right to choose. It assumes that individuals have an intrinsic value and have the right to determine their own destiny. It is the opposite of slavery.)
-THE PRINCIPLE OF BENEFICENCE (The concept that an action or decision should not inflict harm on another, should prevent or remove harm, or promote good to another.)
-THE PRINCIPLE OF JUSTICE (The concept that like cases should be treated alike. This principle is difficult to use because individuals are not alike and often do not desire to be treated alike.)

The Principle of Beneficence or "FIRST DO NO HARM" is at the core of the Hippocratic Oath. There is seldom conflict over the importance of this principle. However, there can be conflict between it and the respect for autonomy, as cited above. For example, a woman may desire to have an implant for breast enlargement even though there is evidence that doing so involves risk, i.e. it "may not be harmless". The burden of informing the person of the extent of risk is shared by the manufacturer and the surgeon. Each may assess the potential for "harm or risk" quite differently. Quantification of risk for an individual is difficult since risk is a statistical concept [4]. Conflicts can and do occur in such situations.

The Principle of Justice leads to the most uncertainty and conflict. This principle maintains that like problems should be treated alike. The theoretical ideal of complete equality is impossible to achieve because individuals are not alike. Also, people may choose, for reasons such as taste or religion, to be treated differently. Respect for autonomy requires that preferences be honored. Problems arise when the autonomous rights of an individual or group limit the implementation of justice to others. Problems also arise when individuals believe falsely that their problem is equivalent to another, when in fact it is different.

Since the concept of ideal justice cannot be achieved in a real world, the formal principle of justice is usually implemented in terms of what are called "The Material Principles of Justice". Table 2, based on Beauchamp and Walters,[3] summarizes alternatives for making decisions regarding distribution of material goods or health care. Most countries have evolved a complicated mixture of these alternatives. For example, the availability of certain implants to a deaf person, such as intracochlear multi channel electrical stimulation, often depends upon family wealth because of the high cost of the device and the lack of insurance for such devices. In developing countries, even simple implants may not be available to the majority of the population due to cost and lack of surgical facilities. The rapidly accelerating cost of health care in all countries may eventually lead to governmental restrictions for many implants with distribution based

Table 2. Material Principles of Justice.

Which include as alternatives:
1) To each person an equal share.
2) To each person according to individual need.
3) To each person according to acquisition in a free market.
4) To each person according to individual effort.
5) To each person according to societal contribution.
6) To each person according to merit.
7) To each person according to age.

Ref: T. L. Beauchamp and L. Walters, *Contemporary Issues in BioEthics (3rd edition)* (Wadsworth Publishing Co., Belmont, CA, 1989)

more and more on personal resources.[5,6] Such restrictions create conflicts between all three ethical principles with little basis for resolving the conflicts.

CONSEQUENCE OF THE THEORETICAL PROBLEM

Beauchamp and Walters summarize[3] the biggest problem facing ethicists and moral philosophers at the present time as: *"The problem of how to value or weigh different moral principles remains unresolved in contemporary moral theory."*

This means that when the Principle of Respect for Autonomy is in conflict with either the Principle of Beneficence or Principle of Justice there is no acceptable means of resolving the conflict.

The most important, practical consequence of this theoretical problem is that it leads to uncertainty in assessing an ethical response in individual cases. Guidelines of ethical behavior can be developed for large population groups but they may not be accepted by individuals within the group. Individuals often consider general guidelines or restrictions to be unjust if they are excluded. Thus, uncertainty in the relative importance of the three ethical principles leads to conflict between individuals and between individuals and the group.

For example, consider the situation when an implant fails. Conflict may arise between the patient and the surgeon, hospital, or manufacturer, or all three. Why is there conflict? The patient chose to have the implant, risks were reviewed and informed consent was obtained; thus the patient's autonomy was respected. The individual case history indicated to the surgeon that the implant and procedure selected had a high probability of success, thereby fulfilling the principle of beneficence. Conflict results, however, when the patient perceives that the principle of justice has been violated. The patient expects not only equal treatment, but also equal results. The patient, and his/her family, do not care about statistics and that 85 or 95% of similar cases treated the same way succeeded. They care only that their case failed.

SOURCE OF CONFLICT

The conflict is due to an unjustified expectation of equal consequences of an act instead of equal performance of the act. The Principle of Justice specifies only that "like cases be treated alike". However, because individuals are different results can be different even when treatment is the same. The difference in results versus expectations can be perceived, wrongly, as unjust.

What are the reasons for unjustified expectations of implant success? Three factors, at least, are involved: human nature, technology, and greed. It is human nature to want the same things as others. This expectation feeds our market-driven economy.[5] The same is true for implants. People do not desire to live with pain as they become older. This is reasonable. They learn from the media or friends that certain implants eliminate pain, therefore, they want an implant if they have painful joints. They do not hear, or are unwilling to accept, that there is a finite risk associated with the surgery and a finite probability of failure of the implant. It is human nature to hear only what you want to hear. This results in unjustified expectations and a conclusion of receiving unjust treatment if difficulties arise.

Technology amplifies the problem. New developments in implants are promoted as superior even when long term data for large populations of patients are not available. We live in a technological age where most people want and expect the "latest", be it electronics, cars, or implants. Along with the "latest" comes the expectation that the latest is best. This is often unjustified but the perception still exists.

Rapid changes in technology also lead to a proliferation of choices. The surgeon and patient no longer are limited to one decision, "should a hip joint be replaced with a prosthesis?" Instead, a series of decisions must be made with regard to: type of stem, type of cup, type of fixation, etc. The statistical basis for risk assessment and beneficence becomes progressively more uncertain the greater the options. The patient may well equate more options with greater expectations of success. This is often false. In fact, the reverse may be the case; i.e., success in a large population decreases as the number of options increases.

Greed can feed on the above factors. As more people want and receive implants the potential for profit increases proportionally. As more options become available it is more likely that products will be promoted for the sake of novelty and image rather than for well established improvements in beneficence.

Economic pressures build to introduce new implant products with only minimal standards of testing in order to have something "new" to offer. Tests which show that problems may occur are undesirable in this context and therefore are avoided unless required by regulatory pressures. Research to obtain solutions to long term reliability problems is often not done because to do so is to admit that long term reliability is a problem. Thus, the implant field grows in volume but not necessarily proportionally in beneficence to the larger number of patients. One consequence is an ever increasing escalation in health care costs which in the United States are now 13% of the Gross National Product.[6]

SPECIFIC ETHICAL CONCERNS IN BIOMATERIALS

As a field we need to promote testing to avoid long-term complications and implant revisions. We need to achieve > 85-90% success of implants over 10-20 years. Failure analysis needs to be done for all implants in use or proposed for use in order to provide a statistical basis for establishing beneficence for the patient. Figure 1 illustrates the type of analysis needed.

Clinical results of implants can be classified as those that result in High Beneficence (top curves) or Low Beneficence (bottom curves). Moderate Beneficence lies between. The ethical Principle of Beneficence requires that an implant meet the high standard of the upper curve because otherwise the principle "first do no harm" may be violated. In other words, any implant that performs in limited trials for 2-3 years as indicated in the lower curve should not be put into general use. Any implant that is in general use with results similar to those in the lower curve should be removed from use. Implants of Moderate Beneficence should be sujected to regulatory monitoring to

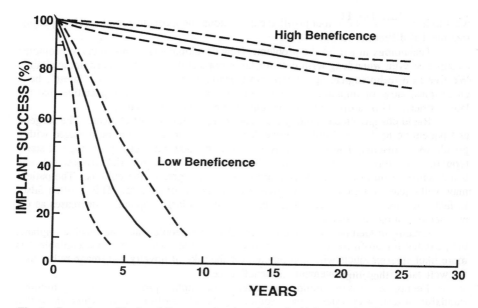

Fig. 1. Comparison of implant failures as a function of time for high vs low beneficence to the patients.

determine the reasons for failures and research funded to improve the performance until the upper curve behavior is achieved.

Figure 1 also illustrates that there exists a distribution of results which broadens with time. Statistical data that can be used to generate such curves needs to be compiled by the professional societies for all the prostheses now in clinical use. Patients need to be informed of their expected benefit with respect to this distribution of results. This is one of the few ways to counter false expectations of results.

Also, more controlled center testing of new products needs to be encouraged in order to produce data for plots such as Fig. 1. A few examples where additional controlled pre-market testing would have been desirable are listed below:

Intrauterine Devices (IUD) for Birth Control
Silicone Injections
Silicone Breast Prostheses
PTFE Powder Injections for Urinary Incontinence
PTFE Powder for Vocal Cord Rehabilitation
PTFE/Carbon Fibre Composites for TMJ Repair
Dense Hydroxyapatite Cones for Edentulous Ridge Maintenance
Plastic and Metallic Prostheses for Ossicular Replacement
Porous Bead Coatings on Orthopedic Implants
Hydroxyapatite Coatings on Orthopedic and Dental Devices

SPECIFIC NEEDS TO MINIMIZE ETHICAL CONCERNS

Clinical results from a number of these implants listed above produce the Low Beneficence curves in Fig. 1. To avoid this type of performance, we need standard in-vivo and in-vitro tests to compare alternative biomaterials under equivalent conditions. Other areas of need to maximize beneficence are:

Long-term predictive tests for biomechanical performance.

Predictive *in vitro* tests to determine biochemical factors in tissue response.

Minimization of animal testing by more effective use of in-vitro tests, as discussed by Saha and Saha.[7]

Avoidance of extensive "me-too" development of "new" biomaterials that are only derivative in nature.

Elimination of extensive, repetitive short term animal testing of unloaded, non-functional devices. This must be replaced with functional device testing under simulated clinical conditions.

Balance the requirement of scientists to generate publications and research dollars with the desire to conduct scientific research that can improve long term performance.

Balance management's often short term outlook with the long term welfare of the patient and society.

Balance the corporate goal of generation of profit with the need for unbiased product quality.

SUMMARY

Many researchers, clinicians and manufacturers of implants have been exposed to the consequences of ethical conflicts which can arise when the Principles of Beneficence and Justice are not reconciled.

The expectations of the population with respect to implant success will continue to rise. Thus, implants must have increased long-term reliability. Failure to ensure this will result in negative consequences for individual patients. Failure will also produce increased governmental regulations and controls. These controls will increase development costs and produce a negative spiral in which fewer manufacturers will be able to afford to produce fewer products and will develop fewer new materials and applications. This negative scenario can be avoided by a concerted effort to improve long-term performance of all types of implants.

The over-riding principle "Respect for Autonomy" must be of concern to all of us. An individual depends on information from all sources to make crucial decisions. Such information should always be the best and most complete we have to offer and be unsullied by commercial or personal preferences. Only by these means can the best decision be made for specific clinical problem.

The new generation of bioceramics described in this book have been developed to improve beneficence for the patient. The alumina heads of total hip joints perform as indicated in the high beneficence curves of Fig. 1. However, as yet there is insufficient data to generate the curves of beneficence (Fig. 1) for many types of bioceramic implants. Until the data is available the implants must be considered developmental and be carefully monitored by the surgeon and the manufacturer.

ACKNOWLEDGEMENT

The author acknowledges Dr. Ross Mackenzie of the Chautauqua Institution who stimulated my interest in this subject and discussions and contributions from Dr. June Wilson and Dr. David Greenspan.

REFERENCES

1.   J. M. Cauwels, *The Body Shop: Bionic Revolutions in Medicine* (C. V. Mosby, St Louis, Mo., 1986).
2.   B. Hilton, *First Do No Harm: Wrestling with the New Medicine's Life and Death Dilemmas* (Abingdon Press, Nashville, TN, 1991).
3.   T. L. Beauchamp and L. Walters, *Contemporary Issues in BioEthics (3rd Edition)* (Wadsworth Publishing Co., Belmont, CA, 1989).
4.   The British Medical Association, *The BMA Guide to Living with Risk* (Penguin Books, New York, NY, 1990).
5.   D. Callahan, *What Kind of Life: The Limits of Medical Progress* (Simon and Schuster, New York, NY, 1990).
6.   *Journal of the American Medical Association*, May 15, 1991.
7.   P. S. Saha and S. Saha, *Journal of Long-Term Effects of Medical Implants*, **1[2]** (1991) 127-134.

Chapter 21

SUMMARY AND FUTURE DIRECTIONS

Larry L. Hench
University of Florida
Advanced Materials Research Center
Alachua, FL   32615

OVERVIEW

One of the great challenges facing the multidisciplinary field of biomaterials is the development of a new generation of implant materials which will last as long as the lifetime of the patient. This is often 15-30 years, double or triple the expected lifetime of many spare parts in use today. A large number of factors influence the long-term clinical performance of a biomaterial. The objective of this chapter is to summarize these factors, discuss their relative importance, and outline future directions for research and development of bioceramics that may yield the increase in implant lifetimes needed. The elements of a "general theory of biomaterials" are reviewed. Implications of this generalization lead to the recommendations presented.

In 1975 Hench and Ethridge proposed as a general theory *"An ideal implant material must have a dynamic surface chemistry that induces histological changes at the implant interface which would normally occur if the implant were not present."*

This theory is supported by the behavior of bioactive ceramics discussed in many chapters of this book. Bone-bonding bioactive ceramics create an environment compatible with osteogenesis, with the mineralizing interface developing as a natural bonding junction between living and non-living materials. Implants that form a hydroxycarbonate apatite (HCA) layer very rapidly incorporate collagen fibrils within the growing HCA agglomerates and thereby create a natural collagenous bonding junction between the implant and soft tissues. Resorbable calcium phosphate ceramics function as a mineralized framework for bony remodeling, with a final "natural" result that is similar to autogeneous bone grafts.

Implant materials that behave differently than is required by the above theory, elicit non-adherent fibrous capsules, which leads to problems in interfacial stability.

A suitable surface chemistry is necessary, but not sufficient, to prevent formation of non-adherent fibrous capsules at an implant-tissue interface. Table 1 summarizes the many factors involved in creating a stable vs unstable implant interface. The nature of the surgery, post-implantation healing, and biomechanical conditions determine not only whether a fibrous capsule will form but also its shape and thickness, all may be independent of implant surface chemistry. Continuous movement at an implant-tissue interface will always produce a capsule. The more extensive the motion and the longer

Table 1. Factors That Influence the Histological Response to an Implant.

Non-Adherent Fibrous Capsule	Bone Bonding	Soft Tissue Bonding
Extent of tissue necrosis	Periosteal invasion	Rapid formation of
Infection	Adherent to cells	HCA layer
Toxic leachables	Adherent to acellular	Adherent to cells
Wear or degradation	constituents	Adherence to
Movement	Tight fit	collagen
Inflammation	Vascularity	Tight fit
	No inflammation	No inflammation
	No wear debris	

the duration, the thicker the capsule. Infection, release of toxic leachables, continuous wear and release of wear particles, or uncontrolled degradation of a surface leads to a fibrous capsule that isolates the implant from normal tissues. It is futile to look for an artificial material or surface chemistry that will overcome these natural limitations of surgery, healing, and implant function.

The revised general theory that evolves from consideration of the factors listed in Table 1 was expressed by Hench and Ethridge in 1982 as:

General Theory of Biomaterials Behavior
An ideal implant material performs as if it were equivalent to the host tissue.

Two axioms follow:

Axiom 1. The tissue at the interface should be equivalent to the normal host tissue.

Axiom 2. The response of the material to physical stimuli should be like that of the tissue it replaces.

These axioms are interdependent. In order for an equivalent physical response to be realized, a stable interfacial bond between tissue and implant must be achieved. Conversely, in order for a stable interface to be produced it is necessary for physical stimuli to be controlled during the repair process. Factors that inhibit one of these relationships hinder the other as well.

Because of the critical interactions of these axioms, the focus of research and understanding of long-term performance of devices must be on creating an implant-tissue interface which is simultaneously histologically and biomechanically stable. However, it is the short term response of tissues following implantation which determines the eventual long-term response.

Many of the surface-chemical features essential for formation of a stable interfacial bond with tissues have been identified. However, the details are yet to be determined. In bone, the central issue seems to be the relative competition between fibrogenesis and osteogenesis at the interface. Many factors favor proliferation of fibroblasts and capsule formation, whereas very specific conditions must be satisfied for osteogenesis to occur. This mimics the situation in the natural repair of bone, as implied by the General Theory.

At the cellular level, fibroblasts are favored over osteoblasts if osteoblast progenitor stem cells: 1) are not present, 2) cannot attach, 3) cannot differentiate, 4) cannot divide, 5) cannot generate bone matrix, or, 6) matrix mineralization is inhibited.

Following bone surgery it is likely that stem cells capable of becoming osteoblasts, given the right environment, will be available at the implant interface along with fibroblasts since both types of stem cells originate from blood and tissues invaded during surgery. Damage to bone results in large changes in: pH, oxygen tension gradients, local electric potentials, concentrations of chemicals, enzymes, and acellular proteins, such as bone growth factors. These local environmental changes lead to differentiation of stem cells into osteoblasts. If an implant perturbs this environment sufficiently to prevent differentiation of osteoblasts then fibroblasts proliferate and the gap between the implant and undamaged tissue is closed with a fibrous capsule.

Davies and colleagues (1991) have shown that attachment of osteogenic stem cells to a substrate and generation of mineralizable bone matrix is necessary for differentiation to proceed and osteoblasts to proliferate. The surface chemistry of a substrate affects these processes. Vrouwenwelder et al. (1992) have demonstrated that primary osteoblast-like cells divide more rapidly on bioactive glass substrates than they do on inactive materials like stainless steel and titanium. In contrast, Seitz et al. (1982) showed that bioactive glass surfaces of the same composition (45S5) slowed down the attachment, spreading, and growth of fibroblast cell lines. Addition of fibronectins to the bioactive glass substrate surface eliminated the substrate effect on the fibroblasts.

These cell culture experiments suggest that highly chemically specific attachment mechanisms control cell morphology, influence the structure of cell membranes, and activate intracellular functions or transport required for differentiation. Differentiation of osteoblasts apparently requires different and more chemically specific cell membrane-attachments than do fibroblasts. Subsequent expressions of osteoblast phenotype and evidence of differentiation, such as alkaline phosphatase production, are a consequence of the structural relationships established at the stage of cell attachment (Davies, 1991). In order for these membrane-specific processes to occur, and osteoblasts to appear, fibroblast proliferation must be impeded. One of the features of bioactive ceramics is to favor proliferation of osteoblasts over fibroblasts. This results in the time dependent differences in relative proportion of soft tissue to bone at the interface of bioactive implants discussed in Chapter 3 (Hench, 1988 and Gross et al, 1988). Bioactive implants that develop a bone-bonding interface the most rapidly *in vivo* also have the largest influence on the growth of fibroblasts and osteoblasts *in vitro*.

The important difference in *in vivo* and *in vitro* behavior of bioceramics is related to their surface reaction kinetics in physiological solutions, as discussed is Chapters 3, 5 and 9. Figure 1 summarizes the sequence of eleven reactions which apparently occur on the surface of a bioactive glass as a bond with bone is formed. However, not all bioactive ceramics possess all eleven steps or surface reaction kinetics as rapid as indicated in Fig. 1. The level of understanding of the chemical processes for Stages 1-5 but is sparse for Stages 6-11. The limitation in several stages, such as 8 and 9, is the uncertainty of the biological processes that control the genetic expression of highly differentiated cells, like osteoblasts.

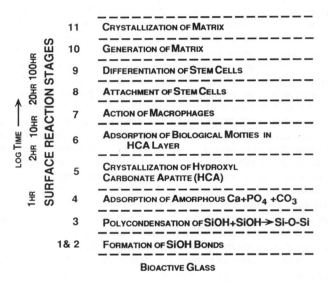

Fig 1. Sequence of interfacial reactions involved in forming a bond between tissue and bioactive ceramics. (Reprinted from L. L. Hench, "Bioceramics: From Concept to Clinic," *J. Amer. Ceram. Soc.*, **74[7]** (1991) 1487-570, with permission.)

The time scale for the surface reactions of a glass with high bioactivity is also shown in Fig. 1. Glasses that bond to bone rapidly undergo surface reaction Stages 1-5 very rapidly. A polycrystalline hydroxyl carbonate apatite (HCA) layer is formed (Stages 4 and 5) on 45S5 Bioglass® within three hours both *in vitro*, in simulated body fluids, and *in vivo*. In contrast, glass or glass-ceramic compositions with intermediate levels of bioactivity require two to three days to form an HCA layer on the material. Compositions that are not bioactive and do not form a bond with bone do not form an HCA layer even after three to four weeks in solution.

The differences in rate of change of the inorganic phases on the surface of the material apparently alters which biological species are adsorbed (Stage 6 in Fig. 1), although the nature and concentration of the species affected is not known. The rates and types of interfacial cellular responses (Stages 7-11) are also altered by the differences in inorganic phases developed on the surface, but the reasons are also uncertain. Research at the molecular chemical level, such as described in the workshop proceedings edited by Davies, 1991, is attempting to answer these questions about the implant-tissue interface.

OPPORTUNITIES FOR THE FUTURE:
STRUCTURAL TAILORING OF COMPOSITES

Two parallel directions of research offer the most promise for developing implants with enhanced lifetimes:

1) Molecular tailoring of compositions and surfaces to match the biochemical requirements of diseased and damaged tissues.

2) Microstructural design of composites that will match the biomechanical requirements of natural host tissues.

Table 2, prepared by T. Kokubo, summarizes the compositions and properties of various bioactive composites. These data and those in Chapter 5 show that the potential design versatility of bioactive materials is beginning to be realized. It is proven that bioactivity is retained in multiphase glass-ceramics and composites. Strength and toughness of bioactive implants have been increased greatly since the first generation of bioactive glasses were reported over 20 years ago (Yamamuro et al., 1990).

The composition of a bioactive matrix phase can be optimized for chemical rates to match physiological requirements. An inactive or less-active reinforcing phase can be optimized with respect to size and distribution, thereby maximizing strength and toughness. A/W glass-ceramic, developed by Yamamuro and Kokubo (1990), utilizes this design concept and the very small apatite and wollastonite crystals reinforce the silicate glass matrix resulting in superior mechanical properties compared with a single phase bioactive glass. Further optimization of microstructure has been done by Kasuga et al. (1990) where transformation toughened zirconia particles are added to a bioactive A/W glass-ceramic prior to hot pressing. The product is a very tough bioactive composite with K_{1C} values of 4 MP.m$^{1/2}$ and bend strengths of 703 MPa.

The disadvantage of this approach is that the composite usually has a higher modulus of elasticity than bone (Table 2) and therefore does not eliminate the problem of stress shielding. A composite based on a polymer matrix with a lower modulus of elasticity, such as polyethylene, has mechanical properties much closer to those of bone (Table 2 and Chapter 16) (Bonfield, 1988). When the reinforcing phase is a bioactive ceramic (HA) or a bioactive glass the composite is also bioactive. The level of bioactivity is proportional to the volume fraction of bioactive phase dispersed within the composite. Machining or etching of the surface exposes the bioactive phase and results in bonding to bone at the interface. The composite can be loaded to about 0.35 volume

Table 2.   Mechanical Properties of Bioactive Composites.
(Prepared by Professor T. Kokubo)

	G-Fe[1]	HAp-Fe[2]	HAp-ZrO$_2$[3]	HAp-TiO$_2$[4]	GC-ZrO$_2$[5]	GC-Ti[6]	PE-HAp[7]
Matrix	Bioglass®	Hydroxyapatite	Hydroxyapatite	Hydroxyapatite	A-W-type glass-ceramic	Ceravital®-type glass-ceramic	Polyethylene
Dispersed Phase	Stainless Fibers	Fe-Cr-Al Fibers	ZrO$_2$(Y$_2$O$_3$) Particles	TiO$_2$ Particles	ZrO$_2$(Y$_2$O$_3$) Particles	Ti Particles	Hydroxyapatite Particles
Content of Dispersed Phase (Vol%)	60	30	10-50	15	30	30	45
Bending Strength (MPa)	340	224	450	252	600	60 (Tensile)	22-26 (Tensile)
Young's Modulus (GPa)	65	142		238		100	6
Fracture Toughness K$_{Ic}$(MPa·m$^{1/2}$)		7.4	3.0	3.0	3.0		3.0

1.   P. Ducheyne and L. L. Hench, "The Processing and Static Mechanical Properties of Metal Fiber Reinforced Bioglass®," *J. Mater. Sci.* 17 (1982) 595-606.

2.   G. DeWith and A. T. Corbjin, "Metal Fibre Reinforced Hydroxy-apatite Ceramics," *J. Mater. Sci.* 24 (1989) 3411-3415.

3.   N. Tamari, I. Kondo, M. Mouri and M. Kinoshita, "Effect of Calcium Fluoride Addition on Densification and Mechanical Properties of Hydroxyapatite-Zirconia Composite Cramics," *J. Ceram. Soc. Japan* 96 (1988) 1200-1202.

4.   J. Li, S. Forberg and L. Hermansson, "Evaluation of the Mehcanical Properties of Hot Isostatically Pressed titania and titania-Calcium Phosphate Composites," *Biomaterials* 12 (1991) 438-440.

5.   T. Kasuga, K. Nakajima, T. Uno and M. Yoshida, "Preparation of Zirconia-Toughened Bioactive Glass-Ceramic Composite by Sinter-Hot Isostatic Pressing," *J. Am. Ceram. Soc.* 75 (1992) 1103-1107.

6.   Ch. Müller-Mai, H-J. Schmitz, V. Strunz, G. Fuhrmann, Th. Fritz and U.M. Gross, "Tissues at the Surface of the New Composite Material Titanium/Glass-Ceramic for Replacement of Bone and Teeth," *J. Biomed. Maters. Res.* 23 (1989) 1149-1168.

7.   W. Bonfield, "Hydroxyapatite-Reinforced Polyethylene as an Anagolous Material for Bone Replacement," in *Bioceramics: Material Characteristics Versus In Vivo Behavior*, eds. P. Ducheyne and J. E. Lemons (Academy of Science, New York, 1988) pp. 173-177.

fraction of the second phase and still retain the ductile characteristics of the matrix which results in high fracture toughness and a low modulus of elasticity.

This approach can also be used to make anisotropic composites with gradients in elastic modulus and bioactivity. The ideal is to have high bioactivity on the surface with a moderately low modulus with a gradient towards the interior of the device with less bioactivity and greater stiffness. Tailoring the microstructure and bioactivity through gradients offers the most promising route towards optimizing biochemical and biomechanical performance of an implant.

OPPORTUNITIES FOR THE FUTURE:
MOLECULAR TAILORING OF SURFACE CHEMISTRY

The surface chemistry of implants needs to be optimized to meet the requirements of aged, diseased and damaged tissues. Most biomaterials in use today were developed by trial and error. There is very little understanding of the effects of disease states, such as osteoporosis or arthritis, on interfacial reactions or the biomechanical behavior of implant-tissue interfaces. Some principles have been established to guide the development of new bioceramics. As indicated in Fig. 1 and discussed in Chapters 3, 5, and 9 the relationships between phases and compositions of bioceramics with their surface reaction kinetics has been determined. The effect of surface reactions on *in vivo* behavior is also known in the most general sense. Details of the cellular responses are being established by tissue and cell culture experiments. Results from these studies will make it possible to modify compositions to optimize their behavior with respect to Stages 6-11 in Fig. 1. Compositions with optimized surface behavior can be used as the bioactive phase in composites with the volume fraction and distribution varied to achieve optimal biomechanical behavior. A synergistic combination of structure and composition of a multiphase material comes much closer to matching the nature of real tissues than any material in clinical use today.

Two new directions of research hold promise for improving the scientific basis for tailoring surface reactions of bioceramics. One is the discovery that sol-gel derived glasses have a much expanded compositional range of bioactivity over glasses made by traditional melting and casting processes (Li et al., 1991). Sol-gel processing is one of the most important new methods for production of new, chemically derived materials (Brinker and Scherer, 1990; Hench and West, 1990). The low temperatures of sol-gel processing, as illustrated in Chapter 1, make it possible to control surface chemistry of the resulting materials with greater flexibility than high temperature melting and casting of glasses or sintering or hot pressing of ceramics. Details of the seven processing steps in making bioactive gel-glasses are discussed in Li et al. and Hench and West. Advantages of sol-gel processing of inorganic biomaterials include: new compositions, greater homogeneity, higher levels of purity, net-shape casting of monoliths, low temperature coating of substrates, control of powder size distribution, control of surface chemistry of the gel-glasses, expanded ranges of glass formation, control of pore networks at a nanometer scale in addition to commercial advantages such as low cost, lower energy consumption and nearly zero environmental impact.

Figure 2 shows the extended range of compositions in the SiO_2-CaO-P_2O_5 system that are bioactive when made by alkoxide based sol-gel processing compared with bioactive compositions made by melting and casting, from studies by Professor Kokubo's group. Gel-derived glasses with as much as 88 % SiO_2 develop hydroxyl carbonate apatite (HCA) layers whereas the limit for melt-glasses is 60%. This is a large shift in compositional limit. Melt-glasses with >55% SiO_2 require several days to form a polycrystalline HCA layer whereas gel-glasses do so in only a few minutes. The chemical origin of these important differences appears to be the large concentration of silanols on the surface of the gels after processing temperatures of 500-800°C.

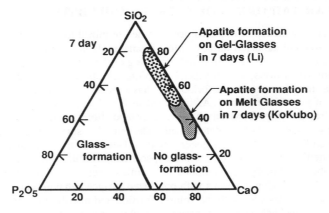

Fig. 2. Apatite formation at 7 days on gel-glasses and melt glasses.

Semi-empirical molecular orbital (MO) calculations, using AM-1 and Extended Huckel methods, show that metastable silica clusters formed from a condensation reaction of neighboring silanols (Stage 3 in Fig. 1) can act as heterogeneous nucleation sites for hydroxyl carbonate apatite crystals (Stages 4 and 5) (Hench and West, 1992). The metastable silica clusters can also act as preferential adsorption sites for amino acids, such as alanine (West and Hench, 1993). These calculational results indicates that the surface reactions of the inorganic material (Stages 1-5) can lead to biologically specific binding sites for protein molecules. The MO calculations show differences in specific adsorption on the inorganic surface which depend on different binding sites on the protein molecules. This may lead to an understanding of the selective adsorption of proteins that act as growth factors or enzymes (Stages 6-8). Such studies may also aid in the interpretation and optimisation of new hybrid inorganic-biological systems, such as alkaline phosphatase enzymes (Avnir et al., 1992) or other optically active organic molecules (Ellerby et al., 1992) trapped within sol-gel silica porous glass matrices.

CONCLUSIONS

Results from molecular orbital modelling calculations, combined with experimental investigations of the adsorption of biological growth factors and other biological species, should make it possible to design a new generation of bioactive phases or use techniques such as ion beam sputter coating to make surface modified layers that enhance the rates of interfacial bonding of even aged or diseased tissues. Such surface enhancement methods may also be incorporated within the design of composites that have high toughness and a modulus of elasticity that will prevent stress shielding of bone. The potential solutions to lifetime problems of prostheses lie in the creative use of materials chemistry. Millions of people will benefit if this potential can be realized. The research direction to follow is finally apparent after many years of trial and error.

ACKNOWLEDGEMENT

The author gratefully acknowledges the financial support of the Air Force Office of Scientific Research Division of Chemical and Materials Science.

READING LIST

L. L. Hench and E. C. Ethridge, *Biomaterials: An Interfacial Approach* (Academic Press, New York, 1982).

J. E. Davies, J. E., ed., *The Bone-Biomaterial Interface* (University of Toronto Press, Toronto, Ontario, Canada, 1991).

Vrouwenvelder, Groot, and de Groot, *J. Biomed. Maters. Res.* (1992).

T. L. Seitz, K. D. Noonan, L. L. Hench, and N. E. Noonan, *J. Biomed. Maters. Res.* **16[3]** (1982) 195-207.

L. L. Hench, in *Bioceramics: Materials Characteristics Versus In-Vivo Behavior*, eds. P. Ducheyne and J. Lemons (Annals of New York Acad. Sci., New York, 1988) Vol. 523, pp 54.

U. Gross, R. Kinne, H. J. Schmitz, and V. Strunz, *CRC Critical Reviews in Biocompatibility* **4** (1988) 2.

T. Yamamuro, L. L. Hench, andl J. Wilson, eds., *Handbook on Bioactive Ceramics, Vol I: Bioactive Glasses and Glass-Ceramics* (CRC Press, Boca Raton, FL, 1990),

T. Yamamuro, L. L. Hench, and J. Wilson, eds. *Handbook of Bioactive Ceramics, Vol. II: Calcium Phosphate and Hydroxylapatite Ceramics* (CRC Press, Boca Raton, FL, 1990).

T. Kasuga, K. Nakajima, T. Uno, and M. Yoshida, in *Handbook of Bioactive Ceramics*, eds. T. Yamamuro, L. L. Hench, J. Wilson (CRC Press, Boca Raton, FL, 1990) Vol. I, pp 137-142.

W. Bonfield, in *Bioceramics: Materials Characteristics vs In-Vivo Behavior*, eds. P. Ducheyne and J. E. Lemons (Annals New York Academy of Sciences, New York, 1988) Vol. 523, pp 173-177.

R. Li, A. E. Clark, and L. L. Hench, *J. Appl. Biomaterials* **2** (1991) 231-239.

C. J. Brinker and G. W. Scherer, *Sol-Gel Science* (Academic Press, San Diego, CA, 1990).

L. L. Hench and J. K. West, *Chem. Rev.* **90** (1990) 33-72.

J. K. West and L. L. Hench, "Reaction Kinetics of Bioactive Ceramics Part V: Molecular Orbital Modeling of Bioactive Glass Surface Reactions," in *Bioceramics 5*, eds., T. Yamamuro, T. Kokubo and T. Nakamura (Kobonshi Kankokai, Inc., Kyoto, Japan, 1992) pp. 75-86.

J. K. West and L. L. Hench, "Adsorption of Alanine on a Silica Surface: MO Calculations," to be published.

D. Avnir, S. Braun and M. Ottolenghi, in *Supramolecular Architecture*, ed. T. Bein (Am. Chem. Soc., Washington, DC, 1992) Chapter 27.

L. M. Ellerby, C. R. Nishida, F. Nishidsa, S. A. Yamanaka, B. Dunn, J. S. Valentine, and J. J. Zink, *Science* **255** (1992)

Appendix

ASTM STANDARDS FOR BIOCERAMICS

Jack E. Lemons* and David Greenspan**
*School of Medicine
The University of Alabama at Birmingham
Birmingham, Alabama 35294
**U.S. Biomaterials Corporation
Alachua, Florida 32615

Consensus standards and the associated development process involves a combination of affiliated interest groups (manufacturers, users and general) working together within a matrix of rules set forth by the standards organization.[1] In general, the overall process for any standard starts with a proposal establishing a need for a consensus document on a material, a test method, or a product performance. The need and appropriateness must be confirmed starting at the task force and section levels followed by approvals from the appropriate committee (F-4 for medical implants) and the parent society. Since consensus standards are for materials, tests and applications that already exist (a product with more than one manufacturer), areas where products are at a research and development phase are not normally considered for standard development. When the proposal is approved, a section chair identifies a task force chairperson and two task force meetings are scheduled each year for document development. Participation in the process is not restricted, however formal voting requires membership within the standards organization.

The process of discussing, writing, rewriting, balloting and approving a standard cannot be completed if any society members vote an objection and the objection is not resolved. Balloting is normally done at subcommittee, main committee, and society levels, and the overall process usually takes two to five years. When completed and approved, the standard is then published annually within the parent organizations' standard publications. At five year intervals, each existing standard is reviewed for either revision or reapproval.

The standards for Bioceramics currently available within the ASTM committee F-4 include: Polycrystalline Alumina (F603), Tricalcium Phosphate (F1088) and Particulate Hydroxylapatite (F1185).[2] Requests for these standards were made by ASTM members within the past 10 years and each existing document is currently being reviewed for revision and reapproval. (Task Force F04.02.03.07, F04.02.03.09 and F04.02.03.13 respectively.)

A number of new task force committees have been established within the past five years. These include: Hydroxyapatite (F04.02.03.03), Single Crystal Aluminum Oxide (F04.02.03.04), Ceramic Coatings (F04.02.03.05), Glass and Glass-Ceramic Biomaterials

(F04.02.03.06), Ca·PO$_4$ Crystallographic Characterization (F04.02.03.08), Ca·PO$_4$ Physical Requirements (F04.02.03.10), Ca·PO$_4$ Mechanical Requirements (F04.02.03.11), Ca·PO$_4$ Environmental Stability, (F04.02.03.12), Zirconia Ceramics (F04.02.03.14), and Anorganic Bone (F04.02.03.15).[3] The task force working on crystallographic, physical, mechanical and environmental stability properties may require augmentation when extended to classes of materials (solid, porous, coatings, etc.). This may also be true for test methods not currently available with ASTM, and for medical or dental applications.

Consensus standard documents must represent known characteristics of materials from commercially available products. The standards are not intended to exclude materials, but rather to provide basic property values including known information on physical, mechanical, chemical, electrical and biological properties from peer reviewed or established and accepted published documents. Since some data are normally not available, test methods may be developed and round-robin testing conducted to establish property limits. An attempt is made to include all available information. For example, in general, material standards include biomaterial chemical analyses, physical and mechanical properties and surface finish requirements for the various biomaterials. These documents as standards, therefore, provide a baseline reference for comparisons of existing and new biomaterials.

The following is a listing of the published ASTM Standards for bioceramic materials, along with pertinent data and requirements for the materials contained in the document. Following the published ASTM Standards are those documents for bioceramic materials and test methods which are still in draft form. Since it is not possible to publish specific information about the data contained in these draft documents, the listing is followed by a brief description of the contents.

F 603 - 83 Standard Specification for High-Purity Aluminum Oxide for Surgical Implant Application

Chemical Requirements - Chemical analysis shall indicate 99.5% aluminum oxide (Al_2O_3) or greater. Silicon dioxide (SiO_2) and alkali oxides must be less than 0.5%.

Physical Requirements - Minimum bulk density of 3.90 g/cm^3 is required (when tested using ASTM C20).

Mechanical Requirements - Flexural Strength (room temperature) shall be 400 MPa (58,000 psi) when tested using ASTM C674. Minimum elastic modulus (room temperature) shall be 380,000 MPa (55.1 x 10^6 psi) when tested using ASTM C674.

Note: This Specification is currently under revision, and is being balloted by the Main Committee of F-4.

F 1088 - 87   Standard Specification for Beta-Tricalcium Phosphate for Surgical Implantation

Chemical Composition - Elemental Analyses of calcium and phosphorus will be consistent with the expected stoichiometry of tricalcium phosphate. Maximum allowable limit of heavy metals are: Lead - 30 ppm, Mercury - 5 ppm, Arsenic - 3 ppm, Cadmium - 5 ppm. The maximum allowable limit of all heavy metals determined as lead will be less than 50 ppm.

Crystal Structure - X-ray diffraction will be used to confirm beta phase of tricalcium phosphate, and establish a minimum phase purity of 95%.

F 1185 - 88   Standard Specification for Composition of Ceramic Hydroxyapatite for Surgical Implants

Description of Terms Specific to this Standard - *hydroxyapatite* - the chemical substance having the empirical formula $Ca_5(PO_4)_3OH$.

Chemical Requirements - Elemental analysis for calcium and phosphorus will be consistent with the expected stoichiometry of hydroxyapatite. A quantitative x-ray diffraction analysis shall indicate a minimum hydroxyapatite content of 95%. The concentration of trace elements shall be limited to the following: Lead - 30 ppm, Arsenic - 3 ppm, Mercury - 5 ppm, Cadmium - 5 ppm. The maximum allowable limit of all heavy metals determined as lead will be 50 ppm.

The following are draft ASTM Standards for various bioceramic materials and test methods for evaluating some of those materials. The numerical designation used along with the Standard title refers to the particular task group within ASTM responsible for drafting the document.

F04.02.03.03 - Draft Specification for Calcium Phosphate Coatings for Implantable Materials

This draft Standard defines a number of terms related to calcium phosphate coatings, specifies chemical and/or crystallographic requirements of the coating, outlines the various physical and mechanical test methods to be used to characterize the coatings, and, where applicable, relates these techniques to existing ASTM Standards.

F04.02.03.04 - Draft Specification for Single Crystal Aluminum Oxide Corundum for Surgical Implant Application

This draft specification covers the material requirements for single crystal aluminum-oxide corundum to be used for surgical implant applications. The document contains requirements for chemical purity, physical properties, and mechanical strength.

F04.02.03.05 - Draft Specification for Calcium Phosphate Powders for Implant Coatings

This draft specification covers the requirements for the calcium phosphate used as raw materials in the fabrication of biocompatible coatings for surgical implants. The document defines a number of different calcium phosphate compounds for standardization of terminology, and specifies chemical compositions for each of the defined compounds. The document also sets specification for particle size analysis and method for certification of the material.

F04.02.03.06 - Draft Specification for Glass and Glass-Ceramic Biomaterials

This draft specification defines the material requirements and characterization techniques for glass and glass-ceramic biomaterials intended for use as bulk, porous, or powdered surgical implants, or as coatings on surgical devices. The document defines and differentiates between glass, glass-ceramic, and bioactive glass and glass-ceramic biomaterials. General chemical requirements and test methods are specified, along with physical and mechanical characterization techniques.

F04.02.03.08 - Draft Specification for Crystallographic Characterization of Calcium Phosphate Ceramics used for Surgical Implants

This draft specification defines terminology to be used in analyzing the crystallinity of calcium phosphate ceramics, such as crystallinity index, impurity phases, and crystalline phase. The document proposes a standard sample configuration to be used for testing, along with standard testing procedures.

F04.02.03.10 - Physical Requirements for Calcium Phosphate Coatings

This draft specification deals with the physical requirements for calcium phosphate ceramic coatings, both porous and non-porous, used for surgical implant applications. This draft excludes crystallographic structures and characterization, mechanical properties and environmental stability. The document proposes standards for thickness of coating, surface roughness and uniformity, porosity, density, color and morphology. Standard test specimen preparation is included.

F04.02.03.11 - Draft Standard for Mechanical Requirements of Calcium Phosphate Coatings

This draft document currently outlines three different test methods to use to determine the mechanical properties of calcium phosphate coatings used for surgical implant applications. The test methods are: 1) Tension Test, 2) Shear Test, and 3) Bending and Shear Fatigue Test. All three test methods detail the significance of the test proposed, appropriate apparatus to be used for the test, method adhesive bonding materials to be tested, test specimen fabrication and preparation, the test method to be followed, calculation of the stress, and a standardized report format.

F04.02.03.12 - Environmental Stability of Calcium Phosphate Coatings

This draft specification covers the *in vitro* evaluation of the dissolution rate of a calcium phosphate coating used for surgical implant applications. No correlation of the results of in-vitro tests to *in vivo* performance is implied. The draft specifies preparation of test specimens, apparatus in which to conduct the tests, media to use for the dissolution test, procedures for conducting the test, analysis methods and parameters, and a reporting format.

F04.02.03.14 - Draft Specification for Zirconia Ceramics for Surgical Implant Applications

This draft specification covers the chemical requirements, allowable impurity levels, physical and mechanical requirements for zirconia ceramics used for surgical implant applications.

F04.02.03.15 - Draft Specification for Anorganic Bone

This draft specification covers the processing, chemical and trace element analysis, physical and mechanical properties, and environmental stability of anorganic bone.

READING LIST

ASTM Guidelines and Rules of Operation.
Annual Book of ASTM Standards, 13.01, Medical Implants (American Society for Testing and Materials, Philadelphia, 1992).
Proceedings of the Ceramics Section ASTM FO4.02.03.

INDEX

386

three-point bending tests 285
thromboresistance 261
Ti-6Al-4V 13, 35, 167, 210, 236, 257, 266, 294
tissue ingrowth 4, 25, 165, 181, 189-191, 210
titanium metal 13, 35, 85, 106, 236, 257, 294
transmission electron microscopy (TEM) 54, 76, 80, 109-112, 157, 162, 240
tricalcium phosphate 2, 6, 146, 185, 202, 229, 377
tris-buffer solutions 45
Tübingen implant 32
tumor 70, 101, 305

U

ultralow-temperature isotropic (ULTI) carbon 261-280
ureteral reflux 71

V

viscosity 15, 41, 251, 283, 306
vitrification 18, 19, 319
vocal cords 68, 71

W

wear debris 5, 31, 366
Weibull factor 159
wollastonite 75,-85, 89-104

X

x-ray diffraction (XRD) 50, 80, 147-171, 211, 213, 229, 264, 323
x-ray photoelectron spectroscopy (XPS) 233
xerogels 22

Y

Young's modulus 12, 13, 26, 27, 33, 76, 128, 129, 145, 158, 188, 241, 267, 269, 285, 299, 300, 370

yttrium aluminosilicate (YAS) glasses 305

Z

zeta potential measurement 331
zirconia 2, 13, 33, 78, 85, 167, 270, 369, 379